Aspects on Fundaments and Applications of Conducting Polymers

Aspects on Fundaments and Applications of Conducting Polymers

Editor

Eckbert Freund

scitus academics

Aspects on Fundaments and Applications of Conducting Polymers

Edited by **Eckbert Freund**

ISBN: 978-1-68117-319-1
Library of Congress Control Number: 2016937438

SCITUS Academics LLC,
www.scitusacademics.com
Box No. 4766, 616 Corporate Way,
Suite 2, Valley Cottage,
NY 10989

Printed in United States of America on Acid Free Paper

Preface

A wide range of basic and applied research including various polymers, copolymers, blends, mixtures and composites has been carried out since the establishment of the conducting properties of intrinsic conductive polymers. Until about 30 years ago, all carbon-based polymers were considered rigidly as insulators. The idea that electricity could be generated by plastics was thought to be absurd. In fact, plastics have been widely used for this property by the electronics industry. They are used as inactive packaging and insulation materials. This very narrow perspective rapidly changes when a new class of polymers is discovered, called intrinsically conductive polymers or electroactive polymers. Consequently, the ultimate understanding of the physical and chemical properties of these materials has been pursued, while the facets applied have progressed very rapidly, crossing the boundaries between disciplines.

Conductive polymers are organic polymers which generate electricity. Such compounds may be either metal or semi-conductive. The main advantage of conducting polymers is their processability, primarily through dispersion. Conducting polymers are not thermoplastics in general. However, organic materials, such as polymers, are used. They can offer high electrical conductivity, but they do not have mechanical properties similar to those of other polymers.

This book, Aspects on Fundamentals and Applications of Conducting Polymers, provides information on the development of fundamental principles and certain polymer applications.

Table of Contents

1

CHAPTER

SMART POLYMERS AND COATINGS OBTAINED BY IONIZING RADIATION: SYNTHESIS AND BIOMEDICAL APPLICATIONS

H. Iván Meléndez-Ortiz[1], Gustavo H. C. Varca[2], Ademar B. Lugão[2], Emilio Bucio[3*]

[1]*Departamento de Materiales Avanzados, Centro de Investigación en Química Aplicada, Saltillo, México*

[2]*Instituto de Pesquisas Energéticas e Nucleares—IPEN-CNEN/SP, Cidade Universitária, São Paulo, Brasil*

[3]*Departamento de Química de Radiaciones y Radioquímica, Instituto de Ciencias Nucleares, Universidad Nacional Autónoma de México, Circuito Exterior, Ciudad Universitaria, México DF, México*

ABSTRACT

Gamma radiation has been shown particularly useful for the functionalization of surfaces with stimuli-responsive polymers. This method involves the formation of active sites (free radicals) onto the polymeric backbone as a result of the exposition to high-energy radiation, in which a proper microenvironment for the reaction among monomer and/or polymer and the active sites takes place, thus leading to propagation which forms side chain grafts. The modification of polymers using high-energy irradiation may be performed by the following methods: direct or simultaneous, pre-irradiation oxidative and pre-irradiation. The most frequent ones correspond to the pre-irradiation oxidative method and the direct one. Radiation-grafting has many advantages over conventional methods considering that it does not require catalyst nor additives to initiate the reaction, and in general, no changes on the mechanical properties with respect to the pristine polymeric matrix are observed. This chapter focused on the synthesis of smart polymers and coatings obtained by the use of gamma radiation. In addition, diverse applications of these materials in the biomedical field are also reported.

Keywords: Smart Polymers, pH Responsive, LCST, Biomedical Applications, Radiation-Grafting

1. INTRODUCTION

Stimuli-responsive polymers are polymers that respond sharply to small changes in physical or chemical conditions with relatively large phase or property changes [1] . The stimuli in which smart polymers respond to are commonly classified into three categories: physical, chemical or biological [2] . Temperature and pH stimuli- responsive macromolecular materials have attracted great attention due to their obvious applications in biomedicine and biotechnology [3] . Some stimuli sensitive monomers exhibit separation from solution and solidification above certain temperatures; thermally-responsive polymers undergo a coil-globule transition in aqueous solution at temperature values which are known to lower critical solution temperature (LCST) [4] . In the case of sensitive polymers, the molecules are soluble below the LCST. Above this transition, they become increasingly hydrophobic and insoluble, and thus lead to gel formation. In contrast, hydrogels that are hydrophobic below critical temperature and hydrophilic above it typically present an upper critical solution temperature (UCST) [5] .

Stimuli-responsive polymers are of particular interest due to their ability to undergo well defined changes in their properties [6] . Taking into account that most grafting reactions occur at the surface or in the bulk polymeric matrix, polymer grafting stands as a suitable method to synthesize such materials. Acrylic acid (AAc) is one of the most popular monomers that have been grafted onto different polymeric matrices to develop pH responsive materials. In this case the carboxyl groups offer sites for the coupling with amino groups terminated poly (ethylene oxide) (PEO) where, via carbodiimide activation method, they are directly used for the interaction with heparin. In general, the carboxylic group has the capability to undergo further chemical reactions to produce new functional groups [7] . The thermosensitive characteristics of poly (N-isopropylacrylamide) (NIPAAm) have been widely studied in the fields of drug delivery and tissue engineering [8] . The concept of a "smart" coating is more recent and has been applied to functional coatings that are able to respond to certain stimuli generated by intrinsic or extrinsic events [9] .

Radiation synthesis of stimuli-responsive polymers and adsorbents is a wide field of important applications of radiation technology [10] . Radiation graft polymerization has been widely used for over 50 years throughout polymer chemistry. Typically, it involves the creation of reactive sites (radicals) on the polymer surface followed by covalent linkage of a preformed polymer or, more frequently by the polymerization of a monomer from those radical sites [11] . Grafting techniques present advantages over conventional approaches with regard to the development of thin functional films for several reasons. They include easy and controllable introduction of polymer chains with a high surface density, precise localization of the chain at the surface and long stability of the grafted layers among others [12] .

Pharmaceutical and biological therapeutics are often limited by short half-lives, poor bioavailability, and physical or chemical instability [13] . Stimuli-responsive (self-regulated) drug delivery systems have been investigated for their applications in pulsatile delivery of certain hormone drugs. Among all systems studied, temperature and pH-responsive delivery systems in particular have drawn much attention considering that some disease states manifest themselves by a change in temperature and/or pH [14] . One relevant application corresponds to the development of delivery devices based on a novel squeezing concept, which relies upon specific swelling-deswelling characteristics under specific temperatures, or temperature/pHs [15] .

2. IONIZING RADIATION AND POLYMERS

Radical polymerization has received the greatest amount of attention among all polymerization methods. About 60% of all available polymers are still obtained by this method [16] . This highlighted position of free radical polymerization is due to its many attractive characteristics including, but not limited to [17] - [19]

a) Its applicability to the polymerization of a wide range of vinyl monomers such as methacrylates, methacrylamides, styrene, butadiene, vinyl acetate and water-soluble monomers such as acrylic acid, the hydroxyacrylates and N-vinyl pyrrolidone;
b) Its ability to provide an unlimited number of copolymers;

c) Its tolerance to a wide range of functional groups (e.g. -OH, $-NR_2$, $-COOH$, $-CONR_2$) and reaction conditions (bulk, solution, emulsion, mini-emulsion, suspension);
d) Its tolerance to water or other impurities in contrast to the great sensitivities of ionic polymerization and of coordination polymerization;
e) The possibility to perform the process under convenient, mild reaction conditions and wide temperature range.
f) The simple and low cost implementation if compared to other technologies.

2.1. Principles of Radiation Chemistry

Radiation chemistry implies chemical effects of interactions of high energy ionizing radiation with materials. The result of energy absorption is the rupture or rearrangement of chemical bonds. One of the most obvious changes produced by ionizing radiation is the formation of ions. However, a significant amount of the radiation energy is also used to excite molecular electrons without extracting them from their parent molecule. Once such primary ion and excited molecules are formed they may undergo various secondary reactions including electron capture, charge neutralization, inter- and intra-molecular energy transfer processes, and molecular dissociation. Precise knowledge of the nature of the primary fragments resulting from the dissociation of molecules of the irradiated medium is of key relevance to the full understanding of radiation-chemical changes. Another specific feature of primary radiation-chemical processes lies in the fact that the primary events are not distributed randomly throughout the irradiated medium but occur in highly localized regions close to the tracks of the ionizing particles [20] [21] .

2.2. Radiation Techniques

Radiation processing has many advantages over conventional methods. When using radiation for materials processing, neither catalysts nor additives are required in order to initiate the reaction, considering the energy source being either gamma-ray, UV or electrons [22] . The main advantage of these radiation-assisted methods is that they are relatively simple.

Surface grafting of polymers has attracted great interest in the past few decades as it allows the tailoring of surface properties of various polymer materials and thus, specific functionalities useful for many applications may be effectively produced [23] .

When polymers are exposed to ionizing irradiation, trapped radicals and peroxides or hydroperoxides are formed as a result. These radicals are capable of initiating polymerization reactions. Various polymerization approaches have been developed, including direct graft polymerization of vinyl monomers that already contain desirable functional groups [20] .

High-energy radiation, such as gamma-irradiation and electron beam lead to activation of polymers in a way that specific monomers may be grafted. To the extent of modification, i.e., the grafting degree may be easily controlled by careful variation of the radiation exposure and reaction conditions. The grafting method by gamma-irradiation is extremely attractive considering that the modified biomaterial is obtained in the purest form possible, no chemical initiators are required and it may be applicable to almost all polymer-monomer combinations and thus allowing enormous possibilities to generate materials with improved physical-chemical and biological characteristics [24] . Another advantage of gamma-irradiation is that it offers much higher penetration into the polymer if compared to the electron beam [25] .

2.3. Types of Polymers Obtained by Ionizing Radiation

The irradiation of macromolecules is known to cause homolytic fission and thus form free radicals on the polymer. The medium is of key relevance in this case, e.g. if irradiation is carried out in air peroxides may be formed on the polymer. The lifetime of the free radical, on the other hand, relies upon the nature of the polymer backbone. The radiation grafting of various water-soluble monomers such as acrylic acid, methacrylic acid, acrylamide, N,N-dimethylacrylamide and N-vinyl-2-pyrrolidone on ultra-high molecular weight polyethylene were synthesized. A large effort is being devoted to the production of membranes via grafting of various hydrophilic monomers onto polyethylene, polypropylene and fluorinated polymers, as base polymers. As an example, the irradiation grafting of dimethylaminoethymethacrylate was studied in several polymers such as polyethylene, polypropylene, polytetrafluoroethylene, silicone and polycarbonates.

2.3.1. Radiation Induced Graft Copolymerization

Radiation-induced graft copolymerization is a well established technology dating back almost 50 years [20] [26] . If the polymer substrate is exposed to ionizing radiation (for instance, electron beams, gamma-rays, X-rays) the active sites randomly produced along polymer chain may act as macroinitiators and initiate free radical polymerization of the monomer as a consequence. This method is applicable for a lot of polymer/monomer combinations and unlike chemically initiated grafting, there is no contamination from initiators. The extent of polymerization of monomers on substrates is expressed in terms of grafting yield, also known as grafting degree. In principle, there are three methods for radiation grafting [27] [28] which are outlined below:

1) Radiation-grafting by simultaneous or mutual method

The so called radiation-grafting by simultaneous or mutual method is perhaps the simplest one. In this case, both polymer and monomers are introduced into a reactor and simultaneously irradiated. The polymer substrate is immersed in the monomer, which may be a liquid, vapor or solid diluted in a suitable solvent, in presence or absence of additives, and then exposed to ionizing radiation. In addition to the formation of the grafted polymer a homopolymer will invariably be formed, considering that both substrate and monomer are exposed to radiation. This technique should be carried out only if the radiochemical yield of the polymer $G_{(R)P}$ is higher than the monomer $G_{(R\cdot)M}$ in order to avoid homopolymer formation, which stands as its main disadvantage [20] . The addition of monomers in vapor phase or at low concentrations in a suitable solvent, and the use of polymerization inhibitors such as Mohr's salt may reduce such formation [29] .

2) Radiation-grafting by pre-irradiation method (post-irradiation grafting).

The pre-irradiation method (post-irradiation grafting) is characterized by pre-irradiation of the polymer under vacuum or inert gas. When radiation reaches the sample, the molecules become ionized and excited species are formed. These species form relatively stable free radicals which build up the active sites. The grafting polymerization reaction is usually initiated at elevated temperatures for monomers introduced in liquid, vapor or as solution in a suitable solvent. The main advantage of this technique is that since monomers are not exposed to radiation, the homopolymer formation is low, whereas its disadvantages are attributed to the short life time of the generated radicals, the need of inert atmosphere and the low grafting yield. The grafting degree strongly depends on reaction conditions such as pre-irradiation dose, monomer concentration, temperature, reaction time, material thickness and the solvent in cases where it is applied [20] [30] .

3) Radiation-grafting by pre-irradiation oxidative or alkylperoxy/hydroperoxy method

The pre-irradiation oxidative method, also known as alkylperoxy/hydroperoxy method, consists in the irradiation of samples in air or oxygen. In a more specific way the free radicals present in the polymer backbone quickly react with oxygen and then form alkylperoxides and hydroperoxides. In a second and sub-sequent step, the monomer or monomer solution is introduced and heated in absence of oxygen. Both peroxides and hydroperoxides are homolytically broken and then become available to react with the monomer. In this case, the grafting degree depends on the above mentioned conditions. The main advantage of this technique is that pre-irradiated samples may be stored at low temperature [20] [30] , while the disadvantages are related to the need of elevated temperatures, long reaction time and solvents in some cases. Another drawback to be considered is related to the formation of hydroxyl radicals which may trigger homopolymerization and then lead to lower grafting yields as a consequence if compared to the simultaneous method [31] , in which this reaction is absent.

Each of these methods (direct, pre-irradiation, and pre-irradiation oxidative) holds advantages and disadvantages. The existence of these methods is particularly useful when the access to a radiation source is limited. The simultaneous method in particular, requires the use of the radiation source during the entire grafting process; Whatsoever, grafting yields are generally higher when compared to the pre-irradiation methods due to reduced radical loss, which are originated from decomposition reactions present in the latter method [32] . In all cases, the rate and efficiency of the initiation is dependent upon the type of radiation, dose rate, and the radiation sensitivity of the material involved [33] .

2.3.2. RAFT-Graft Polymerization

The RAFT-graft polymerization may be achieved by the use of gamma radiation [34] . Controlled radical polymerization, also called living free-radical polymerization is a technique that combines the advantages of free radical polymerization with living polymerization in order to obtain well-defined macromolecules [35] . Reversible addition-fragmentation chain transfer polymerization (RAFT) on the other hand, involves a reversible chain transfer rather than chain termination. In this process initiating radicals are generated by the exposition of the polymer, or monomers in solution to gamma irradiation. These radicals initiate the propagating chains which are capable of binding to RAFT agents and form intermediate radicals. In pre-equilibrium the leaving groups are released and capable of re-initiating propagation [36] .

3. SMART POLYMERS OBTAINED BY IONIZING RADIATION

The temperature sensitive behavior of NIPAAm grafted by electron beam onto polypropylene (PP) and by the gamma pre-irradiation oxidative method was studied by Bucio [37] . NIPAAm grafted onto poly (vinyl chloride) tubes were synthesized by the gamma pre-irradiation method [38] and low LCST was found at 30°C with maximum graft yields near 40% and low swelling percentage of about 19%. The production of temperature sensitive membranes by radiation grafting composed of NIPAAm grafted on a polyvinylidene fluoride (PVDF) porous membrane was studied by Kang [39] . The influence of radiation and reaction conditions on the grafting yield was investigated in details and the results showed that the membranes exhibited the temperature-sensitive LCST of NIPAAm. NIPAAm was also grafted by the use of the pre-irradiation method onto an ethyl-vinyl acetate copolymer (EVA), and it was found that the surface of the copolymer grafted with NIPAAm showed thermosensitivity similarly to partially crosslinked PNIPAAm gels [40] . The thermo and pH sensitivity behavior of 2-(di- methylamino) ethyl methacrylate and 4-vinyl pyridine binary grafted onto PP films was studied by Burillo [41] . AAc is one of the most popular monomers that have been grafted onto different polymeric matrices. Its polymer or copolymers with pH sensitive response and carboxylic groups from poly acrylic acid of pH sensitive response have a capability to undergo further chemical reactions to produce new functional groups [42] .

3.1. Temperature-Responsive Polymers

Temperature-sensitive polymers exhibit LCST behavior where phase separation is induced at a certain temperature threshold (Figure 1). Polymers of this type undergo thermally induced reversible phase transition. They are soluble in aqueous solvents (water) at low temperatures but become insoluble as the temperature rises above the LCST. It is possible to increase the functionality of microgel particles by finding the right balance of hydrophobic and hydrophilic co-monomers or by tuning to a desired temperature range by copolymerization using more hydrophilic (which raises the LCST) or more hydrophobic (which lowers the LCST) co-monomers [43] [44] .

Heskins and Gillet were the first to report an endotherm observed at the LCST upon heating an aqueous solution of PNIPAAm [45] . Grinberg [46] have studied the volume phase transition in responsive polymers using high-sensitivity differential scanning calorimetry (DSC) and as well as the swelling behavior of the polymers at different scanning rates. It was possible to measure the dependence of the transition parameters on the heating rate. The DSC measurements, by heating at different rates (from 1 to 10°C/min), provide results nearly approximating equilibrium and LCST or UCST. The transition temperature, enthalpy, and entropy of this thermosensitive behavior as well as the transition LCST are parameters that may be estimated.

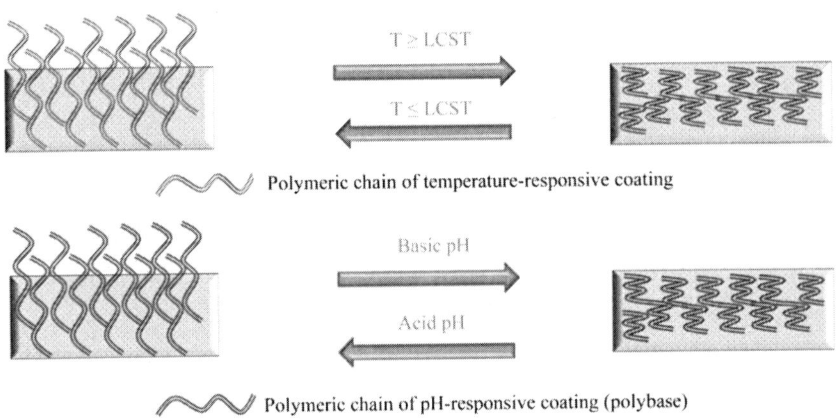

Figure 1. Representative scheme of Thermo and pH-responsive polymeric coatings.

3.2. pH-Responsive Polymers

pH-responsive polymers (Figure 1) consist of ionizable pendants capable of accepting and donating protons in response to environmental changes in pH such as carboxylic acids and basic amino alkyl moieties [47] . The change in the charge of pendant groups causes an alteration of the hydrodynamic volume of the polymer chains [48] . Then, the transition from collapsed state to swollen state is caused by the osmotic pressure generated by mobile counterions which neutralize the charges [49] . The phase transition of pH-sensitive polymers is nominated critical pH. In the human body pH variations are present along the gastrointestinal tract and in problematic sectors like tumor areas and surrounding tissues [50] . Drug delivery in this specific area makes relevant the need for the development of pH-sensitive systems with fast response to changes in environmental stimuli. A fast response of a polymer and a repetitive function of another polymer may be combined using different functionalization techniques such as γ-irradiation. Most grafting reactions which take place on the surface or in the bulk polymeric matrix allow the design of clinically effective controlled drug delivery systems, supported by the concept of achieving optimized combinations between monomer and the polymer itself.

3.3. Light-Responsive Polymers

The functionalization of polymers using light-responsive materials confers the ability to respond to stimuli triggered by the exposition to light sources such as UV, Vis or infrared. Apart from the nature of the induced changes, e.g. swelling or deswelling, these light induced transitions at molecular level assure a wide range of applications, including cell culture, photonics and drug delivery systems among others [51] . Common light- responsive materials may be achieved by the addition of photo-sensitive groups [51] such as azopyridine [52] , spiropyrans [53] , stilbenes [54] , Triphenylmethane-leuco derivatives, bis(4-dimethylamino)phenylmethyl leucocyanide, including modified polyacrylamides [55] and azobenzene materials [56] .

Light allows proper and specific control of the trigger mechanism, as well as assures precise control of the stimulus. A combination of these properties makes light-responsible systems very promising materials for drug delivery purposes. Inconsistent responses to light, uncontrolled structural changes, and slow response correspond to typical drawbacks of these materials [55] . Although these systems may hold very relevant properties for a wide variety of applications in different fields, the use of ionizing radiation for the synthesis and development of light responsive polymers and coatings is slim to none. The main reason for this is related to the high sensitivity to high energy irradiation by the monomers or functional groups to be inserted into the polymer. This makes the use of irradiation techniques not very promising for the design of such materials.

3.4. Magnetic-Responsive Polymers

The so-called magnetic-responsive polymers are result of the combination of polymers with magnetic particles, at nano or micro level, that hold the ability to undergo changes in response to magnetic fields, also known as magnetoelastic or magnetostrictive polymeric composites [57] [58] . On this account, a wide variety of magnetic nanoparticles have been incorporated/added to polymers by distinct methods in order to combine properties of both materials into a stable system, e.g. colloidal formulations [57] [58] . Typical particles are related but not limited to metal particles, ferromagnetic particles [59] and nickel powders [60] . These systems are more applicable and promising towards the development of advanced materials for drug targeting or site specific delivery, including magnetic devices, artificial muscles, micromanipulators and so on [57] .

Composites are not the main focus of this chapter, whatsoever for further reference irradiation mediated pathways abound in literature concerning the development of both magnetic nanoparticles [61] and the composites [62] [63] . An additional feature of the use of high energy irradiation for such purpose is related to the possibility to perform both, nanoparticle synthesis and composite development, simultaneously or not.

4. IONIZING RADIATION AND COATINGS

During the past 50 years, radiation processing of materials has been increasingly applied to improve quality, efficiency, energy savings and for the manufacture of new products with specific properties. Currently, radiation technology continues to enter the marketplace with innovations in materials processing. Radiation technologies applied to materials processing comprise different radiation types and sources for diverse applications. This technology has been used in multi-layer tubes, aerospace application, and is under investigation in numerous other areas such as structural parts for use as automotive panels, for electro-optical devices, healthcare pro- ducts, and in many other areas [64] . Regarding the coating industry this technology has been mainly applied in the curing of diverse materials being UV radiation and electron beam the most used radiation sources [65] .

Radiation curing refers to the process where the energy of UV-light or electron beams is used to induce crosslinking of polymer chains by radical induced mechanism in coatings or in adhesives. This process involves the transformation of a liquid substance into a non-tacky solid within a few seconds [66] . This technique has grown over the last 3 decades and shifted from novelty to a firmly established alternative method for many of the traditionally cured systems [67] . Curing by radiation techniques hold some advantages over traditional methods, mainly attributed to a quick process, the possibility to be carried out at room temperature, no need of solvents and additives (initiators and/or catalysts), the possibility to link the coating to the substrate resulting in more durable coatings, and so on. However, this technique has some disadvantages. For instance, the radiation curing frequently needs inert atmospheres to prevent inhibition of curing, and the high initial cost for the acquisition of the irradiation equipment.

Electron-beam curing has come into extensive use particularly in surface coatings and printings [68] - [71] . Electron beam cured coatings are usually uniformly cured due to the homogenous penetration of the electrons. Such coatings adhere well to a variety of surfaces including metals, wood, textiles, paper, polymers, and glass. Electron beam curing has numerous advantages over conventional thermal curing methods including the ability to combine various materials and functions in a single operation, no substrate heating, low energy consumption, immediate further treatment of converted products without post curing, high efficiency, no emission of organic solvents into the environment, and exact repeatability of process conditions due to a high dose accuracy.

Depending on the applications electron accelerators of different energies may be selected:

1) Low-energy (from 150 keV to 300 keV): extensively applied for the curing of coatings for paper, wood, CDs, DVDs, metals and polymers, surface grafting for membranes, adhesives for plastics and printing inks, films and multilayer packaging, paints, varnishes, and disinfestations of seeds.

2) Mid-energy (from 300 keV to 5 MeV): particularly useful for the irradiation of natural polymers, ion- exchange membranes for fuel cells application, composite materials, wire and cable insulation, biodegradable blends, and fiber modification.

3) High-energy (from 5 MeV to 10 MeV): commonly used for the sterilization of medical devices and some pharmaceutical and biological products including the preservation and disinfestations of foods and agricultural products, treatment of lignocelluloses materials to produce ethanol biofuels, removal of organic compounds from wastewater, and decontamination of pesticide packing [72] .

Alternatively, gamma radiation has been widely applied for the development of coatings from various materials, especially metals and polymers used as biomaterials. As an example, Jin et al. [73] coated the surface of stainless steel (SS) by grafting 2-hydroxyethylmethacrylate via mutual or simultaneous radiation method under ^{60}Co gamma-ray irradiation. Before radiation grafting the surface of SS was modified with dopamine. They demonstrated by water contact angle measurements and platelet adhesion assays that the hydrophilicity and hemocompatibility of the coated SS were improved. On the other hand, Yang et al. [74] grafted acrylamide (AAm) onto titanium metal sheets by using the direct method via gamma irradiation. Results showed that the wettability of the coated titanium surface was greatly improved if compared with the pristine titanium sheet. Also, the hydrophilic coating of polyacrylamide showed good stability because the titanium surface was pretreated using dopamine and presented good adhesive ability.

Coated PP materials have also been prepared by the use of ionizing radiation. An interpenetrating polymer network made of two stimuli-responsive polymers was grafted onto PP films with the aim of developing medicine containing coatings for medical devices. Drug-loaded PP films sustained the delivery of vancomycin, an antimicrobial agent, for several hours and provided release rate values adequate for killing bacteria attempting to adhere to the surface of the PP films [75] . García-Vargas et al. [76] modified separately PP monofilament sutures with glycidyl methacrylate (GMA) and AAc by means of pre-irradiation method using a ^{60}Co source. PP- g-AAc sutures exhibited pH-dependent swelling and remarkable high capability to host vancomycin. Some AAc-functionalized sutures were able to inhibit bacterial growth after successive challenges with fresh lawns. These antimicrobial agent-loaded PP sutures could be used as device combo products that minimize the risk of infections.

5. BIOMEDICAL APPLICATIONS

5.1. Smart Polymers

As above mentioned, smart or stimuli-responsive polymers are high-performance polymers (Figure 2) that change their properties when exposed to external stimuli [47] [77] - [82] . In other words, this nomenclature is directly related to their ability to respond to external or internal environmental changes. The most common stimuli correspond to temperature, pH, electromagnetic field and light, where among these, temperature and pH responsive systems are by far the most important systems concerning biomedical research. Stimuli-responsive polymers have been used in various forms such as graft and block copolymers, hydrogels, polymeric micelles, coatings, and interpenetrating networks [83] - [85] .

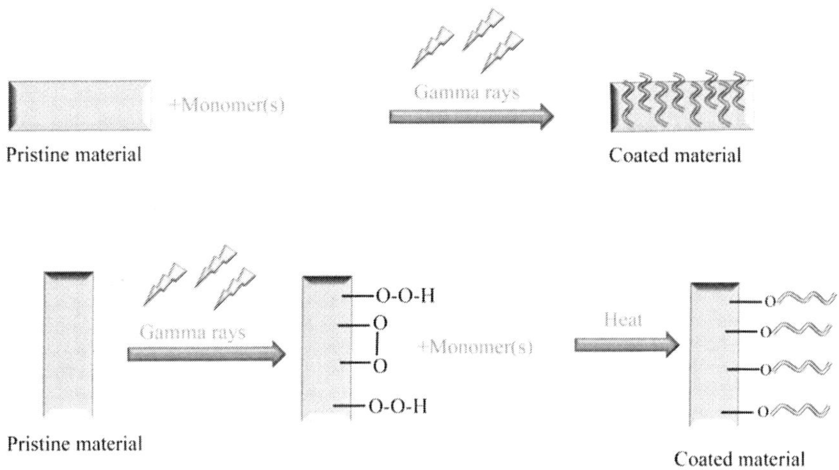

Figure 2. Pristine based thermo-responsive coatings obtained by irradiation grafting for biotechnological purposes.

Temperature-responsive polymers and hydrogels exhibit a volume phase transition at a defined temperature which causes a sudden change in the conformation of the polymeric chains and solvation state as a consequence. The change in temperature is not only relatively easy to control, but also easily applicable to both in vitro and in vivo conditions. Below the LCST, the polymeric chains expand and become hydrophilic and water soluble, while above the LCST the polymeric chains shrink and thus become hydrophobic and water insoluble [47] . LCST and UCST systems are not restricted to aqueous solvent environments. Whatsoever, only the aqueous systems are of interest for biomedical applications. An example of temperature-responsive polymer widely known for a long time is the poly(N-isopropylacrylamide) (PNIPAAm). PNIPAAm dehydrates and its structure collapses at temperatures above its LCST (32°C in water), and becomes hydrated and swells at temperatures below its LCST.

On the other hand, the pH is an important environmental parameter for biomedical applications, taking into account that pH changes do occur in many specific sites of the human body. The pH-responsive polymers contain ionizable groups that can accept and donate protons in response to an environmental change in pH. Upon ionization the electrostatic repulsions of the generated charges (anions or cations) cause a dramatic extension of coiled chains. The ionization of the pendant acidic or basic groups on polyelectrolytes can be partial due to the electrostatic repulsion from other adjacent ionized groups [47] . By means of swelling properties, in the case of polymers with weakly acidic groups, swelling index increases as the external pH increases. The opposite is observed if the polymer contains weakly basic groups. Most anionic pH-sensitive polymers used in drug delivery are based on poly(acrylic acid), poly(methacrylic acid), poly(ethylene imine) and poly(l-lysine).

Currently the most prevalent use for smart polymers in biomedicine is specifically related to targeted drug delivery systems [86] - [88] . However, other biomedical applications such as gene and tissue engineering, chemotherapy, biochromatography, and biomembrane technology have also been reported [89] - [93] . The most relevant inconvenient with the use of smart polymers in biomedicine is related to the toxicity or incompatibility of artificial substances with the human body. Nevertheless, stimuli-responsive polymers hold enormous potential in biomedical applications if these obstacles may be overcome.

Ionizing radiation-induced polymerization is an economical, clean, and convenient technique for the synthesis, grafting or modification of biomaterials [94] - [97] . Ionizing radiation (gamma radiation, electron beam or X-ray) allow the creation of interesting polymers with new architectures. Specifically, the use of gamma radiation has some advantages including the possibility to perform grafting at room temperature in solid, liquid or gaseous state, no need of additives (catalyst or initiators), high penetrability, much higher grafting yields if compared to electron beam radiation, and surface selectivity by tuning reaction conditions, etc. Burillo et al. [97] prepared a temperature and pH-responsive interpenetrating network by gamma radiation from a ^{60}Co gamma source and electron beam using a Van de Graaff accelerator. Interpenetrating networks of temperature sensitive PNIPAAm and pH sensitive PAAc were prepared by two consecutive steps. Cross-linking yields were higher for samples irradiated with gamma radiation than those treated with electron beam. The pH sensitivity from PAAc was conserved while the LCST of PNIPAAm shifted to 28°C. Ruiz et al. [83] synthesized a smart polymer by grafting an interpenetrating polymer network onto PP films. Three consecutive steps were used: graft copolymerization of PAAc onto PP films by gamma radiation, crosslinking of PP-g-PAAc to form the first network, and synthesis of a second network of PNIPAAm. They found that the interpenetrating network grafted onto PP retained the thermal and pH sensitivity from PNIPAAm and PAAc respectively. Contreras et al. [98] grafted N, N'-dime- thylacrylamide and NIPAAm using pre-irradiation method with gamma-rays. They studied the effect of absorbed dose, monomer concentration and reaction time on the degree of grafting. In addition, grafting of stimuli- responsive polymers onto polymeric matrices of biomedical interest has been reported [99] - [103] . PP films have been modified with temperature-responsive polymers to load nalidixic acid, and non-steroidal anti-inflammatory drugs [103] [104] . Stimuli-responsive bacterial cellulose-g-poly-(acrylic acid) hydrogels were prepared for potential use as oral delivery systems of protein-based drugs using electron beam irradiation [93] . These hydrogels exhibited excellent mucoadhesive potential and cytocompatibility.

5.2. Smart Coatings

Conventional and advanced technological applications demand materials with well-defined surface properties. A versatile route employed to provide new properties to certain materials is the surface functionalization. Surface functionalization is the act of modifying the surface of a material by bringing physical, chemical or biological characteristics different from the ones originally found on the surface of a material. A functional surface comprises new and/or modified chemical groups and/or morphological features that determine the interaction with the surrounding environment and subsequently the material properties and behavior [105] - [107] .

The surfaces of many materials can be functionalized by applying coatings. Coatings, whether organic, inorganic or hybrid, are a class of materials that may be applied to the surface of certain materials in order to decorate or modify it. Coatings may partially or completely cover the surface of materials and in addition, they may be tailored for many applications in which they should be able to perform well-defined variety of functions. Coatings do change the surface properties of the substrate such as wettability, adhesion, corrosion or wear resistance, and so on. There are innumerous functional coatings present in our daily life including laptops, paints, lacquers, solar panels, home furniture, cars and even in more advanced applications or systems such as invisible paints, medical devices, or orthopedic implants. A major consideration for most coating processes is that the coating must be applied at a controlled thickness. The thickness of the coating system and the number of layers depends on the final application of the coated material ranging from a few up to several hundreds of micrometers. Generally, each layer is aimed for specific functionalities that include adhesion to the bare metal or adhesion between different layers, corrosion inhibition, anti-fouling, water repellence, and wear resistance, among others [108] - [112] .

The concept of "smart", "intelligent" or stimuli-responsive coatings is relatively recent and has been applied to functional coatings that change their properties in response to environmental stimulus. Smart coatings cover a wide-range of characteristics but the main performance is related to durability. However, concerning industrial coatings, energy consumption of the processes is the most urgent current focus. The applicability of reactive smart coatings to various substrates, like polymers, metals, or composites, generates a fairly universal platform without relying on the broad chemical alteration of the bulk material. As a consequence, this minimizes contaminations and promotes a biocompatible process free of harmful solvents, initiators, or catalysts. Therefore, smart coatings compose a class of materials with enormous technological interest. Intelligent coatings have been commonly used for various applications, including but not limited to the biomedical ones [113] - [115] . Depending upon the applications traditional functional coatings exhibit significantly different physicochemical properties. Polymer-based smart coatings are expected to undergo a reversible phase transition which causes changes in polymer conformation and polymer surface energy. Responses to mechanical deformations, temperature, pH, ionic strength, electric and magnetic fields, surface and interfacial phenomena, and the development of nanostructures are the most common targets when designing coating polymeric films [116] [117] . However, the external stimuli—temperature, pH or ionic strength of solvent, light or electromagnetic fields—often require fine- tuning of surface properties. Coatings in biomedical engineering have emerged as an important topic and play a critical role in biological applications [118] . Coating strategies target the spatially-controlled design of interfaces as well as the development of smart surfaces. In addition, in order to truly mimic a biological interface, biomaterials must not only provide excellent control over the distribution and orientation of biomolecules, but they may also overcome the intrinsic static character of pristine materials. A promising route towards this goal is based on the coating of biomaterials using stimuli-responsive polymers [119] - [121] . These intelligent coatings will provide a dynamic control of material properties to direct, for example, cell or biomolecule adhesion, cell detachment, controlled drug delivery, antimicrobial properties, etc.

Materials can be modified with smart coatings by grafting polymers with reactive end groups and architectures onto their surface. One strategy is the use of radiation techniques such as gamma radiation and electron beam which have proved to be useful in the development of coatings for materials with biomedical use in regenerative medicine and controlled drug delivery [75] [76] [82] [86] [100] - [104] [122] - [124] .

Polycaprolactone, a compound frequently used as the base polymer in scaffolds engineered towards tissue engineering applications was modified with pH-responsive PAAc by means of gamma radiation [124] . The introduced -COOH groups could permit the subsequent attachment of biomolecules and thus making them useful templates for tissue engineering applications. On the other hand, stable coatings of temperature-responsive poly(vinyl methyl ether) have been immobilized onto polystyrene surfaces by low energy electron beam irradiation [125] . Based-polymer coatings have been used to coat biomaterials for their use in drug-eluting medical devices with anti-fouling properties. Chitosan and chitosan hydrogels were immobilized on high density polyethylene tubes previously modified with AAc by electron beam irradiation [126] . The chitosan hydrogel layers were further modified with heparin to improve antibacterial and blood compatibility properties. On the other hand, silicone rubber (SR) has been modified using gamma rays, with some stimuli-responsive polymers to improve its ability to host either antimicrobial or anti-inflammatory drugs [38] [127] [128] . The grafting of these polymers has improved the performance of SR for biomedical applications. Commercial polyurethane catheters (Tecoflex®, TFX) were irradiated with gamma rays and then coated with stimuli-responsive polymers to obtain vancomycin-eluting implantable medical devices [129] . Authors found that TFX-based insertable medical devices were able to inhibit the growth of S. aureus acting as a drug delivery system. In addition, cotton gauzes have been coated via one-step grafting (direct method) using ^{60}Co gamma-rays with the thermo-pH sensitive 2-(di- methylamino)ethyl methacrylate (DMAEMA) [130] . DMAEMA grafting makes gauzes more hydrophilic with enhanced blood absorption capability and favored collagenase activity. Cotton-g-DMAEMA gauzes were subsequently quaternized with methyl iodide in order to make them less prone to be colonized by bacteria (Figure 3).

6. CONCLUSIONS AND FUTURE PROSPECTS

Ionizing radiation stands as an important and efficient technique to modify biomaterials with stimuli-responsive coatings and/or stimuli-responsive monomers/functional groups. In addition to conventional methods, the use of irradiation based techniques allows the development of covalent-linked coatings which are desirable for many applications, including biomedical, automotive, textile and biotechnological industry among others.

From a development point of view the design and the appropriate coating properties should be defined by the applications in which these materials are intended for. On this account, the irradiation source and method, as well as polymer and monomer selection should be based on this premise. On the other hand, from a general perspective the combination of stimuli-responsive coatings and biomaterials led to emerging technology platforms suitable for the development of advanced materials able to mimic biological systems as well as respond to biological changes, such as pH or temperature variations.

Future developments of this technology promise to greatly improve the performance of biomaterials concerning specific biomedical and biotechnological applications including drug-eluting medical devices, anti-fouling biomedical devices and tissue engineering.

Novel and recent approaches in this particular field have been delivered over the last decades by several scientists which include the radiation grafting inside the nanopores of polymeric matrices. This technique may not exclusively hold the ability to confer distinct responsive properties to materials but assure localized changes in the nano/microstructure of the systems and thus contribute to great and novel perspectives in a near future.

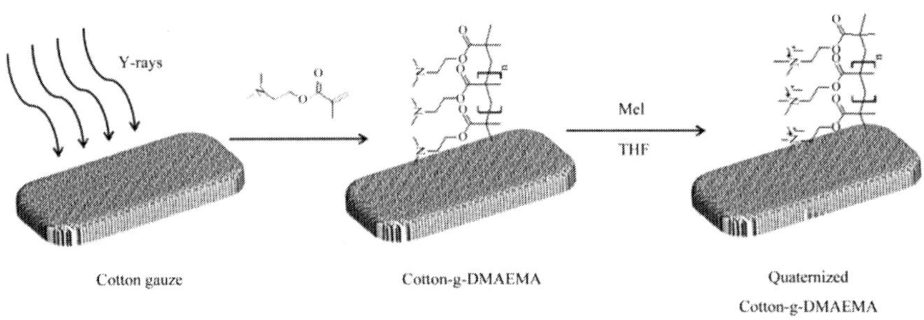

Figure 3. Schematic representation of the cotton-g-DMAEMA synthesis and subsequent quaternization with methyl iodide.

ACKNOWLEDGEMENTS

H.I. Melendez-Ortiz is grateful to the program Cátedras-CONACyT (Mexico). This chapter was supported by DGAPA-UNAM Grant IN200714, CONACYT-CNPq Project 174378 (Mexico) and CNPq project number 490200/2011-7 (Brazil). The authors would like to acknowledge M.L. Escamilla, M. Cruz, and E. Palacios from ICN-UNAM and B. Puente from CIQA for technical assistance.

REFERENCES

1. Hoffman, A.S. (2013) Stimuli-Responsive Polymers: Biomedical Applications and Challenges for Clinical Translation. Advanced Drug Delivery Reviews, 65, 10-16.

2. Peppas, N.A., Hilt, J.Z., Khademhosseini, A., et al. (2006) Hydrogels in Biology and Medicine: From Molecular Principles to Bionanotechnology. Advanced Materials, 18, 1345-1360.

3. Zhou L., Yuan W., Yuan J., et al. (2008) Preparation of Double-Responsive SiO2-g-DMAEMA Nanoparticles via ATRP. Materials Letters 62, 1372-1375.

4. Medeirosa, S.F., Santos, A.M. and Fessi, H. (2011) Stimuli-Responsive Magnetic Particles for Biomedical Applications. International Journal of Pharmaceutics 403, 139-161.

5. Klouda, L. and Mikos, A.G. (2008) Thermoresponsive Hydrogels in Biomedical Applications. European Journal of Pharmaceutics and Biopharmaceutics, 68, 34-45.

6. Ulbricht, M., Ozdemir, S. and Geismann, C. (2006) Functionalized Track-Etched Membranes as Versatile Tool to Investigate Stimuli-Responsive Polymers for "Smart" Nano-and Microsystems. Desalination, 199, 150-152.

7. Adem, E., Avalos-Borja, M., Bucio, E., Burillo, G., Castillon, F.F. and Cota, L. (2005) Surface Characterization of Binary Grafting of AAc/NIPAAm onto Poly(tetrafluoroethylene) (PTFE). Nuclear Instruments and Methods in Physics Research Section B, 234, 471-476.

8. Yang, J.M. and Lin, H.T. (2004) Properties of Chitosan Containing PP-g-AA-g-NIPAAm of NIPAAm/β-CD-g-PP Bigraft Nonwoven Fabric for Wound Dressing. Journal of Membrane Science, 243, 1-7.

9. Montemor, M.F. (2014) Functional and Smart Coatings for Corrosion Protection: A Review of Recent Advances. Surface and Coatings Technology, 258, 17-37.

10. El-Mohdy, H.L. and Safrany, A. (2008) Preparation of Fast Response Superabsorbent Hydrogels by Radiation Polymerization and Crosslinking of N-Isopropylacrylamide in Solution. Radiation Physics and Chemistry, 77, 273-279.

11. Ebara, M., Hoffman, J.M., Stayton, P.S. and Hoffman, A.S. (1999) Surface Modification of Microfluidic Channels by UV- Mediated Graft Polymerization of Non-Fouling and "Smart" Polymers. Radiation Physics and Chemistry, 76, 1409-1413.

12. Uhlmann, P., Ionov, L., Houbenov, N., Nitschke, M., Grundke, K., Motornov, M., et al. (2006) Surface Functionalization by Smart Coatings: Stimuli-Responsive Binary Polymer Brushes. Progress in Organic Coatings, 55, 168-174.

13. Honey, P.J., Rijo, J., Anju, A. and Anoop, K.R. (2014) Smart Polymers for the Controlled Delivery of Drugs—A Concise Overview. Acta Pharmaceutica Sinica B, 4, 120-127.

14. Zhang, K. and Wu, X.Y. (2004) Temperature and pH-Responsive Polymeric Composite Membranes for Controlled Delivery of Proteins and Peptides. Biomaterials, 25, 5281-5291.

15. Gotowska, A., Bark, J.S., Kwon, I.C., Bae, Y.H., Cha, Y. and Kim, S.W. (1997) Squeezing Hydrogels for Controlled Oral Drug Delivery. Journal of Controlled Release, 48, 141-148.

16. Sebenik, A. (1998) Living Free-Radical Block Copolymerization Using Thio-Iniferters. Progress in Polymer Science, 23, 875-917.

17. Moad, G. and Solomon, D.H. (2006) The Chemistry of Radical Polymerization. Second Edition, Elsevier, Oxford.

18. Rizzardo, E., Chiefari, J., Chong, B.Y.K., Ercole, F., Krstina, J., Jeffery, J., et al. (1999) Tailored Polymers by Free Radical Processes. Macromolecular Symposia, 143, 291-307.

19. Moad, G., Rizzardo, E. and Thang, S.H. (2005) Living Radical Polymerization by the RAFT Process. Australian Journal of Chemistry, 58, 379-410.

20. Chapiro, A. (1962) Radiation Chemistry of Polymeric Systems. Interscience Publishers, New York.

21. Chapiro, A. (1977) Radiation Induced Grafting. Radiation Physics and Chemistry, 9, 55-67.

22. Dennis, G.R., Garnett, J.L. and Zilic, E. (2003) Cure Grafting—A Complementary Technique to Preirradiation and Simultaneous Processes? Radiation Physics and Chemistry, 67, 391-395.

23. Kimura, Y., Asano, M., Chen, J.H., Maekawa, Y., Katakai, R. and Yoshida, M. (2008) Influence of Grafting Solvents on the Properties of Polymer Electrolyte Membranes Prepared by γ-Ray Preirradiation Method. Radiation Physics and Chemistry, 77, 864-870.

24. Clough, R.L. (2001) High-Energy Radiation and Polymers: A Review of Commercial Processes and Emerging Applications. Nuclear Instruments and Methods in Physics Research Section B, 185, 8-33.

25. Gupta, B., Anjum, N., Jain, R., Revagade, N. and Singh, H. (2004) Development of Membranes by Radiation-Induced Graft Polymerization of Monomers onto Polyethylene Films. Journal of Macromolecular Science, Part C: Polymer Reviews, 44, 275-309.

26. Odian, G., Sobel, M., Rossi, A. and Klein, R. (1961) Radiation-Induced Graft Polymerization: The Trommsdorff Effect of Methanol. Journal of Polymer Science, 55, 663-673.

27. Mazzei, R., García, G., Massa, G. and Filevich, A. (2007) Grafting of Polypropylene and Poly(vinylidene fluoride) Films Implanted with Ar+ Ions. Nuclear Instruments and Methods in Physics Research Section B, 255, 314-320.

28. Ramírez-Fuentes, Y.S., Bucio, E. and Burillo, G. (2007) Radiation-Induced Grafting of Stimuli-Responsive Binary Monomers onto Polypropylene Films. Nuclear Instruments and Methods in Physics Research Section B, 265, 183-186.

29. Bucio, E., Cedillo, G., Burillo, G. and Ogawa, T. (2001) Radiation-Induced Grafting of Functional Acrylic Monomers onto Polyethylene and

Polypropylene Films Using Acryloyl Chloride. Polymer Bulletin, 46, 115-121.

30. Alvarez-Lorenzo, C., Bucio, E., Burillo, G. and Concheiro, A. (2010) Medical Devices Modified at the Surface by γ-Ray Grafting for Drug Loading and Delivery. Expert Opinion on Drug Delivery, 7, 173-185.

31. Li, J., Sato, K., Ichiduri, S., Asano, S., Ikeda, S., Iida, M., et al. (2004) Pre-Irradiation Induced Grafting of Styrene into Crosslinked and Non-Crosslinked Polytetrafluoroethylene Films for Polymer Electrolyte Fuel Cell Applications. I: Influence of Styrene Grafting Conditions. European Polymer Journal, 40, 775-783.

32. Kamel, I., Machi, S. and Silverman, J.J. (1972) Radiation-Induced Grafting of Styrene Vapor to Polyethylene. Journal of Polymer Science Part A-1: Polymer Chemistry, 10, 1019-1029.

33. Ratner, D. and Hoffman, A.S. (1974) The Effect of Cupric Ion on the Radiation Grafting of N-vinyl-2-pyrrolidone and Other Hydrophilic Monomers onto Silicone Rubber. Journal of Applied Polymer Science, 18, 3183-3204.

34. Barsbay, M. and Güven, O. (2009) A Short Review of Radiation-Induced Raft-Mediated Graft Copolymerization: A Powerful Combination for Modifying the Surface Properties of Polymers in a Controlled Manner. Radiation Physics and Chemistry, 78, 1054-1059.

35. Chiefari, J., Chong, Y.K., Ercole, F., Krstina, J., Jeffery, J., Le, T.P.T., et al. (1998) Living Free-Radical Polymerization by Reversible Addition-Fragmentation Chain Transfer: The RAFT Process. Macromolecules, 31, 5559-5562.

36. Barner, L., Quinn, J.F., Barner-Kowollik, C., Vana, P. and Davis, T.P. (2003) Reversible Addition-Fragmentation Chain Transfer Polymerization Initiated with γ-Radiation at Ambient Temperature: An Overview. European Polymer Journal, 39, 449-459.

37. Bucio, E., Burillo, G., Adem, E. and Coqueret, X. (2005) Temperature Sensitive Behavior of Poly(N-isopropylacrylamide) Grafted onto EB-

Irradiated Polypropylene. Macromolecular Materials and Engineering, 290, 745-752.

38. Burillo, G., Díaz, A. and Bucio, E. (2006) Characterization of Thermo and pH Sensitivity of Binary Graft Copolymers onto Polytetrafluoroethylene. Journal of the Mexican Chemical Society, 50, 1-4.

39. Kang, Y.L. and Neoh, K.G. (2002) Synthesis and Characterization of Poly(N-isopropylacrylamide)-Graft-Poly(vinylidene fluoride) Copolymers and Temperature-Sensitive Membranes. Langmuir, 18, 6416-6423.

40. Pu, H., Ding, Z. and Ma, Z. (1996) Preparation, Characterization, and Properties of EVA Preirradiation Grafted NIPAAm. Journal of Applied Polymer Science, 10, 1529-1535.

41. Burillo, G., Bucio, E., Arenas, E. and Lopez, G.P. (2007) Temperature and pH Sensitive Swelling Behavior of Binary DMAEMA/4VP Grafts on Polypropylene Films. Macromolecular Materials and Engineering, 292, 214-219.

42. Ferraz, C.C., Varca, G.H.C., Ruiz, J.C., Lopes, P.S., Mahtor, M.B. and Lugao, A.B. (2014) Radiation-Grafting of Thermo- and pH-Responsive Poly(N-vinylcaprolactam-co-acrylic acid) onto Silicone Rubber and Polypropylene Films for Biomedical Purposes. Radiation Physics and Chemistry, 97, 298-303.

43. Yan, L., Zhu, Q. and Kenkare, P.U. (2000) Lower Critical Solution Temperature of Linear PNIPA Obtained from a Yukawa Potential of Polymer Chains. Journal of Applied Polymer Science, 78, 1971-1976.

44. Feil, H., Bae, Y.H., Feijen, J. and Kim, S.W. (1993) Effect of Comonomer Hydrophilicity and Ionization on the Lower Critical Solution Temperature of N-Isopropylacrylamide Copolymers. Macromolecules, 26, 2496-2500.

45. Heskins, M. and Guillet, J.E. (1969) Solution Properties of Poly(N-isopropylacrylamide). Journal of Macromolecular Science, 2, 1441-1455.

46. Grinberg, A., Grosberg, Y. and Tanaka, T. (2000) Studies of the Thermal Volume Transition of Poly(N-isopropylacrylamide) Hydrogels by High-Sensitivity Differential Scanning Microcalorimetry. 2. Thermodynamic Functions. Macromolecules, 33, 8685-8692.

47. Gil, E.S. and Hudson, S.M. (2004) Stimuli-Responsive Polymers and Their Bioconjugates. Progress in Polymer Science, 29, 1173-1222.

48. Siegel, R.A. (1993) Hydrophobic Weak Polyelectrolyte Gels: Studies of Swelling Equilibria and Kinetics. Advances in Polymer Science, 109, 233-267.

49. Tonge, S.R. and Tighe, B.J. (2001) Responsive Hydrophobically Associating Polymers: A Review of Structure and Properties. Advanced Drug Delivery Reviews, 53, 109-122.

50. Stubbs, M., McSheehy, P.M.J. and Griffiths, J.R. (1999) Causes and Consequences of Acidic pH in Tumors: A Magnetic Resonance Study. Advances in Enzyme Regulation, 39, 13-30.

51. Katz, J.S. and Burdick, J.A. (2010) Light-Responsive Biomaterials: Development and Applications. Macromolecular Bioscience, 10, 339-348.

52. Lin, L., Yan, Z., Gu, J.S., Zhang, Y.Y., Feng, Z. and Yu, Y.L. (2009) UV-Responsive Behavior of Azopyridine-Containing Diblock Copolymeric Vesicles: Photoinduced Fusion, Disintegration and Rearrangement. Macromolecular Rapid Communications, 30, 1089-1093.

53. Oh, Y.J., Nam, J.A., Al-Nahain, A., Lee, S., In, I. and Park, S.Y. (2012) Spiropyran-Conjugated Pluronic as a Dual Responsive Colorimetric Detector. Macromolecular Rapid Communications, 33, 1958-1963.

54. Eastoe, J., Dominguez, M.S., Wyatt, P., Beeby, A. and Heenan, R.K. (2002) Properties of a Stilbene-Containing Gemini Photosurfactant: Light-Triggered Changes in Surface Tension and Aggregation. Langmuir, 18, 7837-7844.

55. James, H.P., John, R., Alex, A. and Anoop, K.R. (2014) Smart Polymers for the Controlled Delivery of Drugs—A Concise Overview. Acta Pharmaceutica Sinica B, 4, 120-127.

56. Yager, K.G. and Barrett, C.J. (2009) Azobenzene Polymers for Photonic Applications. In: Zhao, Y. and Ikeda, T., Eds., Smart Light-Responsive Materials: Azobenzene-Containing Polymers and Liquid Crystals, John Wiley & Sons, Inc., Hoboken.

57. Andrade, A., Ferreira, R., Fabris, J. and Domingues, R. (2011) Coating Nanomagnetic Particles for Biomedical Applications. In: Fazel-Rezai, R., Ed., Biomedical Engineering—Frontiers and Challenges, InTech, Rijeka.

58. Chen, J. and Chang, C. (2014) Fabrications and Applications of Stimulus-Responsive Polymer Films and Patterns on Surfaces: A Review. Materials, 7, 805-875.

59. Cuevas, J.M., Alonso, J., German, L., Iturrondobeitia, M., Laza, J.M., Vilas, J.L. and León, L.M. (2009) Magneto-Active Shape Memory Composites by Incorporating Ferromagnetic Microparticles in a Thermo-Responsive Polyalkenamer. Smart Materials and Structures, 18, Article ID: 075003.

60. Zhang, D.W., Liu, Y.J. and Leng, J.S. (2010) Study on the Activation of Styrene-Based Shape Memory Polymer by Medium-Infrared Laser Light. Applied Physics Letters, 96, Article ID: 111905.

61. Abedini, A., Daud, A.R. and Hamid, M.A.A. (2014) Radiolytic Formation of Fe3O4 Nanoparticles: Influence of Radiation Dose on Structure and Magnetic Properties. PLoS ONE, 9, e90055.

62. Turcu, R., Nan, A., Craciunescu, I., Pana, O., Leostean, C. and Macavei, S. (2009) Smart Composites Based on Magnetic Nanoparticles and Responsive Polymers. Journal of Physics, 182, Article ID: 012081.

63. Kharissova, O.V., Kharisov, B.I. and Mendez, U.O. (2012) Radiation-Assisted Synthesis of Composites, Materials, Compounds, and Nanostructures. Wiley Encyclopedia of Composites, 1-26.

64. Parejo Calvo, W.A., Duarte, C.L., Machado, L.D.B., Manzoli, J.E., Geraldo, A.B.C., Kodama, Y., et al. (2012) Electron Beam Accelerators Trends in Radiation Processing Technology for Industrial and Environmental Applications in Latin America and the Caribbean. Radiation Physics and Chemistry, 81, 1276-1281.

65. Salleh, N.G.N., Yhaya, M.F., Hassan, A., Bakar, A.A. and Mokhtar, M. (2011) Effect of UV/EB Radiation Dosages on the Properties of Nanocomposite Coatings. Radiation Physics and Chemistry, 80, 136-141.

66. Bajpai, M., Shukla, V. and Kumar, A. (2002) Film Performance and UV Curing of Epoxy Acrylate Resins. Progress in Organic Coatings, 44, 271-278.

67. Webster, G. (1997) UV and EB Curing Technology and Equipment. Wiley/SITA, Oxford.

68. Sohn, J.Y., Gwon, S.J., Choi, J.H., Shin, J. and Nho, Y.C. (2008) Preparation of Polymer-Coated Separators Using an Electron Beam Irradiation. Nuclear Instruments and Methods in Physics Research Section B, 266, 4994-5000.

69. Youssef, H.A., Ali, Z.I. and Zahran, A.H. (2001) Electron Beam Structure Modification of Poly(vinyl chloride)-Wire Coating. Polymer Degradation and Stability, 74, 213-218.

70. Salleh, N.G., Glasel, H.J. and Mehnert, R. (2002) Development of Hard Materials by Radiation Curing Technology. Radiation Physics and Chemistry, 63, 475-479.

71. Ruiz, C.S.B., Machado, L.D.B., Pino, E.S. and Sampa, M.H.O. (2002) Characterization of a Clear Coating Cured by UV/EB Radiation. Radiation Physics and Chemistry, 63, 481-483.

72. Chmielewski, A.G., Al-Sheikhly, M., Berejka, A.J., Cleland, M.R. and Antoniak, M. (2014) Recent Developments in the Application of Electron Accelerators for Polymer Processing. Radiation Physics and Chemistry, 94, 147-150.

73. Jin, W., Yang, L., Yang, W., Chen, B. and Chen, J. (2014) Grafting of HEMA onto Dopamine Coated Stainless Steel by 60Co-γ Irradiation Method. Radiation Physics and Chemistry, 105, 57-62.

74. Yang, W., Yang, L., Shi, Y., Chen, B. and Chen, J. (2013) Radiation Grafting of Acrylamide onto Surface of Dopamine Functionalised Titanium. Surface Engineering, 29, 667-670.

75. Munoz-Munoz, F., Ruiz, J.C., Alvarez-Lorenzo, C., Concheiro, A. and Bucio, E. (2009) Novel Interpenetrating Smart Polymer Networks Grafted onto Polypropylene by Gamma Radiation for Loading and Delivery of Vancomycin. European Polymer Journal, 45, 1859-1867.

76. García-Vargas, M., González-Chomón, C., Magarinos, B., Concheiro, A., Alvarez-Lorenzo, C. and Bucio, E. (2014) Acrylic Polymer-Grafted Polypropylene Sutures for Covalent Immobilization or Reversible Adsorption of Vancomycin. International Journal of Pharmaceutics, 461, 286-295.

77. Cabane, E., Zhang, X., Langowska, K., Palivan, C.G. and Meier, W. (2012) Stimuli-Responsive Polymers and Their Applications in Nanomedicine. Biointerphases, 7, 1-27.

78. Stuart, M.A.C., Huck, W.T.S., Genzer, J., Müller, M., Ober, C., Stamm, M., et al. (2010) Emerging Applications of Stimuli-Responsive Polymer Materials. Nature Materials, 9, 101-113.

79. Chen, J.K. and Chang, C.J. (2014) Fabrications and Applications of Stimulus-Responsive Polymer Films and Patterns on Surfaces: A Review. Materials, 7, 805-875.

80. Meléndez-Ortiz, H.I. and Bucio, E. (2009) Stimuli-Sensitive Behaviour of Binary Graft Co-Polymers (PP-g-DMAEMA)- g-NIPAAm and (PP-g-4VP)-g-NIPAAm in Acidic and Basic Medium. Designed Monomers & Polymers, 12, 99-108.

81. Zhang, Q., Koa, N.R. and Oh, J.K. (2012) Recent Advances in Stimuli-Responsive Degradable Block Copolymer Micelles: Synthesis and Controlled Drug Delivery Applications. Chemical Communications, 48, 7542-7552.

82. Hu, J., Zhang, G., Ge, Z. and Liu, S. (2014) Stimuli-Responsive Tertiary Amine Methacrylate-Based Block Copolymers: Synthesis, Supramolecular Self-Assembly and Functional Applications. Progress in Polymer Science, 39, 1096-1143.

83. Ruiz, J.C., Burillo, G. and Bucio, E. (2007) Interpenetrating Thermo and pH Stimuli-Responsive Polymer Networks of PAAc/PNIPAAm Grafted onto PP. Macromolecular Materials and Engineering, 292, 1176-1188.

84. Liu, X., Guo, H. and Zha, L. (2012) Study of pH/Temperature Dual Stimuli-Responsive Nanogels with Interpenetrating Polymer Network Structure. Polymer International, 61, 1144-1150.

85. Cheng, W., Gu, L., Ren, W. and Liu, Y. (2014) Stimuli-Responsive Polymers for Anti-Cancer Drug Delivery. Materials Science and Engineering: C, 45, 600-608.

86. Lee, S.M. and Nguyen, S.B.T. (2013) Smart Nanoscale Drug Delivery Platforms from Stimuli-Responsive Polymers and Liposomes. Macromolecules, 46, 9169-9180.

87. Mura, S., Nicolas, J. and Couvreur, P. (2013) Stimuli-Responsive Nanocarriers for Drug Delivery. Nature Materials, 12, 991-1003.

88. Ward, M.A. and Georgiou, T.K. (2011) Thermoresponsive Polymers for Biomedical Applications. Polymers, 3, 1215-1242.

89. Terefe, N.S., Glagovskaia, O., De Silva, K. and Stockmann, R. (2014) Application of Stimuli Responsive Polymers for Sustainable Ion Exchange Chromatography. Food and Bioproducts Processing, 92, 208-225.

90. Ayano, E. and Kanazawa, H. (2006) Aqueous Chromatography System Using Temperature-Responsive Polymer-Modified Stationary Phases. Journal of Separation Science, 29, 738-749.

91. Cai, T., Li, M., Neohab, K.G. and Kang, E.T. (2012) Preparation of Stimuli Responsive Polycaprolactone Membranes of Controllable Porous Morphology via Combined Atom Transfer Radical Polymerization, Ring-

Opening Polymerization and Thiol-Yne Click Chemistry. Journal of Materials Chemistry, 22, 16248-16258.

92. Chu, L.Y., Xie, R. and Ju, X.J. (2011) Stimuli-Responsive Membranes: Smart Tools for Controllable Mass-Transfer and Separation Processes. Chinese Journal of Chemical Engineering, 19, 891-903.

93. Ahmad, N., Amin, M.C.I.M., Mahali, S.M., Ismail, I. and Chuang, V.T.G. (2014) Biocompatible and Mucoadhesive Bacterial Cellulose-g-Poly(acrylic acid) Hydrogels for Oral Protein Delivery. Molecular Pharmaceutics, 11, 4130- 4142.

94. El-Mohdy, H.L.A. (2013) Thermo-Responsive Behavior of Radiation-Induced Poly(N-isopropylacrylamide)/Polyethylene Oxide Nanocomposite. Journal of Polymer Research, 20, 206.

95. Meléndez-Ortiz, H.I. and Bucio, E. (2008) Radiation Synthesis of a Thermo-pH Responsive Binary Graft Copolymer (PP-g-DMAEMA)-g-NIPAAm by a Two Step Method. Polymer Bulletin, 61, 619-629.

96. Meléndez-Ortiz, H.I., Bucio, E. and Burillo, G. (2009) Radiation-Grafting of 4-Vinylpyridine and N-Isopropylacrylamide onto Polypropylene to Give Novel pH and Thermo-Sensitive Films. Radiation Physics and Chemistry, 78, 1-7.

97. Burillo, G., Briones, M. and Adem, E. (2007) IPN's of Acrylic Acid and N-Isopropylacrylamide by Gamma and Electron Beam Irradiation. Nuclear Instruments and Methods in Physics Research Section B, 265, 104-108.

98. Contreras-García, A., Burillo, G., Aliev, R. and Bucio, E. (2008) Radiation Grafting of N,N'-Dimethylacrylamide and N-Isopropylacrylamide onto Polypropylene Films by Two-Step Method. Radiation Physics and Chemistry, 77, 936- 940.

99. Munoz-Munoz, F., Ruiz, J.C., Alvarez-Lorenzo, C., Concheiro, A. and Bucio, E. (2012) Temperature- and pH-Sensitive Interpenetrating Networks Grafted onto PP: Cross-Linking Irradiation Dose as a Critical Variable for the Performance as Vancomycin-Eluting Systems. Radiation Physics and Chemistry, 81, 531-540.

100. Ruiz, J.C., Alvarez-Lorenzo, C., Taboada, P., Burillo, G., Bucio, E., De Prijck, K., Neils, H.J., Coenye, T. and Concheiro, A. (2008) Polypropylene Grafted with Smart Polymers (PNIPAAm/PAAc) for Loading and Controlled Release of Vancomycin. European Journal of Pharmaceutics and Biopharmaceutics, 70, 467-477.

101. Meléndez-Ortiz, H.I., Díaz-Rodríguez, P., Alvarez-Lorenzo, C., Concheiro, A. and Bucio, E. (2014) Binary Graft Modification of Polypropylene for Anti-Inflammatory Drug-Device Combo Products. Journal of Pharmaceutical Sciences, 103, 1269-1277.

102. Meléndez-Ortiz, H.I., Alvarez-Lorenzo, C., Concheiro, A. and Bucio, E. (2015) Grafting of N-Vinyl Caprolactam and Methacrylic Acid onto Silicone Rubber Films for Drug-Eluting Products. Journal of Applied Polymer Science, 132, Article ID: 41855.

103. Contreras-García, A., Bucio, E., Concheiro, A. and Alvarez-Lorenzo, C. (2010) Polypropylene Grafted with NIPAAm and APMA for Creating Hemocompatible Surfaces That Load/Elute Nalidixic Acid. Reactive and Functional Polymers, 70, 836-842.

104. Contreras-García, A., Alvarez-Lorenzo, C., Taboada, C., Concheiro, A. and Bucio, E. (2011) Stimuli-Responsive Networks Grafted onto Polypropylene for the Sustained Delivery of NSAIDs. Acta Biomaterialia, 7, 996-1008.

105. Cole, I.S. (2014) 2—Smart Coatings for Corrosion Protection: An Overview. In: Makhlouf, A.S.H., Ed., Handbook of Smart Coatings for Materials Protection, Woodhead Publishing, New Delhi, 29-55.

106. Liu, X., Li, H., Jin, Q. and Ji, J. (2014) Surface Tailoring of Nanoparticles via Mixed-Charge Monolayers and Their Biomedical Applications. Small, 10, 4230-4242.

107. Kiryuhin, D.P., Kim, I.P., Buznik, V.M., Ignat'eva, L.N., Kuryavyi, V.G. and Sakharov, S.G. (2009) Radiation- Chemical Synthesis of Tetrafluoroethylene Telomers and Their Use for Preparation of Thin Protective Fluoropolymer Coatings. Russian Journal of General Chemistry, 79, 589-595.

108. Manvi, G.N. and Jagtap, R.N. (2013) Radiation Cured Branched Polyurethane for Coatings. Pigment & Resin Technology, 42, 309-316.

109. Skorb, E.V. and Andreeva, D.V. (2013) Surface Nanoarchitecture for Bio-Applications: Self-Regulating Intelligent Interfaces. Advanced Functional Materials, 23, 4483-4506.

110. Schulze, A., Maitz, M.F., Zimmermann, R., Marquardt, B., Fischer, M., Werner, C., Went, M. and Thomas, I. (2013) Permanent Surface Modification by Electron-Beam-Induced Grafting of Hydrophilic Polymers to PVDF Membranes. RSC Advances, 3, 22518-22526.

111. Hidzir, N.M., Lee, Q., Hill, D.J.T., Rasoul, F. and Grondahl, L. (2015) Grafting of Acrylic Acid-Co-Itaconic Acid onto PTFE and Characterization of Water Uptake by the Graft Copolymers. Journal of Applied Polymer Science, 132, Article ID: 41482.

112. Pierna, M., Santos, M., Arias, F.J., Alonso, M. and Rodríguez-Cabello, J.C. (2013) Efficient Cell and Cell-Sheet Harvesting Based on Smart Surfaces Coated with a Multifunctional and Self-Organizing Elastin-Like Recombinamer. Biomacromolecules, 14, 1893-1903.

113. Hogg, A., Uhl, S., Feuvrier, F., Girardet, Y., Graf, B., Aellen, T., Keppner, H., Tardy, Y. and Burger, J. (2014) Protective Multilayer Packaging for Long-Term Implantable Medical Devices. Surface and Coatings Technology, 255, 124-129.

114. Dai, G., Xiao, H., Zhu, S. and Shi, M. (2014) Collagen-Immobilized Poly(ethylene terephthalate)-g-Poly(vinyl alcohol) Fibers Prepared by Electron-Beam Co-Irradiation. Journal of Applied Polymer Science, 131, Article ID: 40597.

115. Yoshida, M., Langer, R., Lendlein, A. and Lahann, J. (2006) From Advanced Biomedical Coatings to Multi-Functionalized Biomaterials. Journal of Macromolecular Science, Part C, 46, 347-375.

116. Urban, M.W. (2006) Intelligent Polymeric Coatings; Current and Future Advances. Journal of Macromolecular Science, Part C, 46, 329-339.

117. Chilkoti, A. and Hubbell, J.A. (2005) Biointerface Science. MRS Bulletin, 30, 175-176.

118. Lahann, J. and Langer, R. (2005) Smart Materials with Dynamically Controllable Surfaces. MRS Bulletin, 30, 185-188.

119. Senaratne, W., Andruzzi, L. and Ober, C.K. (2005) Self-Assembled Monolayers and Polymer Brushes in Biotechnology: Current Applications and Future Perspectives. Biomacromolecules, 6, 2427-2448.

120. Bucio, E. and Burillo, G. (2009) Radiation-Induced Grafting of Sensitive Polymers. Journal of Radioanalytical and Nuclear Chemistry, 280, 239-243.

121. Higa, O.Z., Mendonca Faria, H.A. and de Queiroz, A.A.A. (2014) Polyglycerol Dendrimers Immobilized on Radiation Grafted Poly-HEMA Hydrogels: Surface Chemistry Characterization and Cell Adhesion. Radiation Physics and Chemistry, 98, 118-123.

122. Singh, H.D., Han, S.S., Son, J.H. and Kim, S.C. (2015) Poly(ethylene glycol) Dicarboxylate/Poly(ethylene oxide) Hydrogel Film Co-Crosslinked by Electron Beam Irradiation as an Anti-Adhesion Barrier. Materials Science and Engineering: C, 46, 195-201.

123. Kim, S.J., Kim, W.I., Yamato, M., Okano, T., Kikuchi, A. and Kwon, O. (2013) Successive Grafting of PHEMA and PIPAAm onto Cell Culture Surface Enables Rapid Cell Sheet Recovery. Tissue Engineering and Regenerative Medicine, 10, 139-145.

124. Luk, J.Z., Cooper-White, J., Rintoul, L., Tarane, E. and Grondahl, L. (2013) Functionalised Polycaprolactone Films and 3D Scaffolds via Gamma Irradiation-Induced Grafting. Journal of Materials Chemistry B, 1, 4171-4181.

125. Gramm, S., Teichmann, J., Nitschke, M., Gohs, U., Eichhorn, K.J. and Werner, C. (2011) Electron Beam Immobilization of Functionalized Poly(vinyl methyl ether) Thin Films on Polymer Surfaces—Towards Stimuli Responsive Coatings for Biomedical Purposes. Express Polymer Letters, 5, 970-976.

126. Qu, X., Wirsén, A., Olander, B. and Albertsson, A.C. (2001) Surface Modification of High Density Polyethylene Tubes by Coating Chitosan, Chitosan Hydrogel and Heparin. Polymer Bulletin, 46, 223-229.

127. Vázquez-González, B., Meléndez-Ortiz, H.I., Díaz-Gómez, L., Alvarez-Lorenzo, C., Concheiro, A. and Bucio, E. (2014) Silicone Rubber Modified with Methacrylic Acid to Host Antiseptic Drugs. Macromolecular Materials and Engineering, 299, 1240-1250.

128. Contreras-García, A., Bucio, E., Brackman, G., Coenye, T., Concheiro, A. and Alvarez-Lorenzo, C. (2011) Biofilm Inhibition and Drug-Eluting Properties of Novel DMAEMA-Modified Polyethylene and Silicone Rubber Surfaces. Biofouling, 27, 123-135.

129. Munoz-Munoz, F., Bucio, E., Magarinos, B., Concheiro, A. and Alvarez-Lorenzo, C. (2014) Temperature- and pH- Sensitive IPNs Grafted onto Polyurethane by Gamma Radiation for Antimicrobial Drug-Eluting Insertable Devices. Journal of Applied Polymer Science, 131, Article ID: 39992.

130. Luna-Straffon, M.A., Contreras-García, A., Brackman, G., Coenye, T., Concheiro, A., Alvarez-Lorenzo, C. and Bucio, E. (2014) Wound Debridement and Antibiofilm Properties of Gamma-Ray DMAEMA-Grafted onto Cotton Gauzes. Cellulose, 21, 3767-3779.

2
CHAPTER

THEORETICAL STUDY ON THE ELECTRONIC, STRUCTURAL, PROPERTIES AND REACTIVITY OF A SERIES OF MONO-, DI-, TRI- AND TETRACHLOROTHIOPHENES AS WELL AS CORRESPONDING RADICAL CATION FORMS AS MONOMERS FOR CONDUCTING POLYMERS

Hossein Shirani IL Beigi[1*], Saeed Jameh-Bozorghi[2]

[1]*Young Researchers Club, Islamic Azad University, Toyserkan Branch, Toyserkan, Iran.*

[2]*Department of Chemistry, Islamic Azad University, Toyserkan Branch, Toyserkan, Iran.*

ABSTRACT

In this paper, electrical and structural properties of mono-, di-, tri- and tetrachlorothiophenes and their radical cations have been studied using the density functional theory and B3LYP method with 6-311++G** basis set.

The effects of the number and position of the substituent of chlorine atoms on the properties of the thiophene ring for all chlorothiophenes and their radical cations have been studied. Vibrational frequencies, nuclear chemical shielding constants, spin-density distribution, size and direction of dipole moment vector, ionization potential, electric polarizabilities and NICS values of these compounds have been calculated as well. The analysis of these data showed that double bonds in 3-chlorothiophene are more delocalized and it is the best possible candidate monomer among all chlorothiophenes for the synthesis of corresponding conducting polymers with modified characteristics.

INTRODUCTION

Electrically-conducting polymers such as advanced materials and the new class of '*synthetic metals*' have drawn high attention in the last years [1]. The 2000 Nobel Prize in chemistry recognized the discovery of conducting polymers and over 25 years of progress in this field [2, 3]. There has been growing interest in research on conducting polymer nano structures (nano-rods, -tubes, -wires and -fibers) since they incorporate the advantages of organic conductors with low-dimensional systems and so create interesting physicochemical properties and potentially useful applications [4–7]. The varieties of such studies have been reported in the literature that includes: surfactants [8–11], liquid crystals [12], polyelectrolytes [13], nano-wire seeds [14], thiophene oligomers [15], bulky organic dopants [16–21]. Among intrinsic conducting polymers with conjugated double bonds, polythiophenes (PTHs) and related derivatives are suitable for commercial applications [22].

Polythiophenes are normally produced from non aqueous media because the monomer is more soluble in them. The influence of water on polymerization process of thiophene as well as on the redox switching properties has been studied [23, 24]. It has been recently shown that the attachment of a range of electron-donating or electron-withdrawing groups through a conjugated linker has a dramatic effect on polymerization potential and subsequent photovoltaic performance [25, 26]. Polythiophene has a solubility of 75% (w/w) in chloroform and a conductivity of 220 s cm^{-1} [22].

The effect of temperature on polymerization process of thiophene has been investigated as well [27]. Results show that when polymerization is carried out at 15-20°C, polymers with optimal properties will be obtained. Ultrasonication has been used to improve the efficiency (improved yield, lowering of polymerization potentials) of electropolymerization process of polythiophene [28].

Thiophene is oxidized to form a conducting electro active polymer (CEP), with the best conductivity when it is created from α-α' linkages [22]. In polypyrroles, the mechanism of polymerization involves the formation of radical cations which react with another monomer to develop the polymeric structure (Figure 1) [29].

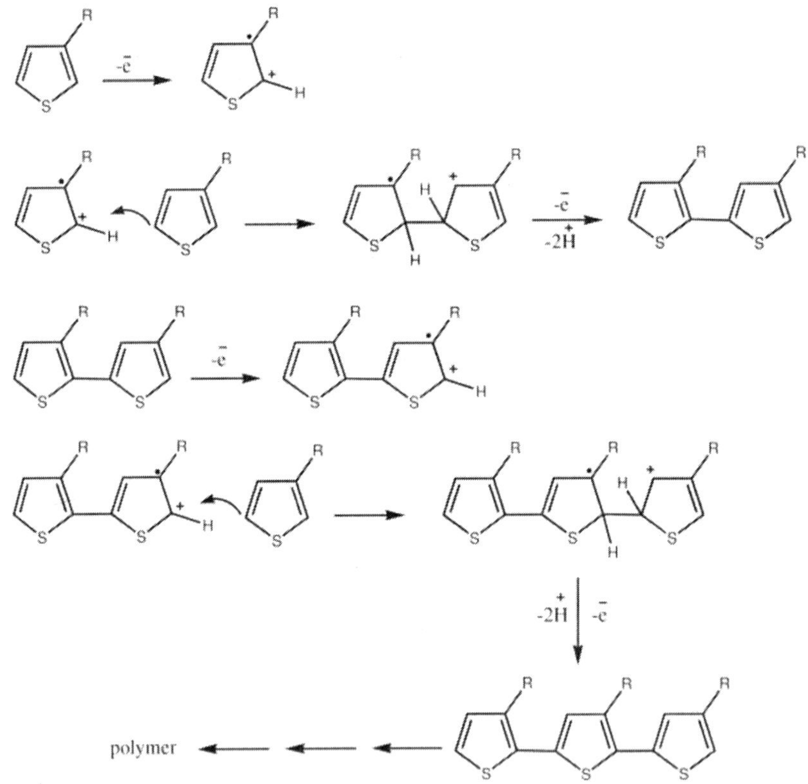

Figure 1 POLYMERIZATION OF THIOPHENE (WHERE R = H OR A SUBSTITUENT).

The objective of the present research is to study electronic and structural properties of all chlorothiophenes (shown in Figure 2) and their radical cations using DFT-B3LYP and DFT-UB3LYP methods.

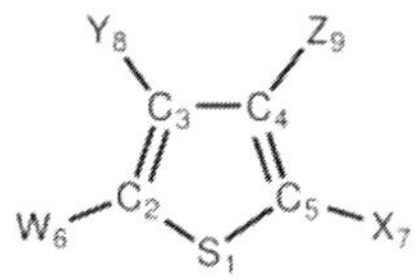

T: W=X=Y=Z=H	thiophene
A: W=Cl, X=Y=Z=H	2-chlorothiophene
B: Y=Cl, W=X=Z=H	3-chlorothiophene
C: W=Y=Cl, X=Z=H	2,3-dichlorothiophene
D: W=Z=Cl, X=Y=H	2,4-dichlorothiophene
E: W=X=Cl, Y=Z=H	2,5-dichlorothiophene
F: Y=Z=Cl, W=X=H	3,4-dichlorothiophene
G: W=Y=Z=Cl, X=H	2,3,4-trichlorothiophene
H: W=X=Y=Cl, Z=H	2,3,5-trichlorothiophene
I: W=X=Y=Z=Cl	2,3,4,5-tetrachlorothiophene

Figure 2 ALL POSSIBLE MONO-, DI-, TRI- AND TETRACHLOROTHIOPHENES STUDIED IN THIS WORK.

The electropolymerization and characteristics of monomers (conductivity and solubility) are related to the stability of their radical cations which it is closely dependent on the type and configuration of the substituent on the thiophene ring [22]. Vibrational frequencies with analyses of their IR intensities and NMR shielding constants will be used to characterize and predict the molecular and spectroscopic properties of fluorothiophenes.

COMPUTATIONAL PROCEDURES

First of all, structures of representative mono-, di-, tri- and tetrachlorothiophenes were optimized using B3LYP/6-311++G** method [30–32]. The geometries of these compounds were optimized using standard gradient techniques with default parameters set in G98W. Vibrational frequencies with their IR intensities are calculated with scale factors, $f = 1.0100$ for low-frequency vibrations (below 1000 cm^{-1}) and $f = 0.9679$ for high-frequency vibrations (above 1000 cm^{-1}) [33]. The vibrational frequencies analysis showed that all structures correspond to local minima in potential energy surface.

RESULTS AND DISCUSSION

Structural Analysis of Chlorothiophenes

In this section, the electrical and structural properties of chlorothiophenes and their related radical cations have been studied. Because of the stabilizing effects of the delocalization of non-bonding *p* electrons of sulfur atom, it is expected that the thiophene molecules possess a plane structure.

In agreement with this expectation, results of the present calculations show that all chlorothiophenes possess planar structures (the optimized values of all dihedral angles of all structures are either 0° or 180°). The optimized structural parameters obtained at B3LYP/6-311++G** level of theory for all chlorothiophenes and UB3LYP/6-311++G** level of theory for all radical cations are listed in Tables 1 and 2.

Table 1 B3LYP/6-311++G** optimized values of bond lengths (Å) and the F_n coefficient for Thiophene and Chlorothiophenes

Molecule	S1-C2	S1-C5	C2-C3	C3-C4	C4-C5	Fn
Thiophene	1.733	1.733	1.366	1.428	1.366	1.045
	1.728	*1.728*	*1.422*	*1.376*	*1.421*	*0.968*
A	1.740	1.737	1.364	1.428	1.364	1.047
	1.760	*1.715*	*1.419*	*1.378*	*1.414*	*0.973*
B	1.732	1.733	1.363	1.424	1.364	**1.044**
	1.681	*1.780*	*1.437*	*1.341*	*1.366*	***0.957***
C	1.741	1.735	1.367	1.427	1.362	1.046
	1.732	*1.743*	*1.445*	*1.395*	*1.383*	*0.987*
D	1.740	1.737	1.362	1.427	1.361	1.048
	1.782	*1.700*	*1.385*	*1.386*	*1.431*	*0.984*
E	1.745	1.745	1.362	1.429	1.361	1.050
	1.750	*1.750*	*1.410*	*1.379*	*1.410*	*0.978*
F	1.730	1.730	1.363	1.433	1.363	1.051
	1.733	*1.734*	*1.413*	*1.387*	*1.412*	*0.982*
G	1.738	1.734	1.367	1.435	1.360	1.052
	1.754	*1.728*	*1.415*	*1.386*	*1.411*	*0.981*
H	1.747	1.744	1.365	1.429	1.360	1.049
	1.734	*1.768*	*1.430*	*1.383*	*1.390*	*0.981*
I	1.745	1.743	1.364	1.437	1.364	1.053
	1.752	*1.752*	*1.413*	*1.388*	*1.413*	*0.982*

Corresponding values for the related radical cations optimized with UB3LYP/6-311++G** are given in the lower row with italic fonts. See Figure 2 for definitions of bond lengths.

Table 2 B3LYP/6-311++G** optimized values of bond angles for Thiophene and Chlorothiophenes

Molecule	C2-S1-C5	S1-C2-W	S1-C5-X	S1-C2-C3	S1-C5-C4	C3-C4-C5	C2-C3-Y	C5-C4-Z
Thiophene	91.5	119.9	119.9	111.5	111.5	112.7	123.4	123.4
	89.7	*120.1*	*120.1*	*113.3*	*113.3*	*111.9*	*122.6*	*122.6*
A	90.7	120.5	119.6	112.5	111.8	111.8	123.4	123.3
	88.0	*120.0*	*119.4*	*114.9*	*114.2*	*112.8*	*122.7*	*122.3*
B	91.7	121.1	119.9	110.6	111.9	111.6	123.4	124.6
	90.6	*123.1*	*118.8*	*111.4*	*113.4*	*110.0*	*120.7*	*125.2*
C	91.0	120.9	120.8	119.6	112.2	112.3	124.2	124.5
	88.2	*122.9*	*118.2*	*113.4*	*115.4*	*110.7*	*120.6*	*124.5*
D	90.9	120.4	120.4	112.8	110.8	114.6	124.5	123.4
	88.5	*118.4*	*121.9*	*115.5*	*112.4*	*115.1*	*125.0*	*120.4*
E	89.9	120.1	120.0	112.7	112.7	112.4	123.4	123.4
	86.3	*119.4*	*119.4*	*115.8*	*115.8*	*111.1*	*122.5*	*122.5*
F	91.9	120.9	120.9	111.3	111.3	112.8	123.6	123.6
	89.9	*122.9*	*121.5*	*113.0*	*112.9*	*112.1*	*123.1*	*123.2*
G	91.3	120.6	120.5	112.1	111.5	113.4	124.4	123.5
	88.3	*121.2*	*120.7*	*114.5*	*114.1*	*112.9*	*123.2*	*122.8*
H	90.3	120.1	120.4	113.0	111.6	113.5	124.5	124.2
	86.7	*121.8*	*118.0*	*114.1*	*116.8*	*109.4*	*120.9*	*124.5*
I	90.6	120.1	120.1	112.2	112.2	112.5	124.2	124.2
	86.9	*120.4*	*120.4*	*115.4*	*115.4*	*111.2*	*123.2*	*123.2*

Corresponding values for the related radical cations are given in the lower row with italic fonts. See Figure 2 for definitions of angles.

To predict the bonding-characteristic behavior of chlorothiophene rings in their corresponding polymer chains and to determine the extent of the π-conjugation character of these polymers, we have used the F_n coefficient defined in Equation 1 for each thiophene ring [34].

$$F_n = \frac{R34}{(R23 + R45)/2} \tag{1}$$

where R34 is the length of C_3-C_4 bond (formally single bond) and [(R23 + R45)/2] is the average length of C_2-C_3 and C_4-C_5 bonds (formally double bonds) in the thiophene ring. The calculated values of the F_n coefficient for thiophene, all chlorothiophenes and their radical cations are listed in Table 1. As shown in this table, F_n coefficient has the following trend for all molecules:

B < Thiophene < C < A < D < H < E < F < G < I

And for all molecular radical cations, values of the F_n coefficient has the following trend:

B < Thiophene < A < E < G = H < F = I < D < C

As this table shows, 3-chlorothiophene has the smallest value of the F_n coefficient. Therefore, it can be suggested that double bonds in 3-chlorothiophene are more delocalized.

Spin-Distribution Analysis

Electropolymerization process for each molecular radical cation is obtained from α-α' (**C_2-C_5**) linkages [29], so it can be said that in a compound, the greater spin density in Positions **C_3** and **C_4** is, the higher electropolymerization rate will be [35].

The spin-density distribution over the ring atoms of chlorothiophene radical cations have been calculated and presented in Table 3. As shown in this table, it is obvious that monomers **B** and **F** have the maximal spin density on positions **C_3** and **C_4**. This means that the electropolymerization of 3-chlorothiophene (**B**) and 3,4-dichlorothiophene (**F**) can be more favorable ratio with respect to other molecules.

Table 3. Distribution of spin density over sulfur and carbon atoms in UB3LYP/6-311++G** optimized structures of Thiophene and Chlorothiophenes radical cations

Molecule	S_1	C_2	C_3	C_4	C_5
Thiophene	-0.131	0.547	0.046	0.046	0.548
A	-0.087	0.377	0.055	0.049	0.456
B	0.041	0.356	**0.289**	-0.179	0.357
C	-0.081	0.358	0.186	-0.108	0.399
D	0.003	0.315	-0.124	0.162	0.425
E	-0.088	0.343	0.061	0.061	0.343
F	0.419	-0.110	**0.236**	0.240	-0.109
G	-0.065	0.384	-0.044	0.014	0.470
H	-0.062	0.349	0.133	-0.064	0.313
I	-0.067	0.362	-0.006	-0.005	0.361

Electric Dipole Moments

Orientations of the polymer chains in the condensed phase are the most important parameters affecting the electric charge transport properties of polymers so that the electrical conductivity of a polymer chain is altered when its orientation and consequently its symmetry and non isotropic interactions are changed.

Chemically synthesized PTHs powders exhibit a partially crystalline structure in which the chains align and pack in a herringbone arrangement (Figure 3) in the crystalline regions [36].

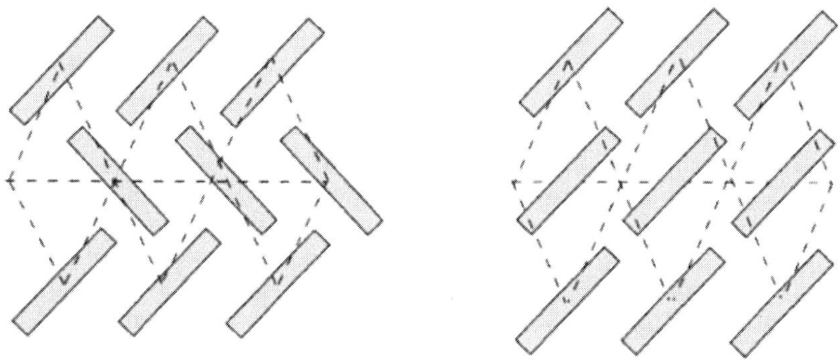

(a) (b)

Figure 3. ILLUSTRATION OF STACKING ARRANGEMENTS PROPOSED FOR POLYTHIOPHENE; (A) HERRINGBONE STACKING ARRANGEMENT, (B) LAMELLAR STACKING ARRANGEMENT.

These overall interactions can be well expressed in terms of interactions between local dipole moments of monomers. The local alignment of monomers' dipole moments in solution with respect to the orientation of the polymer chain determines the electrochemical properties of the polymer formed on electrode surface.

In addition, the orientation of monomer in the double layer of the solution in an electropolymerization cell depends on both size and direction of the dipole moment vectors of both monomers and polymer chain [35].

It is necessary to study the dipole moments of these chlorothiophenes to be able to predict their electropolymerization properties. The size of the dipole moment vector and its components calculated for chlorothiophenes in this study are presented in Table 4.

Table 4 Electric dipole moments, polarizability tensor elements, and ionization potentials calculated at B3LYP/6-311++G** level of theory for the optimized structures of Thiophene and Chlorothiophenes

Molecule	Electric dipole moment (D)			Polarizabilities (Å)				Ionization potentials (ev)
	μ_x	μ_y	μ_{tot}	α_{xx}	α_{yy}	α_{zz}	α_{xy}	IP
Thiophene	0.0000	-0.5169	0.5169	67.093	73.851	41.017	0.012	8.75
A	1.2229	-1.1156	1.6553	83.370	90.474	45.513	0.424	8.93
B	0.4885	-1.3293	1.4162	73.843	99.340	45.713	0.914	**8.77**
C	2.0522	-0.6022	**2.1388**	105.378	99.760	50.761	1.770	8.56
D	-0.0085	-0.6768	0.6768	118.479	89.726	50.349	5.704	8.58
E	-0.0014	-1.1668	1.1668	124.387	85.513	50.238	0.006	8.38
F	1.9947	-0.3062	**2.0181**	113.075	91.086	50.781	-3.497	**9.00**
G	0.2028	-1.8053	1.8166	123.767	114.854	56.216	6.267	8.66
H	-0.3414	-0.6487	0.7331	117.232	125.636	55.339	14.351	8.44
I	0.7756	-0.6432	1.0077	135.731	138.834	61.933	7.424	8.50

μ_z, α_{xz}, and α_{yz} are essentially zero for all compounds because of the planar structure.

μ_z, α_{xz}, and α_{yz} are essentially zero for all compounds because of the planar structure

An analysis of calculated dipole moments shows that the orientation of the dipole moment vector is toward sulfur atom for all chlorothiophenes. The size and direction of the dipole moment vector is more dependent on the position (symmetry) of substituent than its number. It can also be seen in Table 4 that the dipole moments for all compounds have the following trend:

C > **F** > G > A > B > E > I > H > D > Thiophene

The size of the dipole moment vector for monomers **C** and **F** is greater than other compounds. Therefore, compounds **C** and **F** have a high solubility in polar solvents. These characteristics increase the efficiency of electrochemical polymerization processes of these monomers. Moreover it can be seen in Table 4 that thiophene has a low solubility in polar solvents.

Ionization Potentials

The first step in the electropolymerization of conducting polymers is formation of intermediate radical cations from monomers which can be considered as an ionization reaction [37].

$$M \rightarrow M^{+} + \bar{e} \quad (2)$$

where M and $M^{\cdot+}$ denote monomer and its radical cation respectively. So, stability of intermediate radical cations has an important role in the electropolymerization process.

Therefore, it is useful to calculate electronic energy deference between the neutral monomer (as the initial species) and positively charged monomer (as the intermediate). This energy difference is proportional to Ionization Potential (IP) of the monomer. Calculated values of IP for chlorothiophenes based on Equation 2 are presented in Table 4. The trend of electrochemical stability of these monomers is:

$$\mathbf{F} > \mathbf{B} > \text{Thiophene} > G > D > C > A > I > H > E$$

Since ionization potential energies are directly proportional to the electrochemical oxidation potentials of the compounds, it can be said that the electrochemical stability of monomers **F** and **B** are greater than other compounds.

Electric Polarizabilities

Values of Polarizabilities of compounds containing halogen are used to interpret light scattering and intensities of vibrational Raman spectra [34]. Values of exact electrical polarizabilities for all chlorothiophenes have been calculated and listed in Table 4. Based on this table, it can be claimed that with an increase in the number of chlorine atoms on the thiophene ring, polarization values will increase as well.

The zero values obtained for polarizabilities $\mathbf{A}_{xz}$, and $\mathbf{A}_{yz}$ are consequences of the planar structure. The small values of $\mathbf{A}_{xy}$ polarizabilities indicate that the delocalization strength of the π-system of the ring dominates the anisotropic effects of substituent for asymmetric chlorothiophenes.

Vibrational Spectra

The vibrational frequencies for all chlorothiophenes were calculated using optimized structures at B3LYP/6-311++G** level of theory. Calculated IR absorption frequencies and intensities for all chlorothiophenes are presented in Table 5. To facilitate the comparison the frequencies are divided into low-frequency (below 1000 cm^{-1}) and high-frequency groups (above 1000 cm^{-1}). According to analysis of frequencies listed in Table5, it can be suggested that thermal decomposition has the following trend:

Table 5. B3LYP/6-311++G** calculated IR absorption frequencies in cm^{-1} (intensities in km/mol) for Thiophene and Chlorothiophenes

Molecule	Low - frequency range	High - frequency range
Thiophene	**460(0.67) 577 (0.0) 619(0.37) 697(0.0) 727(143) 751(0.41) 842(26.9) 886(1.63) 894(0.30) 928(0.0)**	1017(3.15) 1069(4.45) 1070(0.0) 1234(11.37) 1348(0.71) 1397 (11.2) 1501(0.24) 3093(3.85) 3107(3.7) 3139(0.04) 3142(1.20)
A	208(0.72) 264(0.13) 438(4.5) 471(4.17) 566(0.36) 655(5.28) 655(83.78) 854 (20.46) 920(0.13) 969(50.73)	1033(11.23) 1075(1.96) 1206(9.34) 1329(2.19) 1406(48.15) 1512(8.91) 3100(3.84) 3120(0.21) 3143(0.78)
B	210(0.51) 289(0.11) 426(4.34) 462(0.0) 611(15.0) 642(4.30) 682(3.5) 783 (84.55) 812(1.54) 862(24.13) 886(0.01) 912(44.03)	1061(7.04) 1184(15.57) 1184(8.76) 1342(24.13) 1395(1.59) 1499(29.97) 3119(0.42) 3145(0.39) 3155(3.64)
C	158(0.12) 175(0.04) 240(1.51) 317(1.12) 418(1.45) 476(1.16) 497(5.84) 608 (14.78) 680(6.04) 717(47.74) 815(0.53) 892(64.40) 897(1.62) 991(40.60)	1078(0.73) 1144(15.97) 1327(16.33) 1405(27.92) 1508(27.40) 3118(0.85) 3144(1.22)
D	159(0.05) 189(0.27) 247(0.01) 327(0.52) 393(1.20) 466(6.33) 471(1.12) 505 (11.54) 687(9.9) 732(32.32) 833(35.74) 834(23.82) 883(42.25) 978(38.80)	1066(12.21) 1158(10.25) 1313(31.22) 1409(33.17) 1506(70.38) 3132(1.66) 3154(6.58)
E	117(0.01) 194(0.10) 273(0.0) 318(0.73) 362(1.33) 486(14.08) 527(19.39) 552 (0.0) 672(1.37) 728(3.32) 799(46.01) 884(0.0) 982(7.20) 985(140.11)	1048(0.42) 1189(5.80) 1196(1) 1419(33.77) 1521(38.74) 3111 (0.94) 3122(0.14)
F	170(0.0) 179(0.48) 242(1.90) 335(0.14) 454(4.75) 458(0.65) 467(1.14) 633(0.0) 692(20.02) 707(0.0) 800(67.63) 860(1.4) 863(24.2) 967(62.3)	1104(19.3) 1142(0.1) 1318(58.4) 1491(7.9) 1491(13.2) 3154 (6.8) 3155(3.10)
G	94(0.04) 179(0.02) 187(0.11) 237(1.3) 241(0.19) 363(1.3) 367(0.57) 456(3.83) 481(0.14) 541(3) 658(2.6) 732(33.4) 775(46.6) 856(3.9) 928(71.1)	1009(3.9) 1120(16.22) 1297(79.65) 1410(17.9) 1498(37.1) 3155 (7.6)
H	115(0.0) 178(0.0) 173(0.05) 187(0.0) 279(0.15) 355(1.5) 364(1.5) 420(1.8) 493 (8.6) 582(21) 505(1.8) 695(3.1) 831(20.1) 853(61.7) 972(4.15) 999(85.6)	1123(14.1) 1289(20.6) 1421(13.9) 1516(86) 3130(4.67)
I	88(0.0) 114(0.02) 177(0.0) 181(0.05) 191(0.09) 236(0.91) 271(0.0) 489(4.6) 363 (0.14) 364(1.4) 403(2.6) 456(3.17) 514(2.15) 657(0.0) 671(22.11) 809(45.32) 896(61.9) 972(168)	1025(28) 1270(87) 1424(1.24) 1505(75)

Table 6 Zero-Point vibrational energy in kcal/mol, calculated at B3LYP/6-311++G** level of theory for the optimized structures of Thiophene and Chlorothiophenes

Molecule	ZPE
Thiophene	**41.64**
A	35.85
B	35.81
C	30.01
D	29.92
E	29.97
F	29.93
G	24.07
H	24.09
I	18.20

Thiophene > B > A > F > D > C > E > H > G > I

Thiophene and 3-chlorothiophene have the highest low-frequency values among all these compounds; this means that thiophene and 3-

chlorothiophene have the largest force constants for their bending modes of vibration. Higher values of low frequencies obtained for thiophene and 3-chlorothiophene can be regarded as higher protection from thermal decomposition for these molecules.

The calculated Zero-Point vibrational Energy (ZPE) for thiophene and all chlorothiophenes are listed in Table 6. Having in mind the contents of the table it can be said that ZPE for thiophene is greater than other compounds, as a result the relative thermal stability of thiophene is higher than other compounds.

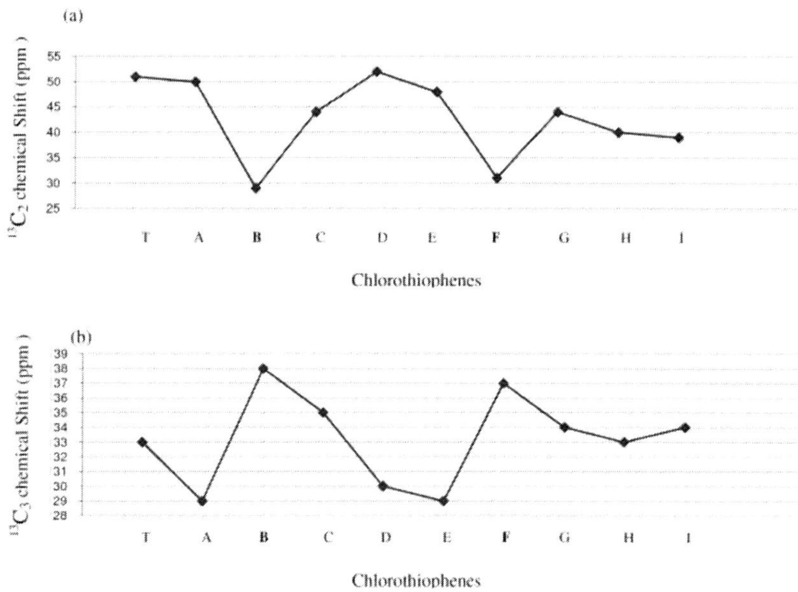

Figure 4. GIAO CALCULATED VALUES OF THE CHEMICAL SHIFTS FOR THIOPHENE AND CHLOROTHIOPHENES (REFERENCED TO TMS); (A) $^{13}C_2$ CHEMICAL SHIFTS, (B) $^{13}C_3$ CHEMICAL SHIFTS.

NMR CHEMICAL SHIFTS

The NMR chemical shifts can be used to predict ring current and estimate aromaticity of all chlorothiophenes [38–40]. ^{13}C chemical shielding have been calculated for DFT-B3LYP/6-311++G** optimized geometries of all chlorothiophenes. The isotropic and anisotropic magnetic shielding have been calculated using GIAO and CSGT methods [41–43].

Values of chemical shifts in Positions C_2 and C_3 for all chlorothiophenes are calculated and presented in Figure 4. It can be seen from this figure that compounds **B** and **F** have the lowest chemical shifts in Position C_2, these molecules have the highest chemical shifts in Position C_3, as well. It should be noted that these compounds had the highest values of spin-density in their Position C_3.

NICS ANALYSIS

The Nuclear Independent Chemical Shift (NICS) at GIAO/B3LYP/6-311++G** and magnetic susceptibilities at CSGT/B3LYP/6-311++G** level of theory [44–49] in center, 0.5 and 1 angstrom upper than thiophene ring have been calculated and listed in Table 7. Negative values of NICS represent the magnetic shielding and electric current in cyclic compounds. NICS results showed that compounds **B** and **F** have the highest NICS values in rings center, 0.5 and 1 angstrom upper than thiophene ring; these compounds have the largest electrical current and most conductivity among all other compounds. Based on this table, it can be claimed that the magnetic susceptibilities of these molecules increase with the number of the chlorine atoms, as well.

Table 7. Calculated NICS values (ppm) at GIAO/B3LYP/6-311++G** and magnetic susceptibilities (cgs-ppm) at CSGT/B3LYP/6-311++G** for Thiophene and Chlorothiophenes

Molecule	NICS(0)	NICS(0.5)	NICS(1)	χ
Thiophene	-11.91	-11.49	-9.28	-51.86
A	-12.30	-11.54	-9.05	-68.66
B	**-13.49**	**-12.22**	**-9.57**	-69.26
C	-12.57	-11.45	-8.67	-86.79
D	-12.77	-11.50	-8.56	-85.45
E	-12.13	-10.91	-8.19	-85.24
F	**-13.51**	**-12.12**	**-9.31**	-86.81
G	- 13.48	-11.87	-8.60	-104.38
H	-12.48	-10.90	-7.81	-103.01
I	-13.31	-11.38	-7.93	-123.23

Conclusion

B3LYP method with 6-311++G** basis set calculations have been carried out successfully to study the structural, electric and spectroscopic properties of all chlorothiophenes. Values of dipole moment, spin density distribution, ionization potential, vibrational frequencies and NICS have been calculated and analyzed. The analysis of the data showed that 3-chlorothiophene has the smallest values of the F_n coefficient. Therefore, it can be suggested that double bonds in 3-chlorothiophene are more delocalized. Electrochemical stability of 3-chlorothiophene and 3,4-dichlorothiophene is greater than other chlorothiophenes. Thermal stability and Zero-Point Energy of thiophene are greater than all chlorothiophenes. The 3-chlorothiophene and 3, 4-dichlorothiophene possesses the largest NICS value; these compounds have the largest electrical current in the thiophene ring. Accordingly, compound **B** (3-chlorothiophene) is the best candidate monomer among all chlorothiophenes for the synthesis of corresponding conducting polymers with modified characteristics. The synthesis feasibility, experimental measurement and thermo chemical stability of these chlorothiophenes are interesting subjects for study.

DECLARATIONS

Acknowledgements

This research was supported by the Islamic Azad University, Toyserkan Branch, Toyserkan, Iran.

Competing interests

The authors declare that they have no competing interests.

AUTHORS' CONTRIBUTIONS

SJ-B proposed and supervised the program, and participated in data analysis and draft reversion, coordinated the study and modified the text. HS participated in data collection and manuscript drafting, completed the molecular design, the arrangement of the work and modification of the manuscript. SJ-B and HS participated in the separation and purification of compounds, acquisition of data, and collection of literature and helped to draft the manuscript. Authors read and approved the final manuscript.

REFERENCES

1. Jameh-Bozorghi S, Shirani ILH: THEORETICAL STUDY ON THE ELECTRONIC, STRUCTURAL, PROPERTIES AND REACTIVITY OF A SERIES OF MONO-, DI-, TRI- AND TETRAFLUOROTHIOPHENES AS WELL AS CORRESPONDING RADICAL CATION FORMS. J Fluor Chem 2011, in press.
2. MacDiarmid AG: SYNTHETIC METALS: A NOVEL ROLE FOR ORGANIC POLYMERS. Chem Int E 2001, 40:2581–2585.
3. Hall AW, Blackwood KM, Milne PE, Goodby JW: NOVEL UV CURED COATINGS AND ADHESIVES BASED ON THE PHOTOINITIATED CYCLOPOLYMERIZATION OF DERIVATIVES OF DIALLYLAMINE. Chem Commun 2003, 2530–2531.
4. Neves S, Gazotti WA, De Paoli M-A: CONDUCTING POLYMERIC NANOMATERIALS. In In Encyclopedia of Nanoscience and Nanotechnology. Volume 2. Edited by: Nalwa HS. Los Angeles: American Scientific Publishers; 2004:133–152.
5. Gangopadhyay R: CONDUCTING POLYMER NANOSTRUCTURES. In In Encyclopedia of Nanoscience and Nanotechnology. Volume 2. Edited by: Nalwa HS. Los Angeles: American Scientific Publishers; 2004:105–131.
6. Wallace GG, Innis PC, Kane-Maguire LAP: INHERENTLY CONDUCTING POLYMER NANOSTRUCTURES. In In Encyclopedia of Nanoscience and Nanotechnology. Volume 4. Edited by: Nalwa HS. Los Angeles: American Scientific Publishers; 2004:113–130.
7. Epstein AJ: INTRODUCTION TO ELECTRONIC POLYMERS: INFLUENCE OF NANOSTRUCTURE ON ELECTRONIC PHENOMENA. In In Organic Electronic Materials: Conjugated Polymers and Low Molecular Weight Organic Solids. Volume 41. Edited by: Farchioni R, Grosso G. Amsterdam: Springer; 2001:3–39.
8. Kolla HS, Surwade SP, Zhang X, MacDiarmid AG, Manohar SK: ABSOLUTE MOLECULAR WEIGHT OF POLYANILINE. J Am Chem Soc 2005, 127:16770–16771.
9. Richard SP, Rajasekhar M, Subramania A: SYNTHESIS OF POLYTHIOPHENE NANOPARTICLES BY SURFACTANT - ASSISTED DILUT POLYMERIZATION METHOD FOR HIGH PERFORMANCE REDOX SUPERCAPACITORS. Int J Electrochem Sci 2009, 4:1289–1301.
10. Bai H, Shi G: GAS SENSORS BASED ON CONDUCTING POLYMERS. Sensors 2007, 7:267–307.

11. Henry D, Tran YW, Julio MD, Richard BK: TOWARD AN UNDERSTANDING OF THE FORMATION OF CONDUCTING POLYMER NANOFIBERS. ACS Nano 2008, 2:1841–1848.

12. Ohkawa S, Ohta R, Kawabata K, Goto H: POLYMERIZATION IN LIQUID CRYSTAL MEDIUM: PREPARATION OF POLYTHIOPHENE DERIVATIVES BEARING A BULKY PYRIMIDINE SUBSTITUENT. Polymers 2010, 2:393–406.

13. Viinikanoja A, Lukkari J, Aaritalo T, Laiho T, Kankare J: PHOSPHONIC ACID DERIVATIZED POLYTHIOPHENE: A BUILDING BLOCK FOR METAL PHOSPHONATE AND POLYELECTROLYTE MULTILAYERS. Langmuir 2003, 19:2768–2775.

14. Grevin B, Rannou P: ELECTROCHEMISTRY: ARRAYS OF POLYMER NANOWIRES. Nature Materials 2004, 3:503–504.

15. Nishinaga T, Yamazaki D, Tateno M, Iyoda M, Komatsu K: SYNTHESIS AND PROPERTIES OF OCTITHIOPHENE DICATION STERICALLY SEGREGATED BY ANNELATION WITH BICYCLE [2.2.2]OCTENE UNITS. Materials 2010, 3:2037–2052.

16. Lipton-Duffin JA, Miwa JA, Kondratenko M, Cicoira FB, Sumpter G, Meunier V, Perepichka DF, Rosei F: STEP-BY-STEP GROWTH OF EPITAXIALLY ALIGNED POLYTHIOPHENE BY SURFACE-CONFINED REACTION. PNAS 2010, 107:11200–11204.

17. Ramanathan K, Bangar M, Yun M, Chen W, Myung N, Mulchandani A: BIOAFFINITY SENSING USING BIOLOGICALLY FUNCTIONALIZED CONDUCTING-POLYMER NANOWIRE. J Am Chem Soc 2005, 127:496–497.

18. Wayne E, Jones Jr, Chiguma J, Johnson E, Pachamuthu A, Santos D: ELECTRICALLY AND THERMALLY CONDUCTING NANOCOMPOSITES FOR ELECTRONIC APPLICATIONS. Materials 2010, 3:1478–1496.

19. Wan MX: CONDUCTING POLYMER NANOTUBES. In In Encyclopedia of Nanoscience and Nanotechnology. Volume 2. Edited by: Nalwa HS. Los Angeles, American Scientific Publishers; 2004:153–169.

20. Maddalena F, Meijer EJ, Asadi K, Leeuw DM, Blom PWM: DOPING KINETICS OF ORGANIC SEMICONDUCTORS INVESTIGATED BY FIELD-EFFECT TRANSISTORS. Appl Phys Lett 2010, 97:0433021–0433023.

21. Yang L, Qi Y, Yuan X, Shen J, Kim J: DIRECT SYNTHESIS, CHARACTERIZATION AND CATALYTIC APPLICATION OF SBA-15 CONTAINING HETEROPOLYACID. Journal of Molecular Catalysis A: Chemical 2005, 229:199–205.

22. Wallace GG, Spinks GM, Kane-Maguire LAP, Teasdale PR: SYNTHESIS AND PROPERTIES OF POLYTHIOPHENES. In In Conductive electroactive polymers: intelligent polymer systems. Queensland: CRC Press; 2009:197–226.

23. Lukkari J, Viinikanoja A, Paukkunen J, Salomaki M, Mervi Janhonen, Aaritalo T, Kankare J: OXIDATION INDUCED VARIATION IN POLYELECTROLYTE MULTILAYERS PREPARED FROM SULFONATED SELF-DOPABLE POLY(ALKOXYTHIOPHENE). Chem Commun 2000, 571–572.

24. Asami R, Fuchigami T, Atobe M: DEVELOPMENT OF A NOVEL ENVIRONMENTALLY FRIENDLY ELECTROPOLYMERIZATION OF WATER-INSOLUBLE MONOMERS IN AQUEOUS ELECTROLYTES USING ACOUSTIC EMULSIFICATION. Langmuir 2006, 22:0258–10263.

25. Chen J, David LO, Jennifer MP, Douglas RMF, Chee OT, Gordon GW: PHOTOELECTROCHEMICAL SOLAR CELLS BASED ON POLYTERTHIOPHENES CONTAINING PORPHYRINS USING IONIC LIQUID ELECTROLYTE. Electrochemical and Solid-State Letters 2005, 8:A528-A530.

26. Silvestri F, Marrocchi A: ACETYLENE-BASED MATERIALS IN ORGANIC PHOTOVOLTAICS. Int J Mol Sci 2010, 11:1471–1508.

27. Oterom TF, Azelain EL: SYNTHESIS AND PROPERTIES OF POLYTHIOPHENES. Polym Comm 1988, 29:21–50.

28. Alvaro AA, Rosa LT: STUDY OF KINETIC FORMATION AND THE ELECTROCHEMICAL BEHAVIOR OF POLYPYRROLE FILMS. J Chil Chem Soc 2009, 54:14–19.

29. Wallace GG, Spinks GM, Kane-Maguire LAP, Teasdale PR: In Conductive electroactive polymers: intelligent polymer systems. Queensland: CRC Press; 2009:77–107.

30. Frisch MJ, Trucks GW, Schlegel HB, Scuseria GE, Robb MA, Cheeseman JR, Zakrazewski VG, Montgomery JA, Startmann RE, Burant JC, Dapprich S, Millam JM, Daniels AD, Kudin KN, Strain MC, Farkas O, Tomasi J, Barone V, Cossi M, Cammi R, Mennucci B, Adamo C, Clifford S, Ochterski J, Petersson GA, Ayala PY, Morokuma QCK, Malik DK, Rabuck AD, Raghavachar K, et al.: GAUSSIAN 98 (Revision A3), Gaussian Inc. Pittsburgh: PA; 1998.

31. Gross EK, Dreiz RM: Density Functional Theory. New York: Plenum Press; 1995.

32. Carlier P, Deora N, Lam P, Harmon J, Williams L: [http://www.gaussian.com] Carlier group gaussian user manual. 2010. Version 1.22

33. Andersson MP, Uvdal P: NEW SCALE FACTORS FOR HARMONIC VIBRATIONAL FREQUENCIES USING THE B3LYP DENSITY FUNCTIONAL METHOD WITH THE TRIPLE- BASIS SET 6–311+G(D,P). J Phys Chem A 2005, 109:2937–2941.

34. Omrani A, Sabzyan H: THEORETICAL STUDY OF CHLOROPYRROLES AS MONOMERS FOR NEW CONDUCTIVE POLYMERS. J Phys Chem A 2005, 109:8874–8879.

35. Sabzyan H, Omrani A: AB INITIO AND DFT STUDY OF ALL MONO-, DI-, TRI- AND TETRAFLUOROPYRROLES AND THEIR CATIONS: PREDICTING STRUCTURAL, SPECTROSCOPIC, ELECTROPOLYMERIZATION AND ELECTROCHEMICAL PROPERTIES. J Phys Chem A 2003, 107:6476–6482.

36. Wallace GG, Spinks GM, Kane-Maguire LAP, Teasdale PR: PROCESSING AND DEVICE FABRICATION. In In Conductive electroactive polymers: intelligent polymer systems. Queensland: CRC Press; 2009:231–245.

37. Nikoofard H, Sabzyan H: DFT STUDY OF MOLECULAR STRUCTURE AND ELECTRONIC PROPERTIES OF FLUOROMETHYLPYRROLE OLIGOMERS INCLUDING DI-, TRI- AND TETRAMER. Journal of Fluorine Chemistry 2007, 128:668–673.

38. Kleinpeter E, Klod S: AB INITIO CALCULATION OF THE ANISOTROPY EFFECT OF MULTIPLE BONDS AND THE RING CURRENT EFFECT OF ARENES-APPLICATION IN CONFORMATIONAL AND CONFIGURATIONAL ANALYSIS. J Chem Soc Perkin Trans 2001, 2:1893–1898.

39. Martin NH, Caldwell BC, Carlson KP, Teague MR: AB INITIO CALCULATION OF THROUGH-SPACE MAGNETIC SHIELDING OF LINEAR POLYCYCLIC AROMATIC HYDROCARBONS (ACENES): EXTENT OF AROMATICITY. J Mol Graph Model 2009, 27:689–692.

40. Chen Z, Wannere CS, Corminboeuf C, Puchta R, Schleyer PVR: NUCLEUS-INDEPENDENT CHEMICAL SHIFTS (NICS) AS AN AROMATICITY CRITERION. Chem Rev 2005, 105:3842–3888.

41. Baldridge KK, Siegel JS: CORRELATION OF EMPIRICAL (TMS) AND ABSOLUTE NMR CHEMICAL SHIFTS PREDICTED BY AB INITIO COMPUTATIONS. J Phys Chem A 1999, 103:4038–4042.

42. Zhang Y, Wu A, Xu Xin, Yan Y: GEOMETRIC DEPENDENCE OF THE B3LYP-PREDICTED MAGNETIC SHIELDING AND CHEMICAL SHIFTS. J Phys Chem A 2007, 111:9431–9437.

43. GAUSSIAN 03 ONLINE MANUAL [http://www.nd.edu/~wschnei1/courses/CBE_547/g03_man/g_ur/k_nmr.htm]

44. Sabzyan H, Omrani A: AB INITIO AND DENSITY FUNCTIONAL STUDY OF ELECTRICAL AND THERMOCHEMICAL PROPERTIES OF MONO-, DI-, TRI- AND TETRAFLUOROPYRROLES AND THEIR CATIONS AND ANIONS. J THEOCHEM 2005, 713:43–49.

45. Schleyer PVR, Manoharan M, Wang Z-X, Kiran B, Jiao H, Puchta R, Hommes NJRVE: DISSECTED NUCLEUS-INDEPENDENT CHEMICAL SHIFT ANALYSIS OF Π-AROMATICITY AND ANTIAROMATICITY. Org Lett 2001, 3:2465–2468.

46. Corminboeuf C, Heine T, Seifert G, Schleyer PVR, Weber J: INDUCED MAGNETIC FIELDS IN AROMATIC [N]-ANNULENẼSINTERPRETATION OF NICS TENSOR COMPONENTS. Phys Chem Chem Phys 2004, 6:273–276.

47. Alkorta I, Rozas I, Elguero J: AN AB INITIO STUDY OF THE NMR PROPERTIES (ABSOLUTE SHIELDING AND NICS) OF A SERIES OF SIGNIFICANT AROMATIC AND ANTIAROMATIC COMPOUNDS. Tetrahedron 2001, 57:6043–6049.

48. Stanger A: NUCLEUS-INDEPENDENT CHEMICAL SHIFTS (NICS): DISTANCE DEPENDENCE AND REVISED CRITERIA FOR AROMATICITY AND ANTIAROMATICITY. J Org Chem 2006, 71:883–893.

49. Fallah-Bagher-Shaidaei H, Wannere CS, Corminboeuf C, Puchta R, Schleyer PVR: WHICH NICS AROMATICITY INDEX FOR PLANAR Π-RINGS IS BEST? Org Lett 2006, 8:863–866.

3

CHAPTER

FLEXIBLE CONDUCTING POLYMER/REDUCED GRAPHENE OXIDE FILMS: SYNTHESIS, CHARACTERIZATION, AND ELECTROCHEMICAL PERFORMANCE

Wenyao Yang, Yuetao Zhao, Xin He, Yan Chen, Jianhua Xu*, Shibin Li, Yajie Yang and Yadong Jiang

ABSTRACT

In this paper, we demonstrate the preparation of a flexible poly (3,4-ethylenedioxythiophene) -poly (styrenesulfonate)/reduced graphene oxide (PEDOT-PSS/RGO) film with a layered structure via a simple vacuum filtered method as a high performance electrochemical electrode. The PEDOT-PSS/RGO films are characterized by scanning electron microscopy (SEM), X-ray diffraction, Raman spectroscopy, and Fourier transform infrared (FT-IR) spectrometry. The results indicate that a layer-ordered structure is constructed in this nanocomposite during the vacuum filtering process. The electrochemical performances of the flexible films are characterized by electrochemical impedance spectroscopy, cyclic voltammetry, and galvanostatic charge/discharge. The results reveal that a

193.7 F/g highly specific capacitance of nanocomposite film is achieved at a current density of 500 mA/g. This flexible and self-supporting nanocomposite film exhibits excellent cycling stability, and the capacity retention is 90.6 % after 1000 cycles, which shows promising application as high-performance electrode materials for flexible energy-storage devices.

Keywords: Flexible PEDOT-PSS Reduced graphene oxide Supercapacitor

BACKGROUND

Supercapacitors, also called electrochemical capacitors, are one of the important energy-storage devices with fast charging/discharging, high power density, high energy density, and long cycle life, which could fill the gap between the conventional capacitors and the batteries [1, 2]. Conductive polymers can store charges not only through pseudocapacitances but also in the electrical double-layer capacitances (EDLCs). The pseudocapacitance is due to faradic charge transfer, while the EDLCs could store energy by an ion adsorption-desorption process in the double electrical layer at the electrode/electrolyte interfaces [3, 4]. As a result, the conductive polymer electrodes present higher specific capacitance than that of pure carbon-based capacitors with EDLCs [5]. One of the most significant problems is that pure conducting polymers are usually mechanically weak or brittle, leading to poor cycling stability during the long cycle charge/discharge process [6, 7]. Coupling conductive polymers to a carbon material has been proven to be an effective approach to overcome this problem [8, 9]. The reduced graphene oxide (RGO), which exhibits high performance of electrical and electrochemical activity combined with favorable mechanical strength, can be obtained by reducing graphene oxide (GO) through chemical treatment [10, 11]. The nanocomposites, based on conducting polymer and RGO, exhibit chemical stability and ideal capacitive behavior when they are used as electrode materials [12–14].

Due to the fast development of flexible electronics, the requirement for high-performance flexible energy-storage devices is bursting. Furthermore, a self-supporting and collector-free supercapacitor electrode with large energy density and great mechanical strength is an important component for a flexible energy-storage device. Recently, much attention has been focused on the conducting polymer poly (3,4-ethylenedioxythiophene)-poly (styrenesulfonate) (PEDOT-PSS), which was used as a flexible electrode due to high conductivity and flexible processability [15–17].

In this paper, we demonstrate the preparation of flexible and self-supporting PEDOT-PSS/RGO films as an electrochemical electrode via a simple vacuum filtration method. In the first step, the GO was prepared by a modified Hummer's method, and the RGO was obtained by a following hydrazine reduction treatment. Then, the PEDOT-PSS/RGO nanocomposites were prepared by an in situ solution polymerization. The PEDOT-PSS was polymerized on the surfaces of RGO sheets, and a sandwich-like PEDOT-PSS/RGO structure was formed. After that, the self-supporting nanocomposite films with a layer-by-layer structure were prepared by vacuum- filtered method. The PEDOT-PSS/RGO nanocomposite films showed better electrochemical properties compared with the pure RGO and PEDOT-PSS films. The combination of flexibility and electrochemical activity makes this self-supporting nanocomposite film attractive for applications in organic and flexible electrode materials.

METHODS

Materials

Three, 4-ethylenedioxythiophene (EDOT) was purchased from Bayer Company (Leverkusen, Germany). Graphite flakes (Grade 3061) and Polystyrenesulfonate (PSS 25 %) were bought from Sigma-Aldrich (St. Louis., MO, USA). Cation exchange resin (001 × 7) and anion exchange resin (201 × 7) were purchased from Bohong Company (Tianjin, China). Ammonium persulphate ($(NH4)_2S_2O_8$, 98 %), Mohr's salt ($(NH_4)_2Fe(SO_4)_2 \cdot 6H_2O$, 99 %) and ethanol ($C_2H_6O$, 99.5 %) were purchased from Kelong Chemicals (Chengdu, China).

Preparation of PEDOT-PSS/RGO self-supporting films

The RGO was synthesized from graphite flakes by the modified Hummer's method and hydrazine reduction method [18, 19]. Deionized water and ethanol-mixed solution (volume ratio 1:1) was used to improve the solubility of EDOT. 100 mg RGO was dispersed in 200-mL mixture solution by sonicating for at least 2 h. In situ polymerization solution of PEDOT-PSS/RGO was prepared by introducing 1.4 mmol PSS and 0.7 mmol EDOT into the 200-mL RGO dispersion. This in situ polymerization solution was vigorously stirred and kept at 5 °C. Then, 1.4 mmol ammonium persulphate and 0.0014 mmol Mohr's salt dissolved in 100 mL deionized water were added slowly while stirring overnight. Finally, the inorganic impurities in the final PEDOT-PSS/RGO dispersion were removed by 001 × 7 cation-exchange resin and 201 × 7 anion- exchange resin.

The PEDOT-PSS/RGO dispersion was sonicated for 2 h and vacuum filtered by PTFE filter membranes (pore size 0.22 μm, diam. 47 mm). Then the films were washed with ethanol and dried at 60 °C for 24 h in vacuum. After that, the dried films were stabilized at 235 °C in vacuum for 30 min and then peeled off from the filter to form self-supporting films.

The PEDOT-PSS/RGO self-supporting films synthesized from different mass ratios were labeled as PS/RG ratio. The PS/RG 1:1 indicates that the mass ratio of EDOT and RGO was 1:1. Meanwhile, the pure PEDOT-PSS and pure RGO self-supporting films were synthesized by a similar synthesis method mentioned above.

Characterization and electrochemical measurements

The morphologies of PEDOT-PSS/RGO films were investigated with S-4800 scanning electron microscopy (SEM). X-ray diffraction (XRD) patterns were performed on an X'Pert PRO diffractometer with Cu Ka radiation ($\lambda = 0.154$ nm). Raman spectra were characterized with a Horiba LabRAM HR system (633 nm laser). Fourier transform infrared (FT-IR) spectra were characterized with a ThermoElectron Nicolet 6700 spectrophotometer. A CHI660D electrochemistry workstation was used to characterize the electrochemical performance of the samples at ambient temperature. Electrochemical impedance spectroscopy (EIS), cyclic voltammetry (CV), and galvanostatic charge/discharge (GCD) were performed with a 1-mol/L Na_2SO_4 aqueous electrolyte. The platinum foils and an Ag/AgCl electrode were used as counter and reference electrodes, respectively. All the measurements were performed at ambient temperature.

RESULTS AND DISCUSSION

The SEM images of pure RGO and PEDOT-PSS/RGO nanocomposite films are shown in Fig. 1. The changes of the microstructures of the films provide important information to reveal the strengthening mechanism of the nanocomposite films. The SEM image in Fig. 1a indicates that the pure RGO film is composed of a large quantity of curved nanosheets. As PEDOT-PSS/RGO nanocomposite films (shown in Fig. 1b), a porous 3D network structure with a few folds is presented. It is clear that the RGO nanosheets are homogeneously coated by PEDOT-PSS, indicating that PEDOT-PSS was successfully polymerized on the surfaces of RGO nanosheets to form sandwich-like structures. The cross-section of PEDOT-PSS/RGO nanocomposite film (shown in Fig. 1c and d) present a layer-by-layer formation, which is probably caused by the flow-assembly effect of RGO sheets during filtration [20–22].

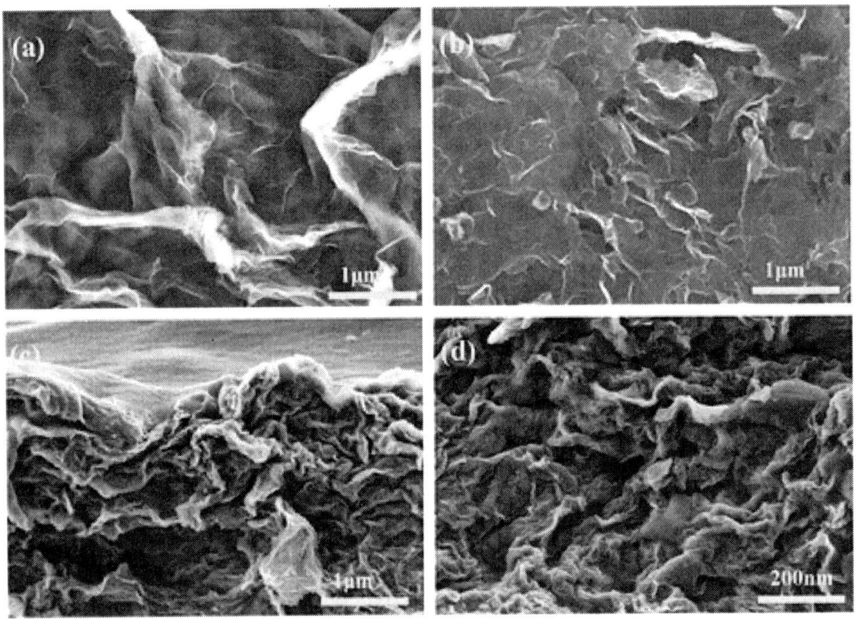

Figure 1. SEM images of flexible films as follows: (**A**) pure RGO, (**B**) PS/RG 1:1 and (**C, D**) cross-section images of PS/RG 1:1

Raman and FT-IR spectroscopy provide powerful tools to further investigate the microstructure of composites. Fig. 2a shows the Raman spectra of pure RGO and PEDOT-PSS/RGO composite films. The Raman spectrum of RGO displays two typical bands centered at 1347 and 1579 cm^{-1}, corresponding to the D-band and G-band, respectively. The D-band indicates the increasing of the sp^3 domains and the decrease of the in-plane sp^2 domains. The G-band, due to the E_{2g} mode, is closely related to the vibration of sp^2-bonded carbon atoms in a 2D graphene layer [23, 24]. The Raman band of PEDOT-PSS/RGO composite films (shown in Fig. 2b) at 985 cm^{-1} is assigned to oxyethylene ring deformation. The band at 1123 cm^{-1} is originated from C-O-C deformation. The bands at 1281 and 1363 cm^{-1} are originated from the C_α-C_α inter-ring and C_β-C_β stretch, respectively. The bands at 1432 and 1501 cm^{-1} are attributed to C = C symmetric stretch. The asymmetric C_α-C_β bond is evidenced by the presence of band at 1561 cm^{-1} [23, 25]. The series of bands indicates the successful formation of PEDOT-PSS/RGO nanocomposites.

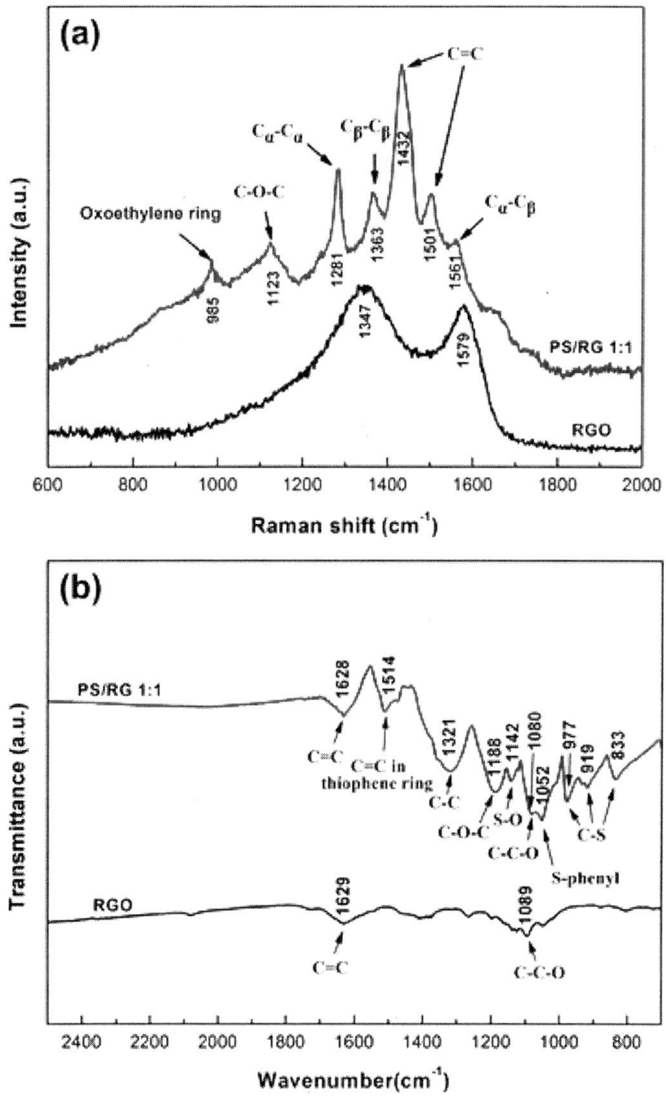

Figure 2. (**A**) Raman spectra of pure RGO and PS/RG 1:1 films; (**B**) FT-IR spectra of RGO, and PS/RG 1:1 films

Figure 2b presents the FT-IR spectra of pure RGO and PS/RG 1:1 composite films. The spectrum of RGO film displays two peaks about at 1629 and 1089 cm^{-1}, attributing to the aromatic C = C and oxygenated bond C-C-O, respectively [26]. We conclude that the oxygenated bond C-C-O arises from residual oxygen containing function groups in the RGO sheets, which improves the solubility of RGO. From the FT-IR spectrum of PS/RG 1:1, the spectrum contains peaks of C = C, C-C and C-O-C bonds in the thiophene ring at 1514, 1321, and 1188 cm^{-1} [8]. The C-S bond in the thiophene ring is evidenced by the presence of bands at about 977, 919 and 833 cm^{-1} [27]. Additionally, the bands at 1142 and 1052 cm^{-1} are attributed to the S-O bond and S-phenyl bond in PSS [28]. Moreover, the above-mentioned characteristic peaks of RGO are reflected in the spectrum of the PS/RG 1:1 film. However, the peak corresponding to the aromatic C = C and oxygenated bonds C-C-O are blue-shifted to 1628 and 1080 cm^{-1}, suggesting interactions between PEDOT-PSS backbone and RGO sheets [29]. Therefore, the series of bands in FT-IR spectra indicates the successful formation of PEDOT-PSS coatings on the surfaces of the RGO.

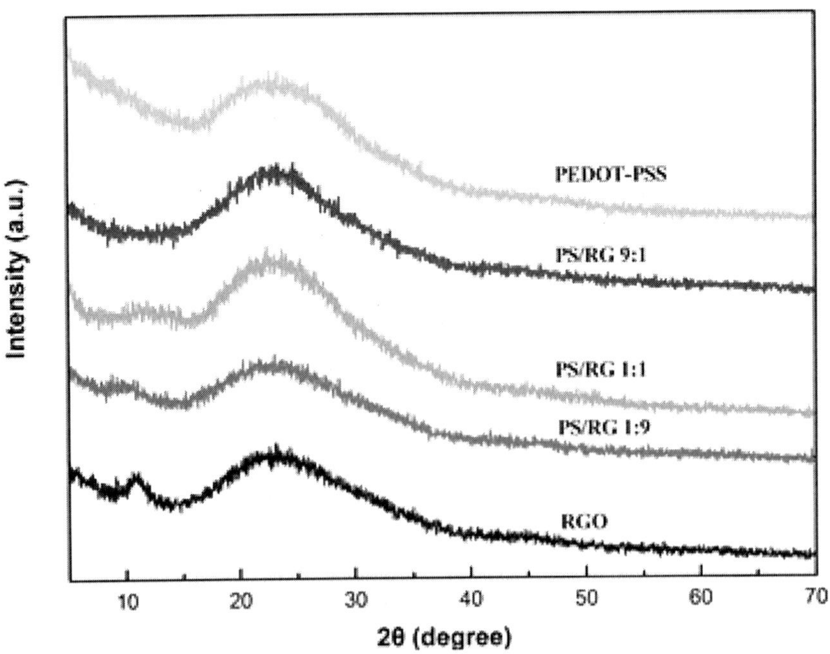

Figure 3. XRD patterns of RGO, PEDOT-PSS, and composite films

Figure 3 shows the XRD patterns of pure PEDOT-PSS, pure RGO, and PEDOT-PSS/RGO nanocomposite films with different RGO contents. The broad peak of PEDOT-PSS pattern ranging from 16.4 to 30.9 ° is attributed to PEDOT-PSS [30]. The pure RGO, PS/RG 1:9, and PS/RG 1:1 exhibits a small peak centered at 10.9 °. We conclude that this small peak arises from residual oxygen containing function groups in the RGO sheets, which improves the solubility of RGO. However, this small peak disappears in PS/RG 9:1 probably due to the low mass ratio of RGO. Compared with pure RGO, the XRD pattern of the PEDOT-PSS/RGO nanocomposites exhibits a broad peak ranging from 15 to 30 °, which is similar with RGO. The intensity of the peaks increases a little, which is ascribed to the reflection of PEDOT-PSS. Therefore, the XRD pattern further confirms the formation of PEDOT-PSS coatings on the surfaces of the RGO, which is consistent with that of the Raman investigations [31].

The electrochemical performances of PEDOT-PSS/RGO films were analyzed by using EIS, CV, GCD, and cycle-life tests. The Nyquist plots of pure PEDOT-PSS, pure RGO, and PEDOT-PSS/RGO nanocomposite films with different RGO contents are showed in Fig. 4. It can be seen that all the plots exhibit the capacitive-type behaviors. It is well known that the first interception of the plots at the Z' axis represents the equivalent series resistance (ESR), and the diameter of the semicircle is ascribed to the charge transfer resistance at the electrode/electrolyte interface [26]. As shown in Fig. 4, the ESR of the PEDOT-PSS, PS/RG 9:1, PS/RG 1:1, PS/RG 1:9, and RGO films are 2.56, 2.15, 1.62, 1.87, and 1.35 Ω, respectively. For the pure PEDOT-PSS and PS/RG 9:1 films, the radii of the semicircles in the Nyquist plot are larger than PS/RG 1:1, displaying a small semicircle characterization in high frequency range. This result indicates that the charge transfer resistance of PS/RG 1:1 film was lower than that of pure PEDOT-PSS and PS/RG 9:1 films. This might be due to the fact that the agglomeration of PEDOT-PSS films during the charge/discharge process decreased the effective surface area between the electrolyte and electrodes. Meanwhile, the inconspicuous semicircles at high frequency of PS/GR 1:9 and RGO films suggest that the interfacial charge transfer resistance is significantly low due to the high electrical conductivity of RGO [10]. The sloped portion closed to 45 ° in these plots at low frequency is typical of Warburg resistance, which is the result of the frequency dependence of ion transport or diffusion in the electrolyte [32, 33]. The Warburg region of RGO, PS/RG1:9, and PS/RG1:1 films are smaller than that of PEDOT-PSS or PS/RG 9:1 films, indicating that the shorter ion diffusion paths are formed during the charge/discharge process. Hence, we deduce that the more contents of RGO in nanocompoistes, the lower Warburg resistance is achieved in electrode materials. At the low frequency, all plots exhibit almost vertical straight lines, especially those films with a high mass ratio of RGO, indicating that these nanocomposite films exhibited an ideal capacitive behavior during the energy storage process. This result also reveals that a low-resistance interface was formed between the PEDOT-PSS and RGO, which is suitable for fast adsorption and desorption of solution ions.

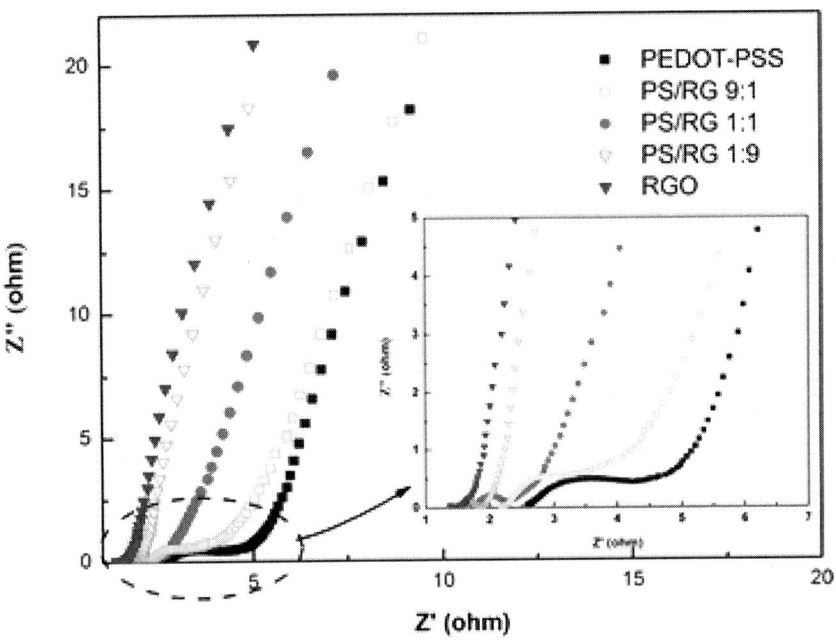

Figure 4. Nyquist plots of varied films. Inset shows plots with enlarged scale

Figure 5a shows the CV curves of composite films within potential range from −0.2 to 0.8 V in a 1-M Na_2SO_4 electrolyte at a scan rate of 20 mV/s. It can be seen that the CV curves of all the films displayed an almost rectangular shape, indicating the good capacitive properties of RGO and PEDOT-PSS nanocomposites. It is clear that PEDOT-PSS/RGO films exhibit larger CV area than that of pure PEDOT-PSS and pure RGO films, revealing higher charge-storage capability of nanocomposites. It is well known that as an electrochemical capacitor electrode, the PEDOT-PSS/RGO nanocomposite films can afford both pseudo capacitance and EDLCs during the electrochemical energy storage process. The enhanced-charge storage capability is attributed to the well dispersion of RGO in PEDOT-PSS, providing a three-dimensional net work for charge transmission. The highly specific surface area of RGO and high pseudo capacitance of conducting polymer lead to excellent capacitance performance of nanocomposite electrode. As shown in Fig. 5a, the PS/RG 1:1 film shows the largest CV area corresponding to the highest specific capacitance, indicating the optimum contents of conducting polymer in nanocomposites. Accordingly, the decreasing of conducting polymer contents in nanocomposites causes lesser contribution of pseudo capacitance, leading to smaller CV area and specific capacitance. It needs to be mentioned that the lower contents of RGO in a composite film will not hinder the conductive polymer from agglomerating effectively during the charge/discharge process, leading to smaller specific capacitance. It has been found that a mass ratio of EDOT to RGO, about 1:1, was the optimum proportion to prepare PEDOT-PSS/RGO nanocomposites with excellent capacity performance. Moreover, the Fig. 5a also suggests that the PEDOT-PSS/RGO film exhibits an excellent electrical synergistic effect, and low contact resistance is formed between electrolyte and electrodes.

A comparison of capacitive performance was performed using galvanostatic charge/discharge curves at a constant current density of 500 mA/g, which is shown in Fig. 5b. It is clearly seen that the shape of the charge/discharge curves of varied films are closely triangular in shape, indicating a good reversibility during the charge/discharge processes. The discharge times of the PEDOT-PSS/RGO nanocomposite films are longer than that of the pure PEDOT-PSS and pure RGO films, suggesting the coating of the PEDOT-PSS on the RGO surface greatly extending the discharge time. Moreover, the PS/RG 1:1 film shows the longest charge/discharge time, indicating the largest specific capacitance of these electrode films. The specific capacitance can be calculated from the capacitance equation:

$$C_m = \frac{i\Delta t}{m\Delta V}$$

Where the C_m, i, Δt, ΔV and m represent the specific capacitance, discharge current, discharge time, discharge potential window, and mass of nanocomposite films, respectively. According to the capacitance equation evaluated from the slopes of the discharge curves, the specific capacitance of pure PEDOT-PSS, PS/RG 9:1, PS/RG 1:1, PS/RG 1:9, and pure RGO films are 106.3, 143.4, 193.7, 129.2, and 81.8 F/g, respectively. The highly specific capacitance in PEDOT-PSS/RGO films results from the synergetic effect between PEDOT-PSS and RGO, which was constructed with a layer-by-layer and porous structure. This special structure also results in improving electrolyte-ion transfer efficiency and better frequency performance of electrode films during the charge/discharge process.

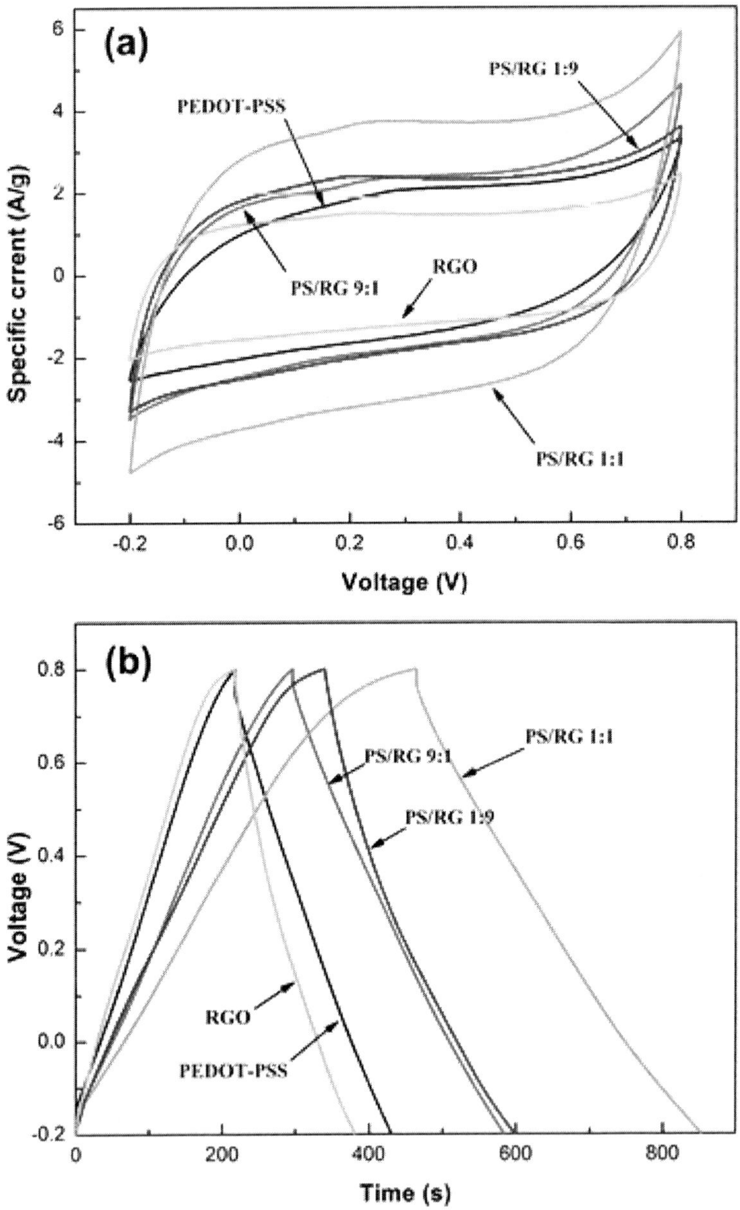

Figure 5. (**A**) Cyclic voltammograms curves of varied films at a scanning speed of 20 mV/s; (**B**) Galvanostatic charge to discharge curves of varied films at a current density of 500 mA/g

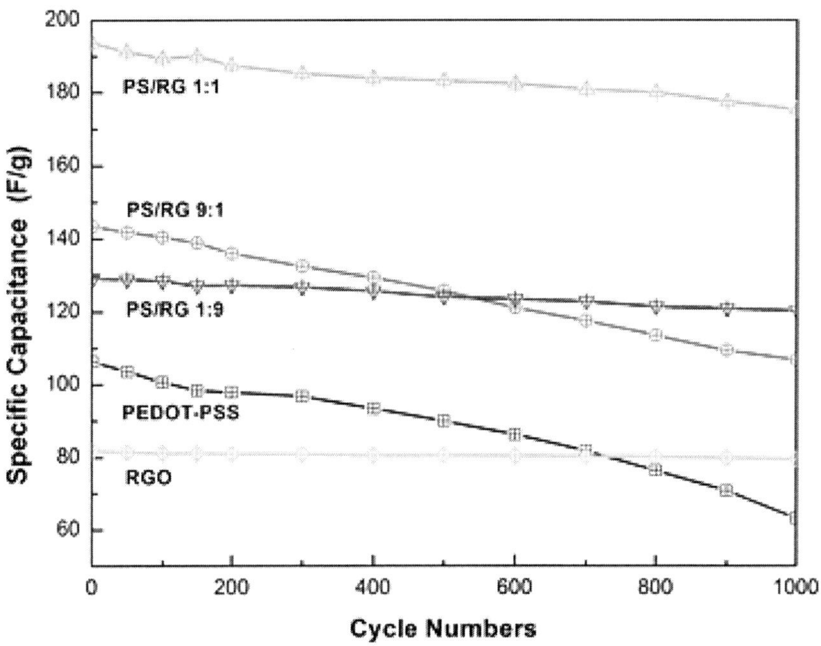

Figure 6. Cycling performances of varied films at a current density of 500 mA/g

The stability of different films as energy storage electrodes was evaluated by charge/discharge cycling tests at a current density of 500 mA/g (Fig. 6). The pristine specific capacitance of RGO is 81.8 F/g, and decreases slightly to 79.5 F/g (97.2 % retention) after 1000 cycles. The specific capacitance loss of RGO film is small, and the carbon nanomaterial shows excellent capacity-retention capability. In contrast, the PEDOT-PSS film shows higher specific-capacitance loss than RGO film with poor cycling performance. The capacitance value of PEDOT-PSS film decreases from 106.3 to 98.4 F/g in the first 150 cycles (7.4 % loss) and decreases slightly to 96.8 F/g during the next 150 cycles (1.5 % loss). An obvious decrease of specific capacitance is observed after the 1000 cycles (40.5 % loss), which is attributed to the poor mechanical capability of conducting polymer during the charge/discharge process, especially at high current density. The specific capacitance of PS/RG 1:1 film can maintained 90.6 % of initial capacitance (175.4 F/g) after 1000 cycles. Obviously, the addition of RGO into PEDOT-PSS improves the mechanical performance of PEDOT-PSS, which results in a better cycling stability of conducting polymer. The PS/RG 9:1 and PS/RG 1:9 films also present better capacity retention performance than PEDOT-

PSS film and keep 74.5 % (106.8 F/g) and 93.1 % (120.3 F/g) of initial specific capacitance over 1000 cycles, respectively. It can be estimated that, under the proper mass ratio of RGO, more RGO in nanocomposites results in better cyclic stability of composite films. This result reveals that the RGO sheets provide a robust mechanical support for PEDOT-PSS, preventing conductive polymer from swelling and shrinking during the long-time charge/discharge process. Therefore, it has been found that an optimum proportion of EDOT to RGO with 1:1 is better to prepare composite electrode with excellent cycling performance, which is consistent with that of CV curves.

CONCLUSIONS

We demonstrated preparation of flexible and self-supporting PEDOT-PSS/RGO nanocomposite films with a layered structure through a simple vacuum-filtered method. The superior capacitive performance of the PEDOT-PSS/RGO nanocomposites was confirmed by the tests of electrochemical properties. The mass ratio of PS/RG showed distinct influence on electrochemical performance of as-prepared composite electrodes. A 193.7 F/g highly specific capacitance was achieved at a current density of 500 mA/g. Furthermore, compared with pure PEDOT-PSS film, this nanocomposite film exhibited better energy-storage stability and can keep 90.6 % of the original specific capacitance after 1000 cycles.

DECLARATIONS

Acknowledgements

The work was supported by the National Science Foundation of China (NSFC) (No.51477026 and No.61471085).

Competing interests

The authors declare that they have no competing interests.

Authors' contributions

WY conceived of the study, carried out the fabrication of PEDOT/RGO composites, and drafted the manuscript. YC, YZ, and XH helped to deposit the films and analyze the data. YJ and JX helped to develop the idea and guided the study. YY and SL helped to draft the manuscript. All authors read and approved the final manuscript.

Authors' information

WY, YC, YZ, and XH are students of a doctor's degree at the School of Optoelectronic Information, University of Electronic Science and Technology of China. YY and SL are associate professors at the School of Optoelectronic Information, University of Electronic Science and Technology of China. JX and YJ are professors at the School of Optoelectronic Information, University of Electronic Science and Technology of China.

REFERENCES

1. Frackowiak E, Béguin F. Carbon materials for the electrochemical storage of energy in capacitors. Carbon. 2001;39:937–50.
2. Burke A. Ultracapacitors why, how, and where is the technology. J Power Sources. 2000;91:37–50.
3. Snook GA, Kao P, Best AS. Conducting polymer based supercapacitor devices and electrodes. J Power Sources. 2011;196:1–12.
4. Yoon SB, Kim KB. Effect of poly(3,4-ethylenedioxythiophene) (PEDOT) on the pseudocapacitive properties of manganese oxide (MnO_2) in the PEDOT-MnO_2-multiwall carbon nanotube (MWNT) composite. Electrochim Acta. 2013;106:135–42.
5. Obreja VN. On the performance of supercapacitors with electrodes based on carbon nanotubes and carbon activated material—a review. Physica E. 2008;40:2596–605.
6. Sonia TS, Mini PA, Nandhini R, Sujith K, Avinash B, Nair SV, et al. Composite supercapacitor electrodes made of activated carbon-PEDOT-PSS and activated carbon-doped PEDOT. B Mater Sci. 2013;36:547–51.
7. Yang YJ, Zhang LN, Li SB, Yang WY, Xu JH, Jiang YD, et al. Electrochemical performance of conducting polymer and its nanocomposites prepared by chemical vapor phase polymerization method. J Mater Sci-Mater EL. 2013;24:2245–53.
8. Han YQ, Shen MX, Wu Y, Zhu JJ, Ding B, Tong H, et al. Preparation and electrochemical performances of PEDOT/sulfonic acid-functionalized graphene composite hydrogel. Synthetic Met. 2013;172:21–7.

9. Sun D, Jin L, Chen Y, Zhang JR, Zhu JJ. Microwave-Assisted in situ synthesis of Graphene/PEDOT hybrid and its application in supercapacitors. Chem Plus Chem. 2013;78:227–34.

10. Yang YJ, Yang XJ, Yang WY, Li SB, Xu JH, Jiang YD. Ordered and ultrathin reduced graphene oxide LB films as hole injection layers for organic light-emitting diode. Nanoscale Res Lett. 2014;9:537.

11. Yang SL, Deng BC, Ge RJ, Zhang L, Wang H, Zhang ZH, et al. Electrodeposition of porous graphene networks on Nickel foams as supercapacitor electrodes with high capacitance and remarkable cyclic stability. Nanoscale Res Lett. 2014;9:672.

12. Österholm A, Lindfors T, Kauppila J, Damlin P, Kvarnström C. Electrochemical incorporation of graphene oxide into conducting polymer films. Electrochim Acta. 2012;83:463–70.

13. Han YQ, Ding B, Tong H, Zhang XG. Capacitance properties of graphite oxide Poly(3,4-ethylene dioxythiophene) composites. J Appl Polym Sci. 2011;121:892–8.

14. Yang WY, Xu JH, Yang YJ, Chen Y, Zhao YT, Li SB, et al. Porous conducting polymer and reduced graphene oxide preparation, characterization and electrochemical performance. J Mater Sci-Mater EL. 2015;26:1668–77.

15. Lim K, Jung S, Lee S, Heo J, Park J, Kang JW, et al. The enhancement of electrical and optical properties of PEDOT:PSS using one-step dynamic etching for flexible application. Org Electron. 2014;15:1845–55.

16. Yang Q, Pang SK, Yung KC. Study of PEDOT–PSS in Carbon nanotube/conducting polymer composites as supercapacitor electrodes in aqueous solution. J Electroanal Chem. 2014;728:140–7.

17. Antiohos D, Folkes G, Sherrell P, Ashraf S, Wallace G, Aitchison P, et al. Compositional effects of PEDOT-PSS/single walled Carbon nanotube films on supercapacitor device performance. J Mater Chem. 2011;21:15987–94.

18. Hummers WS, Offeman RE. Preparation of graphitic oxide. J Am Chem Soc. 1958;80:1339.

19. Rajagopalan B, Chung JS. Reduced chemically modified graphene oxide for supercapacitor electrode. Nanoscale Res Lett. 2014;9:535.

20. Wu Q, Xu YX, Yao ZY, Liu A, Shi GQ. Supercapacitors based on flexible Graphene/Polyaniline nanofiber composite films. ACS Nano. 2010;4:1963–70.

21. Dikin DA, Stankovich S, Zimney EJ, Piner RD, Dommett GHB, Evmenenko G, et al. Preparation and characterization of graphene oxide paper. Nature. 2007;448:457–60.

22. Chen HQ, Müller MB, Gilmore KJ, Wallace GG, Li D. Mechanically strong, electrically conductive, and biocompatible graphene paper. Adv Mater. 2008;20:3557–61.

23. Zhang JT, Zhao XS. Conducting polymers directly coated on reduced graphene oxide sheets as high-Performance supercapacitor electrodes. J Phys Chem C. 2012;116:5420–6.

24. Wan L, Wang B, Wang SM, Wang XB, Guo ZG, Dong BH, et al. Well-dispersed PEDOT:PSS/graphene nanocomposites synthesized by in situ polymerization as counter electrodes for dye-sensitized solar cells. J Mater Sci. 2015;50:2148–57.

25. Elya F, Matsumoto A, Zoetebiera B, Peressinotto VS, Hirata MK, Sousa DA, et al. Handheld and automated ultrasonic spray deposition of conductive PEDOT:PSS films and their application in AC EL devices. Org Electron. 2014;15:1062–70.

26. Weng YT, Wu NL. High-performance poly(3,4-ethylene-dioxythiophene)-polystyrenesulfonate conducting-polymer supercapacitor containing hetero-dimensional carbon additives. J Power Sources. 2013;238:69–73.

27. Chu CY, Tsai JT, Sun CL. Synthesis of PEDOT-modified graphene composite materials as flexible electrodes for energy storage and conversion applications. Int J Hydrogen Energy. 2012;37:13880–6.

28. Xu YF, Wang Y, Liang JJ, Huang Y, Ma YF, Wan XJ, et al. A hybrid material of graphene and poly (3,4-ethyldioxythiophene) with high conductivity, flexibility, and transparency. Nano Res. 2009;2:343–8.

29. Zhou H, Yao W, Li G, Wang J, Lu Y. Graphene/poly(3,4-ethylenedioxythiophene) hydrogel with excellent mechanical performance and high conductivity. Carbon. 2013;59:495–502.

30. Yan D, Liu Y, Li YH, Zhuo RF, Wu ZG, Ren PY, et al. Synthesis and electrochemical properties of MnO_2/rGO/PEDOT:PSS ternary composite electrode material for supercapacitors. Mater Let. 2014;127:53–5.

31. Tai Z, Yan X, Lang J, Xue Q. Enhancement of capacitance performance of flexible Carbon nanofiber paper by adding graphene nanosheets. J Power Sources. 2012;199:373–8.

32. Zhang LL, Zhao SY, Tian XN, Zhao XS. Layered graphene oxide nanostructures with sandwiched conducting polymers. Langmuir. 2010;26:17624–8.

33. Lu XJ, Dou H, Yuan CZ, Yang SD, Hao L, Zhang F, et al. Polypyrrole carbon nanotube nanocomposite enhanced the electrochemical capacitance of flexible graphene film for supercapacitors. J Power Sources. 2012;197:319–24.

4

CHAPTER

STIMULI-RESPONSIVE POLYMERS AND THEIR APPLICATIONS IN NANOMEDICINE

Etienne Cabane , Xiaoyan Zhang , Karolina Langowska ,Cornelia G. Palivan* Wolfgang Meier*

Chemistry Department, University of Basel, Klingelbergstrasse 80, 4056 Basel, Switzerland

ABSTRACT

This review focuses on smart nano-materials built of stimuli-responsive (SR) polymers and will discuss their numerous applications in the biomedical field. The authors will first provide an overview of different stimuli and their corresponding, responsive polymers. By introducing myriad functionalities, SR polymers present a wide range of possibilities in the design of stimuli-responsive devices, making use of virtually all types of polymer constructs, from self-assembled structures (micelles, vesicles) to surfaces (polymer brushes, films) as described in the second section of the review. In the last section of this review the authors report on some of the most promising applications of stimuli-responsive polymers in nanomedicine. In particular, we will discuss applications pertaining to diagnosis, where SR polymers are used to construct sensors capable of selective recognition and quantification of analytes and physical variables, as well as imaging devices. We will also highlight some examples of responsive systems used for therapeutic applications, including smart drug delivery systems (micelles, vesicles, dendrimers ...) and surfaces for regenerative medicine.

E. Cabane, X. Zhang contributed equally to this work.

This article is part of the Topical Collection "In Focus: Nanomedicine".

1 INTRODUCTION

Challenges confronted by medicine today include the increasing demand for sensitive, efficient systems and approaches that will improve responses to pathology. In this respect, for detection purposes, there is a need for new agents that will simultaneously increase sensitivity while their concentrations in the body decrease to avoid accumulation and side-effects. Such agents are intended to efficiently detect pathological conditions in their early stages or distinguish slight changes in areas where surgery has been done, serving to enhance prognoses, especially in complex diseases such as cancer, HIV, and degenerative diseases. The necessity of decreasing doses while increasing efficacy is essential for therapeutic approaches, while decreased side effects will improve a patient's condition, especially in chronic disease or diseases requiring the administration of toxic compounds, for example cancer or HIV. The design of new systems and approaches must meet challenges associated with administration in the body: (i) a simple route of administration, (ii) effective delivery to the desired biological compartment, (iii) response adapted to the pathological event, either rapid or slow, depending on the bio-specificity, and (iv) the use of non-toxic, biocompatible and biodegradable systems. Current know-how in nanotechnology is making possible new ways to fight a number of diseases. As the development of the fast growing field known as nanomedicine employs nanostructures and nanodevices to diagnose, treat, and prevent diseases [1]. In this respect, nanoscience offers novel systems and methods for medical use by providing carriers such as particles, micelles, dendrimers, and vesicles to transport active compounds (drugs, contrast agents, proteins, DNA), and "active" surfaces adapted to biosensing, regeneration and wound healing, An efficient way to improve these systems is to make them stimuli-responsive. A smart response to external or internal stimuli allows: (i) better localization of the system in the desired biological compartment, (ii) controlled release of payload at the location of the pathological event, and (iii) rapidly addressing/imaging the pathological event. In particular, polymers have proven themselves clever options in developing stimuli-responsive systems because their chemistry permits modulating the properties by including responsiveness via sensitive chemical moieties. A large variety of polymers/copolymers has been synthesized to response to physical stimuli (temperature, pH, light), chemical stimuli (various "signaling" molecules), or biological stimuli (enzymes). Stimuli-responsive polymers undergo dramatic and abrupt physical and chemical changes in response to external stimuli [2]. They are also termed 'smart-' [3, 4], 'intelligent-' [5], or 'environmentally sensitive' polymers [6]. One important feature of this type of material is reversibility, i.e. the ability of the polymer to return to its initial state upon application of a counter-trigger. In nature, biopolymers such as proteins and nucleic acids are all basic stimuli-responsive components of living organic systems and often remain stable over wide ranges of external variables but undergo drastic conformational changes abruptly at given critical points [3, 7]. These 'natural' stimuli-responsive polymers have led to the development of numerous synthetic polymers that have been designed to mimic their adaptive behaviours.

By incorporating functional groups that are amenable to a change in character (e.g. charge, polarity and solvency) along a polymer backbone, the resulting relative changes in chemical structure will be amplified synergistically, leading to dramatic transformations in macroscopic material properties. Typically, the 'response' of a polymer in solution alters its individual chain dimensions/size, secondary structure, solubility, or the degree of intermolecular association [8]. In most cases, the present or destruction of secondary forces (hydrogen bonding, hydrophobic effects, electrostatic interactions, etc.), simple reactions (e.g., acid–base reactions) of moieties linked to the polymer backbone, and/or osmotic pressure differences are responsible for this response. Another type of 'response' is due to dramatic alterations in the polymeric structure, such as degradation of polymers upon the application of a specific stimulus by bond breakage in the polymer backbone or at pendant cross-linking groups [8].

Stimuli-responsive systems containing polymers can be designed either with a responsive polymer, or by combining a polymer with a responsive compound, the polymer serving only as a template/carrier for that compound. Here we will focus only on the stimuli-responsive systems involving polymers as smart components, i.e. their properties and structures are changing in response to a specific stimulus. In addition, we are interested to mainly present supramolecular polymers assemblies in solution because they are extensively used both in therapeutic and in detection approaches. Note that the huge chemical diversity of polymers proposed for their stimuli-responsiveness (we will describe in the first part of our review) is dramatically reduced when medical applications are intended due to the biological constraints, we mentioned above. In this particular field it is extremely important to understand the parameters and mechanisms related to the distribution and transport of the nanosystems in the body. Controlling these parameters is necessary to answer the various concerns that will arise regarding environmental risk and side effects associated with the use of nanostructures in the body [9].

In this respect in the last part of the review we will focus on systems that are already used in medical applications, or have possible medical applications.

2 STIMULI-RESPONSIVE POLYMERS

The strategy underlying polymer-containing responsive systems is a dramatic physicochemical change caused by stimuli. At the macromolecular level, polymer chains can be altered in different ways, including changes in hydrophilic-to-hydrophobic balance, conformation, solubility, degradation, and bond cleavage, and these, in turn, will cause detectable behavioral changes to self-assembled structures [10]. Many designs that vary the location of responsive moieties or functional groups are possible. Locations include, but are not limited to: side chains on one of the blocks, chain end-groups, or junctions between blocks. The response may be reversible or not, depending on the strategy employed.

Stimuli are commonly classified in three categories: physical, chemical, or biological (Fig. 1) [11, 12]. Physical stimuli (light, temperature, ultrasound, magnetic, mechanical, electrical) usually modify chain dynamics, i.e. the energy level of the polymer/solvent system, while chemical stimuli (solvent, ionic strength, electrochemical, pH) modulate molecular interactions, whether between polymer and solvent molecules, or between polymer chains [13]. Biological stimuli (enzymes, receptors) relate to the actual functioning of molecules: enzymatic reactions, receptor recognition of molecules [14]. In addition, there are dual stimuli-responsive polymers that simultaneously respond to more than one stimulus.

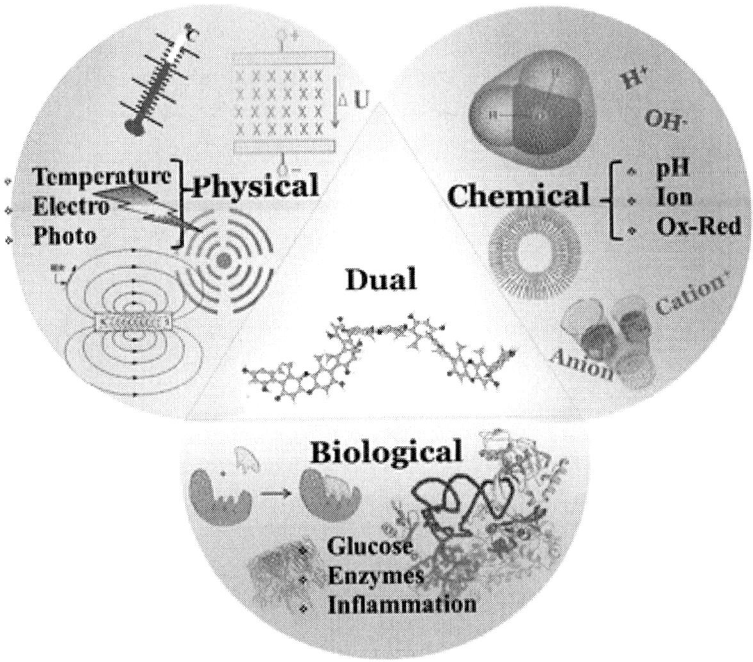

Figure 1. Classification of stimuli of stimuli-responsive polymers

2.1 Physically Dependent Stimuli

Physically dependent stimuli mainly include: temperature, electric field, light, ultrasound, magnetic fields and mechanical deformation. However, in this review we focus only on the stimuli-responsiveness of polymer/copolymer systems, hence, the physical stimuli reported as actively changing their properties/supramolecular structures are temperature, light, and electric field. We mention that magnetic fields and ultrasound have been used only for compounds that have been entrapped/encapsulated in polymer assemblies, and therefore we will not include them here.

2.1.1 Temperature Responsive Polymers

Temperature-responsive polymers have attracted great attention in bioengineering and biotechnology applications, because certain diseases manifest temperature changes [15]. Normally, these copolymers are characterized by a critical solution temperature around which the hydrophobic and hydrophilic interactions between the polymeric chains and the aqueous media abruptly change within a small temperature range. This induces the disruption of intra- and intermolecular electrostatic and hydrophobic interactions and results in chain collapse or expansion (a volume phase transition). Typically, these polymer solutions possess an upper critical solution temperature (UCST) above which one polymer phase exists, and below which a phase separation appears. Alternatively, polymer solutions that appear as monophasic below a specific temperature and biphasic above it generally possess a so-called lower critical solution temperature (LCST). Depending on the mechanism and chemistry of the groups, various temperature-responsive polymers have been reported: poly(*N*-alkyl substituted acrylamides), e.g. poly(*N*-isopropylacrylamide) (PNiPAAm) [16, 17], poly (*N*-vinylalkylamides), e.g. poly(*N*-vinylcaprolactam) (PNVC) [18], and copolymers such as poly(l-lactic acid)-poly(ethylene glycol)-poly(l-lactic acid) (PLLA-PEG-PLLA) triblock copolymers [19], and poly(ethylene oxide)-poly(propylene oxide)-poly (ethylene oxide) (PEO–PPO–PEO) copolymers [20].

2.1.2 Electro-Responsive Polymers

Electrical and electrochemical stimuli are widely used in research and applications, due to their advantages of precise control via the magnitude of the current, the duration of an electrical pulse or the interval between pulses [21, 22]. Typical electrically responsive polymers are conducting polymers, as for example polythiophene (PT) or sulphonated-polystyrene (PSS), which can show swelling, shrinking or bending in response to an external field [23, 24]. There are different effects upon electrochemical stimulation: (a) an influx of counter ions and solvent molecules causes an increase in osmotic pressure in the polymer, resulting in a volumetric expansion, (b) control of the loading/adsorption of polyelectrolyte on to oppositely charged porous materials, (c) formation and swelling of redox-active polyelectrolyte multilayers. For example, when an electrochemical stimulus is applied to multilayer polyacrylamide films, the combined effects of H^+ ions migrating to the region of the cathode and the electrostatic attraction between the anode surface and the negatively charged acrylic acid groups lead to shrinking of the film on the anode side [25, 26].

2.1.3 Photo-Responsive Polymers

Because light can be applied instantaneously and under specific conditions with high accuracy, it renders light-responsive polymers highly advantageous for applications [6]. The light can be directly used at the polymer surface or can be delivered to distant locations using optical fibers. Ideally, the wavelength of the laser is tuned to the so-called biologically 'friendly' window [27], the near-infrared part of the spectrum, which is less harmful and has deeper penetration in tissues than visible light. In this case, the light is both minimally absorbed by cells/tissue and maximally so by the polymers. Most photo-responsive polymers contain light-sensitive chromophores such as azobenzene groups [28, 29], spiropyran groups [30, 31], or nitrobenzyl groups [32, 33]. A variety of azobenzene or spiropyran-containing photo-responsive polymers, as for example PAA [34, 35], PHPMAm [36, 37], and PNIPAM [38, 39], have been reported.

2.2 Chemically-Dependent Stimuli

Chemically-dependent stimuli comprise pH, ionic strength, redox and solvent.

2.2.1 pH-Responsive Polymers

pH is an important environmental parameter for biomedical applications, because pH changes occur in many specific or pathological compartments. For example, there is an obvious change in pH along the gastrointestinal tract from the stomach (pH = 1–3) to the intestine (pH = 5–8), chronic wounds have pH values between 7.4 and 5.4 [40], and tumour tissue is acidic extracellularly [41, 42]. Therefore, unlike temperature changes, this property can be exploited for a direct response at a certain tissue or in a cellular compartment. The key element for pH responsive polymers is the presence of ionisable, weak acidic or basic moieties that attach to a hydrophobic backbone, such as polyelectrolytes [6, 10, 43]. Upon ionization, the electrostatic repulsions of the generated charges (anions or cations) cause a dramatic extension of coiled chains. The ionization of the pendant acidic or basic groups on polyelectrolytes can be partial, due to the electrostatic effect from other adjacent ionized groups [44].

Another typical pH responsive polymer exhibits protonation/deprotonation events by distributing the charge over the ionisable groups of the molecule, such as carboxyl or amino groups [45]. pH induces a phase transition in pH responsive polymers very abruptly. Usually, the phase switches within 0.2–0.3 U of pH [46]. pH responsive polymers typically include chitosan [47], albumin [48], gelatin [49], poly(acrylic acid) (PAAc)/chitosan IPN [50], poly(methacrylic acid-*g*-ethylene glycol) [P(MAA-*g*-EG)] [51, 52], poly(ethylene imine) (PEI) [53], poly(*N*,*N*-diakylamino ethylmethacrylates) (PDAAEMA), and poly(lysine) (PL) [54, 55].

2.2.2 Ion-Responsive Polymers

The responsiveness to ionic strength is a typical property of polymers containing ionisable groups. These polymer systems exhibit unusual rheological behaviour as a result of the attractive Coulombic interactions between oppositely charged species, which may render the polymer insoluble in deionized water but soluble in the presence of a critical concentration of added electrolytes where the attractive charge/charge interactions are shielded [56–58]. Therefore, changes in ionic strength cause changes in the length of the polymer chains, the polymer solubility and the fluorescence quenching kinetics of chromophores bound to electrolytes [57, 59, 60].

2.2.3 Redox-Responsive Polymers

Polymers containing labile groups present an beneficial opportunity to develop redox-responsive biodegradable or bioerodible systems. Acid labile moieties inside polyanhydrides [61, 62], poly(lactic/glycolic acid) (PLGA) [63], and poly(β-amino esters) (PbAEs) [64] induce redox responsiveness. Disulfide groups have also been used to induce redox responsiveness, because they are unstable in a reducing environment, being cleaved in favour of corresponding thiol groups [65, 66]. Polymers with disulfide cross-links degrade when exposed to cysteine or glutathione, which are reductive amino-acid based molecules [67]. Another typical redox responsive polymer is poly(NiPAAm-co-Ru(bpy)$_3$), which can generate a chemical wave by the periodic redox change of Ru(bpy)$_3$ into an oxidized state of lighter colour [68]. This redox reaction alters the hydrophobic and the hydrophilic properties of the polymer chains and results in swelling and deswelling of the polymer.

2.3 Biologically Dependent Stimuli

Biologically dependent stimuli typically involve analytes and biomacromolecules such as glucose, glutathione, enzymes, receptors, and over-produced metabolites in inflammation.

2.3.1 Glucose Responsive Polymers

Precisely engineered glucose sensitive polymers have huge potential in the quest to generate, for example, self-regulated modes of insulin delivery [11, 69]. For glucose responsive polymers, glucose oxidase (GOx) is conjugated to a smart, pH-sensitive polymer. GOx oxidizes glucose to gluconic acid, which causes a pH change in the environment [6]. The pH sensitive polymer then exhibits a volume transition in response to the decreased pH [69]. In this way, drastic changes in the polymer conformation are regulated by the body's glucose level, which, in turn, significantly affects enzyme activity and substrate access.

2.3.2 Enzyme-Responsive Polymers

In nature, bacteria located mainly in the colon produce special enzymes, including reductive enzymes (e.g. azoreductase) or hydrolytic enzymes (e.g. glycosidases) which are capable of degrading various types of polysaccharides, such as pectin, chitosan, amylase/amylopectin, cyclodextrin and dextrin [70–72]. In most enzyme-responsive polymer systems, enzymes are used to destroy the polymer or its assemblies. The biggest advantage of enzyme-responsive polymers is that they do not require an external trigger for their decomposition, exhibit high selectivity, and work under mild conditions. For example, polymer systems based on alginate/chitosan or DEXS/chitosan microcapsules are responsive to chitosanase [73]. And azoaromatic bonds are sensitive to azoreductase [74]. In this respect, they have great potential for in vivo biological applications. However, the main disadvantage is the difficulty of establishing a precise initial response time.

2.3.3 Inflammation-Responsive Polymers

The inflammatory process is initiated by T- and B-lymphocytes, but amplified and perpetuated by polymorphonuclear (PMN) leukocytes and macrophages. Various chemical mediators in the process, including arachidonic acid metabolites, proteolytic enzymes and oxygen metabolites, can cause tissue damage. For inflammation-responsive systems, the reactive oxygen metabolites (oxygen free radicals) released by PMNs and macrophages during the initial phase of inflammation are the stimuli [75]. Such chemical mediators have been successfully used as stimuli for responsive drug delivery. For example, in vivo implantation experiments revealed that hyaluronic acid (HA) cross-linked with glycidylether can degrade in response to inflammation [76].

2.4 Dual-Stimuli

For biomedical applications, a step forward is realized if the smart materials respond simultaneously to more than one stimulus. Therefore,

Table 1. Summarize on stimuli responsive polymers grouped by stimulus–response, and contain information about the synthesis method and application

Table 1 Summarize on stimuli responsive polymers grouped by stimulus–response, and contain information about the synthesis method and application

Type of stimulus–response	Stimulus-responsive polymers	Synthesis method	Application
Physically dependent stimuli			
Temperature-responsive polymers	PNiPAAm [15, 16]	Living radical polymerization	Water soluble polymer sensor, Tissue adhesion prevention material
	PNVC [17]	Living radical polymerization	Thermosensitive hydrogel at any temperature
	PLLA/PEG/PLLA [18]	Ring open polymerization	Potential anti-cancer drug carrier
	PEO–PPO–PEO [19]	Crosslinking the ethoxysilane-cap	Drug carrier
Electro-responsive polymers	PT [23]	Electrochemical Synthesis	Drug release and cancer chemotherapy
	PSS [22]	Emulsion polymerization	Drug carrier
Photo-responsive polymers	Azobenzene or spiropyran-containing		
	PAA [33, 34]	Copolymerization	Photocchromic polymer
	PHPMAm [35, 36]	Copolymerization	Sensor
	PNIPAM [37, 38]	Copolymerization	Photodegradation material
Chemically dependent stimuli			
pH-responsive polymers	chitosan [46]	Biosynthesis	Drug release
	Albumin [47]	Biosynthesis	Enzyme immobilization
	Gelatin [48]	Biosynthesis	Immunoassay
	PAAc/chitosan IPN [49]	UV irradiation	Wound dressing material and drug release
	P(MAA-g-EG) [50, 51]	Free-radical, solution photopolymerization	Controlled insulin delivery
	PEI [52]	Solution polymerization	pH-sensitive controlled release systems
	PDAAEMA		
	PL [53, 54]	Biosynthesis	Vectors for gene delivery
Ion-responsive polymers			
Redox-responsive polymers	Polyanhydrides [60, 61]	Melt condensation polymerization	Potential oral drug delivery systems
	PLGA [62]	Double emulsion solvent evaporation	Controlled delivery systems
	PbAEs [63]	Addition solution polymerization	Efficient carrier for cytotoxic agents
	Poly(NiPAAm-co-Ru(bpy)3) [67]	Living radical copolymerization	Artificial muscles, artificial reptile
Biologically dependent stimuli			
Glucose-responsive polymers	GOx conjugated chitosan [6, 68]	Carbodiimide chemistry	Self-regulated insulin delivery
Enzyme-responsive polymers	DEXS/chitosan [72]	Layer-by-layer assembly	Local and sustained drug release
	Azoaromatic crosslinked hydrogel [73]	Copolymerization	Specific delivery of peptides and proteins
Inflammation-responsive polymers	Glycidylether crosslinked HA [75]	Suspension solution reaction	Implantable drug delivery
Dual-stimuli	PLL block PEG–PLL [76]	Side chain reaction and crosslinking	Enhance gene expression

increasing the efficacy of drug therapies may require polymeric materials, which are responsive to several kinds of stimuli. These will support the diagnosis of patients by monitoring several physiological changes at once. The dual-stimuli responsive approach is ideally suited for theragnostic (a combination of diagnostics and therapy) because some functionalities can provide on-site feedback and diagnostics, while others could initiate curing and therapy. Availability of various physical, chemical and biological stimuli is indispensable for multiple response functions. Therefore, multi-stimuli-responsive polymers, especially dual temperature- and pH-responsive systems, are attracting increasing attention recently for their advantages in biotechnological and biomedical applications. For example, a dual-stimuli-responsive delivery system, using both pH and glutathione-responsive polymeric modules, was developed to therapeutically deliver medicinal molecules [77]. It was possible to tune the release kinetics by systematically varying the composition of the pH-sensitive hydrophobic moiety (butyl acrylate), by modifying the glutathione-responsive moiety (pyridyl disulfide acrylate), or by modifying both of them.

Table 1 summarizes stimuli responsive polymers grouped by stimulus–response, and contain information about the synthesis method and application.

3 STIMULI RESPONSIVE POLYMERS WITH DIFFERENT PHYSICAL FORMS

3.1 Dendrimers

Dendrimers are macromolecules characterized by highly branched structures. Their properties attract attention for their applicability as delivery vessels, carriers of imaging agents, and therapeutically active compounds [78–80].

3.1.1 Temperature Responsive Dendrimers

Various examples of temperature responsive dendrimer systems (with differing architecture and chemical composition) used to encapsulate and release drugs are described in literature: star-shaped poly(ε-caprolactone)-*b*-poly(2-(dimethylamino)ethyl methacrylate) (HPs-Star-PCL-*b*-PDMAEMA) [81], core–shell dendritic poly(ether-amide) (DPEA) modified with carboxyl end-capped linear poly(*N*-isopropylacrylamide) (PNIPAAm–COOH) and carboxyl end-capped methoxy polyethylene glycol (PEG–COOH) [82]. It was shown that the temperature sensitivity of dendrimers can depend on their generation and molecular mass [83]. Dendrimers based on poly(aminoamide) (PAMAM) or poly(propyleneimine) (PPI) were obtained by introducing isobutyramide (IBAM) groups onto the chain ends and, in the case of PAMAM dendrimers, the thermoresponse was further modulated by introducing various peripheral alkylamide groups [84].

3.1.2 Photo-Responsive Dendrimers

Photo-responsive carbosilane dendrimers containing 4-phenylazobenzonitrile units at each terminal end were synthesised for potential applications in conversion of photo-energy into dynamic energy or in drug delivery systems [85]. The molecular size of a dendrimer with azobenzene derivatives depends on the photo- and heat-isomerization abilities of the azobenzene unit. The photoresponse can also be obtained by introducing *O*-nitrobenzyl groups to the surface of hyperbranched polyglycerols (HPGs) for drug release [86]. The presence of a hexa(ethylene glycol) outer-shell instead of the hexene increased the stability of the formed host–guest complexes but resulted in lower guest release. The stability of the host–guest complexes depended on the counterion of the guest molecules. This system offers the opportunity to tune the nanocapsules to control guest binding and release.

3.1.3 pH- and Ion-Responsive Dendrimers

PAMAM (polyamidoamine) and PPI (polypropylene imine) dendrimers are known to be ion- or pH- responsive in an aqueous environment, due to the charge repulsion of the multiple amine groups [53, 87, 88]. Biocompatible acetylated poly(amidoamine) (PAMAM) dendrimers were used for drug delivery, with dexamethasone 21-phosphate (Dp21) as the model drug [89]. Cationic (non-acetylated) and acetylated (acetylation is a convenient strategy to neutralize the peripheral amine group) dendrimers exhibited different pH-dependent micellization, complexation, and encapsulation behaviour. The acetylated dendrimer encapsulated the Dp21 under acidic conditions (pH = 3.0), while the cationic dendrimer encapsulated the drug under both acidic (pH = 3 and pH = 5.0) and neutral conditions (pH = 7.4). In addition, pH-responsive release was different for an acetylated- and a non-acetylated dendritic matrix. Non-acetylated dendrimers showed a much slower release rate than acetylated dendrimers under conditions of lower pH and a much faster release rate from non-acetylated dendrimer as pH values decreased. Degradable 1,3,5-triazaadamantane (TAA) dendrimers were able to be triggered by the addition of HCl [90]. TAAs units are stable under basic conditions but hydrolyze rapidly under acidic conditions to yield basic by-products [tris(amino-methyl)-ethane]. In the polyphosphazene-functionalized diaminobutane poly(propyleneimine) (DAB-PN) dendrimeric system used for hydrophobic drug delivery, release was triggered by sodium chloride ions [91]. Cations such as Na^+, K^+ complexate ethyleneoxy moieties on polyphosphazene chains, which results in the swelling of the polyphosphazene external groups.

3.1.4 Redox-Responsive Dendrimers

Degradable polylysine dendrimers with multiple spermine groups on the surface and non-covalently bound DNA were synthesized via attachment of the spermine by a disulfide linker [92], which was cleaved by mild reducing agents such as glutathione (GSH), therefore causing the release of DNA. Chemically and electrochemically triggered release of dendrimer end groups was obtained, based on different generations of poly(propyleneimine) dendrimers with redox-labile, trimethyl-locked quinone (TLQ) end groups [93]. The TLQ units were released by chemical ($Na_2S_2O_4$) or electrochemical (electrolytic current) redox reaction. Redox-triggered release of dendrimer end groups can be caused by the physiological redox cofactors (e.g., redox proteins, ascorbic acid, thiols).

3.1.5 Enzyme/Protein-Responsive Dendrimers

An interesting example of an enzyme-responsive dendrimer was obtained by the synthesis of dendrimers with a hexyl ester functionality as the hydrophobic part and polyethylene glycol (PEG) as the hydrophilic part [94]. These dendrimers disassembled in response to an enzymatic trigger (enzyme-porcine liver esterase) due to the incorporation of enzyme-cleavable ester moieties at the hydrophobic part of the dendrimers. Enzymatic cleavage of the ester groups caused disintegration of the dendritic structure and release of the guest molecule (Fig. 2). The rate of guest release systematically decreased with an increase in the dendron generation (higher generation dendrimers are more tightly packed, which sterically protects them—the ester functionalities are less accessible for enzymatic degradation). A similar strategy was used for the preparation of dendritic micellar containers [95], based on receptor-ligand binding interactions. PEG was chosen as the hydrophilic part and a decyl chain as the hydrophobic part. In order to disintegrate the dendritic structure, biotyn was incorporated (via click chemistry) as a ligand that bonded to a specific protein-extravidin. The disintegration of the system was caused by the biotin–extravidin interaction, which dramatically changed the hydrophilic–lipophilic balance (HLB) of the dendrimer molecule. The selectivity of this binding and release is based on molecular recognition.

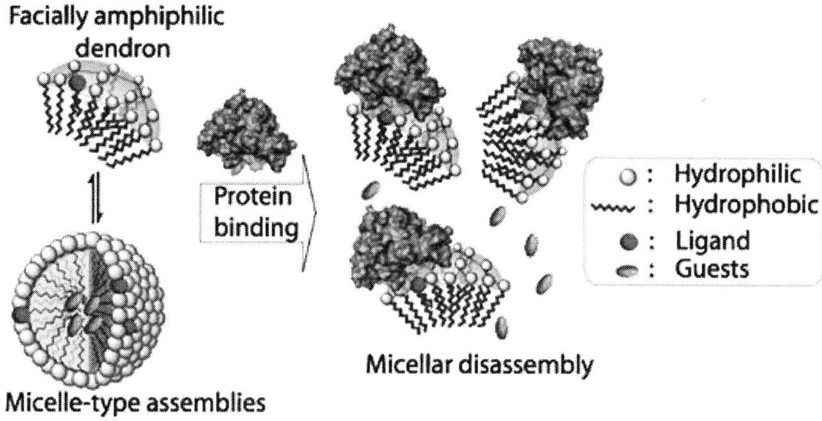

Figure 2. Disintegration of dendrimer-ligand assemblies upon protein-ligand binding [95]

3.2 Micelles

Block copolymer micelles are generally formed by the spontaneous self-assembly of amphiphilic copolymer molecules in an aqueous environment. Usually they are spherically shaped core–shell structures with sizes varying in the range of 10–100 nm. The hydrophobic blocks form the micelle cores, while the hydrophilic blocks form the micelle corona (shells). Lipophilic drugs can be solubilized in the hydrophobic micelle cores, significantly increasing the drug concentration in an aqueous environment.

3.2.1 Photo-, Thermo- and pH-Responsive Micelles

Copolymerization of a spiropyran-containing methacrylate (SPMA) with di(ethylene glycol) methyl ether methacrylate (DEGMMA) resulted in dual-response (photo- and thermo-responsive) PSPMA–PDEGMMA material, which formed micelles and reverse micelles in aqueous solution (Fig. 3) [96]. Upon exposure to UV light, ring-opening isomerization of spiropyran (non-polar, hydrophobic, and colourless under visible light irradiation) occurred, resulting in the coloured, polar, hydrophilic form. The photo-switchable PSPMA block and the thermo-responsive PDEGMMA block, both PSPMA-core and PDEGMMA-core micelles, were obtained by changing the temperature (from 15 to 30°C) of the solution and by photo irradiation. These micelles were used for encapsulation and controlled release and re-encapsulation of the model drug coumarin 102.

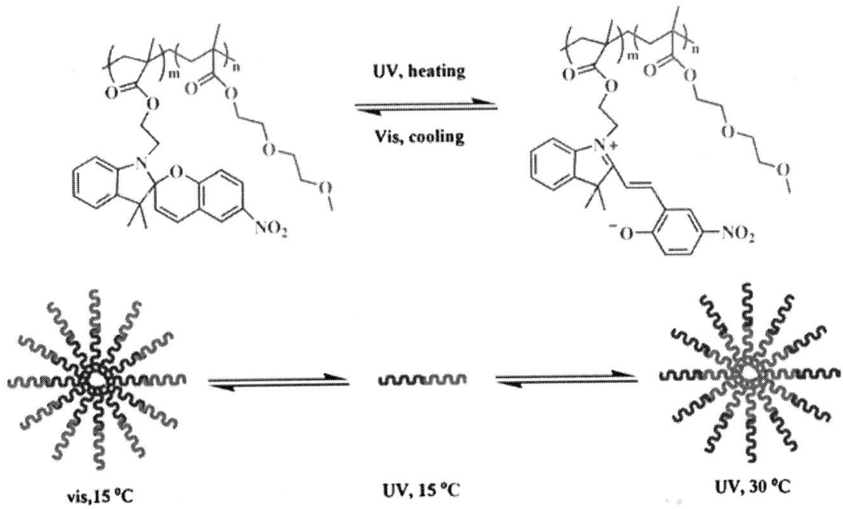

Figure 3. Temperature- and UV-responsive micellar transition of PSPMA-*b*-PDEGMMA copolymer in aqueous solution [96]

Spiropyran-decorated amphiphilic polypeptide-based block copolymers PLGASP-*b*-PEO (poly(l-glutamic acid)-*b*-polyethylene oxide) that form micelles and micellar aggregates also showed conformational changes (from alpha-helix to random coil and vice versa) under UV and visible light, respectively [97]. Because the light used was a medically non-invasive, highly penetrating UV source, these photoresponsive rod-coil block polypeptides could be applied as viable model systems to study photo-induced drug release or light-controlled biomedical applications. Acid labile micelles of a model amphiphilic block copolymer, poly(hydroxyethyl acrylate)-*b*-poly(*n*-butyl acrylate) (PHEA-*b*-PBA) with encapsulated doxorubicin (DOX) demonstrated that hydrolysis of less than half of the cross-links in the core was sufficient to release DOX at acidic pH (5.0) faster than at neutral pH (7.4) [98].

3.2.2 Enzyme-Responsive Micelles

Examples of polymer peptide conjugates, particles of which disintegrated in response to the proteinase K signal [99], are the graft-type polymers (NIPAM–PEP and NIPAM–PEPEP, NIPAM is *N*-isopropylacrylamide, PEP and PEPEP are peptide units) containing a substrate peptide of protein kinase A (PKA) (PKA forms one of the most important intracellular signals in cellular signal transduction). The lower critical solution temperature (LCST) of

NIPAM–PEP was raised from 36.7 to 40°C in response to phosphorylation by activated PKA. The NIPAM–PEPEP containing a different poly(ethylene glycol) unit formed a polymer micelle-type particle above the LCST. These particles disassembled and released drug in response to phosphorylation catalysed by PKA. The micellization of the complex of the polymer poly(potassium acrylate) (PPA) and the surfactant cetyltrimethylammonium bromide (CTAB), using the fluorescent pyrene as a guest molecule, resulted in an enzyme responsive system [100, 101].

3.3 Vesicles

Polymer vesicles, also called polymersomes, are spherical shell structures in which an aqueous compartment is enclosed by a bilayer membrane made of amphiphilic block copolymers. Their advantages compared to liposomes are: greater toughness, greater stability, tunable membrane properties, capacity to transport both hydrophilic and hydrophobic compounds (genes, proteins, imaging agents, anticancer and anti-inflammatory drugs and others), making them good candidates for applications including drug delivery, nanoreactors and templates for micro- or nano-structured materials. They can be used as stimuli-responsive controlled drug release systems [102–104].

3.3.1 pH-, Ion-Responsive Vesicles

The response of polypeptides to pH and ionic strength was used to produce pH-and ion-responsive nanoparticles with controlled sizes and shapes. Amphiphilic poly(butadiene)-*b*-poly(γ-l-glutamic acid) (PB-*b*-PGA) diblock copolymer vesicles underwent reversible coil-helix transition in response to pH and, as a result, the sizes of the particles changed from 100 to 150 nm [105]. Also, peptide based biocompatible polybutadiene-*b*-poly(l-glutamic acid) (PB-*b*-PGA), polyisoprene-*b*-poly(l-lysine) (PI-*b*-PLys) and poly(l-glutamic acid)-*b*-poly(l-lysine) (PGA-*b*-PLys) vesicles demonstrated multi-responsive behaviour [106]. pH-responsive polymer vesicles obtained by the aqueous self-assembly of carboxy-terminated hyperbranched polyesters have the advantage of simple synthesis (a one-step esterification of the commercially available hydroxy-terminated hyperbranched polyester) and the possibility of controlling vesicle size (from 200 nm to 10 mm) by pH changes [107].

The potential of a drug to be released as triggered by pH changes was demonstrated with poly(ethylene oxide)-*b*-poly-(glycerolmonomethacrylate) (PEO-*b*-PG2MA) drug conjugates [108]. At a

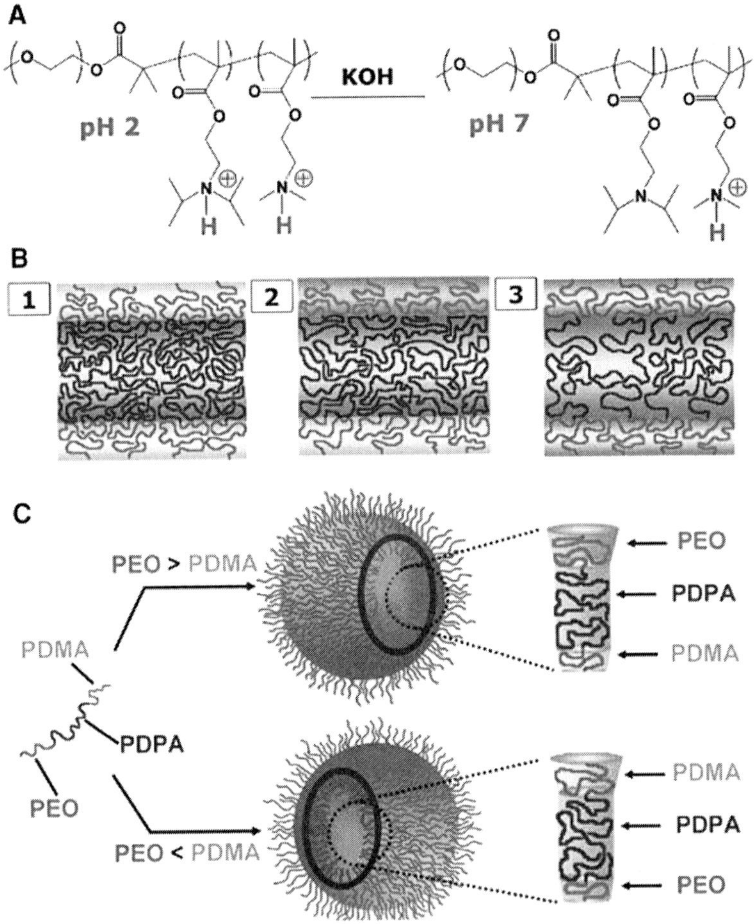

Figure 4. A Effect of solution pH on the degree of protonation of the P and M chains. **B** Three possible membrane structures depending on the block copolymer morphology: **1** AB diblock copolymers form an interdigitated membrane with chemically identical faces; **2** ABC triblock copolymers where the central hydrophobic 'B' block bridges the membrane with segregated 'A' and 'C' interfaces; **3** central 'B' block of ABC triblock copolymer forms a 'loop' within the membrane, with the 'A' and 'C' chains forming a non-segregated membrane. **C** Effect of varying the relative volume fractions of the hydrophilic 'A' and 'C' blocks on the polymersome structure [110]

pH close to neutral, ester-bond linkages were stable and vesicular structures were formed. When pH was lowered to 2.0–3.5, hydrolysis of the ester bond took place and the drug was released. pH-sensitive vesicles made of the copolypeptide polyarginine-*b*-polyleucine (PARG-PLE) were obtained based on the presence of a polyarginine block [109], the properties of which allowed vesicular self-assembly and intracellular delivery.

ABC triblock copolymers (PEO–PDPA–PDMA) [poly(ethylene oxide)-poly(2-(diisopropylamino)ethyl methacrylate)-poly(2-(dimethylamino) ethylmethacrylate] of varying block compositions with asymmetric membranes were used to demonstrate that the surface chemistry of polymersomes plays a crucial role (Fig. 4). PEO and PDMA blocks were hydrophilic and the pH-sensitive PDPA block changed from hydrophilic in acidic solution to hydrophobic at pH 7.0. In vitro cell delivery studies suggest that the vesicles can be either biocompatible or cytotoxic, depending on whether the PEO or PDMA block is at the exterior surface [110].

3.3.2 Temperature-Responsive Systems

Thermo-responsive cross-linked polymer vesicles were formed by self-assembly of the block copolymer poly(2-cinnamoylethyl methacrylate)-*b*-poly(*N*-isopropylacrylamide) (PCEMA-*b*-PNIPAM) and following photo-cross-linking of PCEMA shells, and were used for temperature-(higher than 32°C) triggered release of 4-aminopyridine [111].

Self-assembly of amphiphilic hyperbranched star copolymers with a hydrophobic hyperbranched poly[3-ethyl-3-(hydroxymethyl)oxetane] (HBPO) core and many hydrophilic polyethylene oxide (PEO) arms also showed thermo-sensitive behaviour [112]. The thermo-sensitivity of the vesicles results from the partial dehydration of the PEO vesicle corona.

Diblock copolymer poly(*N*-(3-aminopropyl)methacrylamidehydrochloride)-*b*-(*N*-isopropylacrylamide) (PAMPA-*b*-PNIPAM) vesicles showed not only temperature responsiveness in a narrow range (25–45°C), depending on the length of the building blocks structures of the polymer, but were also "locked" by ionic cross linking of the PAMPA block [113]. Vesicles were stable between pH 0 and 11. However, the particle size was shown to vary with the pH of the solution. At lower pH values, the vesicles were bigger (310 nm at pH 3.0), and increasing the pH value of the solution decreased the size of the vesicles (e.g. 220 nm at pH 10.8).

Thermo-responsiveness can also be obtained by using the synthetic poly(trimethylene carbonate)-*b*-poly(l-glutamic acid) (PTMC-*b*-PGA), diblock copolymer [114]. Temperature induced reversible crystallization/melting of the PTMC-*b*-PGA vesicles in water depended on the vesicle size (membrane thickness). The disruption of the vesicular structure occurred when the temperature was increased above the melting point of the PTMC block (34–35°C).

Dual-response poly[(*N*,*N*-diethylaminoethyl methacrylate)-*b*-(*N*-isopropyl acrylamide)] [P(DEAEMA-*b*-NIPAM)s] systems capable of "schizophrenic" (two or more responsive blocks that can form two different structures triggered by stimuli) aggregation in aqueous solution were controlled by varying the pH and temperature [115].

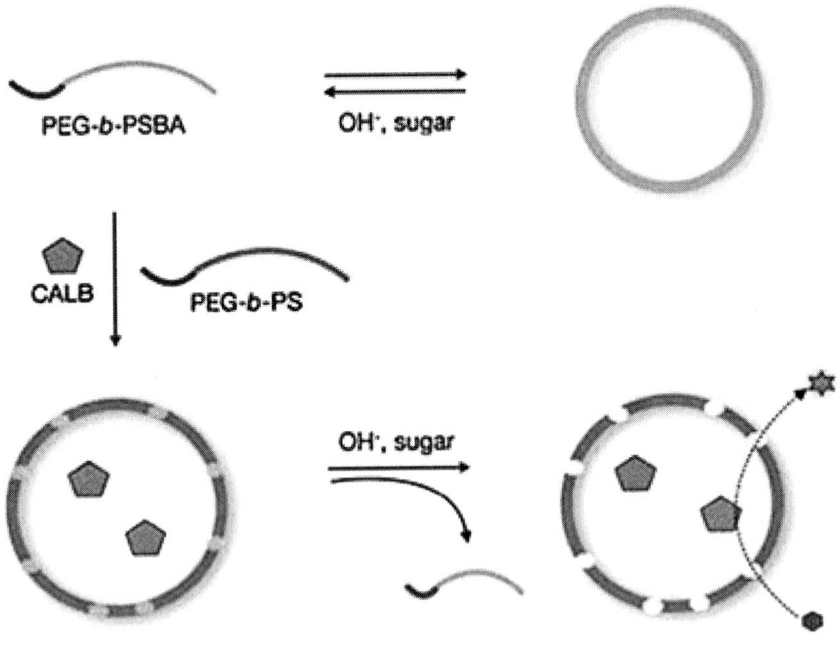

Figure 5. Schematic structure of PEG-*b*-PSBA block copolymers and their equilibrium with d-glucose in a basic aqueous environment, and formation of polymersomes with a permeable membrane induced by the sugar responsiveness of the block copolymers [117]

3.3.3 Glucose-Responsive Systems

Oxidation-responsive vesicles from amphiphilic block copolymers based on ethylene glycol and propylene sulphide (PPS) exposed to oxidative conditions were destabilized [116]. Thioethers in the hydrophobic PPS blocks were changed into hydrophilic sulfoxides, influencing the hydrophilic–lipophilic balance of the amphiphile and inducing its solubilization. A poly(ethylene glycol)-*b*-poly(styrene boronic acid) (PEG-*b*-PSBA) system with boronic acid moieties showed both pH and sugar-responsive behaviour [117]. Disruption of the assemblies occurred after adding 0.5 M NaOH to the vesicle solution (Fig. 5). In addition, in the presence of 200 mM d-glucose, vesicles were also disrupted. The binding of the sugar molecules to the ionized boronic acid increased solubility of the PSBA blocks in water. The polymersomes disassembled completely in the presence of d-fructose (100 mM) in medium of pH 10.

3.3.4 Glutathione-Responsive Systems

Drug release systems based on reversibly crosslinked temperature-responsive nano-sized polymersomes of poly(ethylene oxide)-*b*-poly(acrylic acid)-*b*-poly(*N*-isopropylacrylamide) (PEO–PAA–PNIPAM), were formed in water (no organic solvents), which is important in the delivery of biopharmaceutics [118]. The polymersomes showed high stability in organic solvent, high salt concentrations, and at different temperatures, but in the presence of 10 mM dithiothreitol (DTT) the fast release of encapsulated species was observed. Polymersomes based on hydrophilic poly(ethylene glycol) (PEG) and hydrophobic poly(propylene sulfide) (PPS) connected by a disulfide bridge, PEG_{17}–SS–PPS_{30} were disrupted in the presence of cysteine, at a concentration corresponding to the intracellular level [65]. A similar system, also based on PEG–PPS block copolymers, was reported earlier [119]. This was the first example of the use of oxidation (in the presence of H_2O_2) in order to destabilize PEG–PPS–PEG vesicles and oxidize the central-block sulphide moieties to sulphoxides and finally to sulphones, this oxidation causing an increase in the hydrophilicity of the initially hydrophobic central block.

3.3.5 Light-Responsive Systems

Zhao and coworkers [120] reported the formation of vesicles with PAzo-*b*-P(tBA-AA) copolymers, where PAzo is a hydrophobic methacrylate-based azobenzene containing side-chain liquid crystalline polymer, and p(tBA-AA) stands for the weakly hydrophilic poly-(*tert*-butyl acrylate-co-acrylic acid) polymer. Upon UV-irradiation, the hydrophilicity switch of the PAzo block from hydrophobic to hydrophilic causes a change in the hydrophilic/hydrophobic balance of the copolymer, inducing vesicle dissociation.

Using the same chromophore, Lin et al. [121]. reported a novel photoresponsive polymersome, obtained by self-assembly of a copolymer composed of hydrophilic poly(ethylene oxide) (PEO) and hydrophobic azopyridine containing poly(methacrylate) (PAP). Upon UV-exposure, several morphology changes were observed, and were described as a cycle including transitions from initial vesicles to larger vesicles via fusion, disintegration and rearrangements. These transitions resulted from the deformation of the membrane structure due to the isomerization of azopyridine moieties disturbing the tight packing of the polymer chains in the membrane.

Recently, Mabrouk et al. [122]. reported on a very original light-responsive system. They fabricated polymeric vesicles in the micrometer-size range, with asymmetric membranes composed of inert poly(ethylene glycol)-*b*-polybutadiene (PEG-*b*-PBD), and a liquid crystal-based copolymer, PEG-*b*-PMAazo444 (PAzo). Upon self assembly, the PEG–PBD copolymers are segregated in the inner leaflet of the membrane, while the PAzo copolymers compose the outer leaflet of the membrane, hence forming an asymmetric membrane. When the azo moieties are in the *trans* form, the PAzo polymer adopts a rod-like structure in the membrane. When light is switched on, azo moieties are in the *cis* form, and the PAzo polymers undergo a conformational change to reach a coil conformation. Subsequently, the volume occupied by the PAzo chains increased, leading to a spontaneous change in curvature and to bursting of the giant vesicles by "curling" of the membrane.

3.4 Smart Surfaces: Surface-Supported Polymer Layers and Films

In all nanomedicine studies, a major challenge is determining how nanomaterials will interact with mucosa, tissues, and targeted cells. New modulation systems that control the surface properties or solubility of materials in response to an external signal are designed using the stimuli-responsive polymers on a material surface, or by modifying the surface with bioactive substances, such as enzymes. Indeed, smart surfaces that respond to specific chemical and biological species have been the basis for the fabrication of highly sensitive, reagent-less, re-usable biosensors [22].

Surface grafted polymers can be defined as long chain polymer molecules that are attached to a surface through one or a few anchor sites [123]. Two primary covalent attachment techniques, i.e. "grafting-to" and "grafting-from", have been reported to create polymer brushes. In the "grafting-to" technique, a pre-formed end-functionalised polymer in a solution reacts with a suitable substrate surface to form a tethered polymer brush. In the "grafting-from" method, also called the surface-initiated polymerization method, monomers are polymerised from surface-anchored initiators generally immobilised by the self-assembled monolayer technique (SAM) [124, 125]. SAMs offer ease of preparation and versatile surface chemistry, while polymer brushes can be produced by surface-initiated polymerization techniques with improved control of surface coverage, thickness and composition.

Stimuli-responsive polymer films can be prepared on substrate surfaces using several deposition techniques of differing complexities and applicability, such as spin coating, chemical vapour deposition, laser ablation, plasma deposition, and chemical or electrochemical reactions [126–128]. The choice of deposition methods depends on the physicochemical properties of the polymer material, the film quality requirements and the substrate being coated.

3.4.1 Temperature-Responsive Surfaces

The most widely studied temperature-controlled films are built from PNiPAAm, a thermo-responsive polymer that has an LCST of 32°C in aqueous solution [129]. PNiPAAm chains present a widespread hydrogen bonding network between the amide groups and water molecules. Above LCST, PNiPAAm films undergo a phase transition, from a hydrated swollen state to yield a collapsed morphology (solvent is forced out) [130–132]. The reversible volume phase transition of PNiPAAm films can be utilised to develop thermo-responsive culture media for cells [133–135].

Surface attached stimuli-responsive polymers do not aggregate to form a separate phase, but the conformational transition from the hydrophilic to hydrophobic state endows the surface with regulated hydrophobicity. For example, when PNiPAAm was end-grafted to solid substrates, it provided the surface with thermally controlled wettability and thickness [136]. At low temperatures, the composition profiles are approximately parabolic and extend into the solvent, while at temperatures above the LCST, the polymer profiles are collapsed near the surface. Moreover, nano-patterned thermo-responsive poly(2-(2-methosyethosy)-ethyl methacrylate) brushes demonstrate switching of both the thickness and the topography under temperature stimuli [137].

3.4.2 Electro-Responsive Surfaces

Height changes of polyelectrolyte brushes in response to the presence of ions of different sizes and charge were recently actively explored. When polymer chains bond with counter ions, the swelling and the hydrophilic/hydrophobic properties of the polymer layer change, while patterned brushes with two oppositely charged polyelectrolytes provide reversible switching of wettability, charge, and topography in an inverse manner. For example, by employing the electrochemical reaction in which aromatic nitro (NO_2) groups can be chemically modified by a redox process to amino (NH_2) groups, a surface can be functionalized by site-selective and reaction-controlled immobilisation of DNA [138, 139], and protein [140]. Also, by using the electroactive *O*-silyl hydroquinone moiety to tether the RGD peptide ligand to the monolayer, electroactive functionalised surfaces based on the hydroquinone–quinone redox couple have been shown to allow real-time control of molecular interactions that mediate peptide attachment and consequently the adhesion of cells [141].

On the basis of reversible doping of conducting polymers, a variety of anions have been electrostatically entrapped in conducting polymer films and released by electrical stimulus in a controlled way. As an example of this, positive charged neurotransmitter dopamine was successfully released from a conducting composite polymer, poly(*N*-methyl pyrrolylium)/poly(styrene sulfonate), prepared by anodic polymerization [142]. In its reduced state, this film was able to bind dopamine cations, which were then released by oxidizing the polymer film. Another example is polypyrrole films that can reversibly change their oxidation state, and consequently their properties and surface binding characteristics [143].

3.4.3 Photo-Responsive Surfaces

As described previously, there are mainly two types of photo-responsive molecules that may be used for a photo-triggered response. Spiropyran derivatives can transform from a hydrophobic spiro conformation to a polar hydrophilic zwitterionic merocyanine conformation under UV light, and can reversibly change with visible light [144, 145]. This change from the hydrophobic to the hydrophilic state upon isomerisation has been applied to demonstrate UV light-induced modification of surfaces [145]. The second type is azobenzene molecules that can change from the stable *trans* form to the *cis* state under UV light irradiation (300–400 nm), and reverse the isomerisation by irradiation with visible light [146–148].

A photo-responsive copolymer monolayer combining PNiPAAm and spiropyran chromophores has been used to tailor cell-adhesion by switching light on or off [149]. Change in surface hydrophilicity was obtained by irradiation with 365 nm light and 'reset' by visible light irradiation (400–440 nm) [144]. Additionally, a surface that can be photo-activated for spatio-temporal control of cell adhesion has also been developed by the release of nitric oxide from 2-nitrobenzyl ester-terminated monolayer [150, 151]. The 2-nitrobenzyl groups were selectively removed and consequently the protein and polymer dissociated from the surface.

3.4.4 pH-Responsive Surfaces

Polyelectrolyte brushes are pH-responsive materials that undergo structural changes at interfaces when their chains are charged and/or discharged because of the protonation/dissociation of acid/base groups [152]. As a result, upon an alteration in pH, polyelectrolyte brushes transform from the swollen state to a shrunken state in which the polymer chains collapse [153]. For example, surfaces grafted with an Os-complex redox unit modified poly(4-vinyl pyridine) [154]. Another type of surface was obtained from a mixed polyelectrolyte brush consisting of poly(2-vinylpyridine) and poly(acrylic acid) that had switchable permeability for both anions and cations [155]. When the ambient pH was acidic (pH < 3), the poly(2-vinylpyridine) chains were positively charged and permeable to the anionic probe. However, the redox process for the cationic probe was prevented, resulting in a lack of transport for positively charged ions.

3.4.5 Dual-Stimuli Responsive Surfaces

A smart and stable polymer brush interface based on PNiPAAm, PAA and poly(*N*-isopropylacrylamide-co-acrylic acid) was able to reversibly respond to temperature, ionic strength and pH, independently or simultaneously [156]. The reversible change in hydrogen bonding between the two components (NIPAm and AAc) and water, and the ionization of carboxylate groups under different environmental condition resulted in the dual-stimuli response.

Chitosan based PNiPAAm films possessing both thermal and pH sensitivity were prepared by blending chitosan with PNiPAAm and PEG [157]. The resulting film had an LCST at around 32°C, due to PNiPAAm, and showed pH responsiveness due to the amino groups of chitosan component. Poly(vinylidene fluoride) (PVDF) hydrophobic films grafted with PAA via radiation grafting demonstrated convective permeability that changed significantly with the pH and/or the salt concentration of the surrounding fluids [158].

3.5 Polymer–Protein and Polymer–Drug Conjugates

Polymers conjugated with therapeutic agents have been extensively investigated over the past 30 years. Conjugation of polymers to therapeutic molecules resulted in macromolecular systems that synergistically combined the individual properties of the components. Drug solubilization, protein efficacy and stability are increased by conjugation, while immunogenicity and toxicity are lowered.

3.5.1 Temperature-Responsive Conjugates

Azido-terminated poly(*N*-isopropylacrylamide) (PNiPAAm–N_3) was conjugated to bovine serum albumin (BSA) [159]. When the temperature increased above the PNiPAAm lower critical solution temperature (LCST), the PNiPAAm–BSA bioconjugates formed stable nanoparticles composed of dehydrated polymer and hydrophilic protein. As an alternative to this systems, protein–polymer conjugates are based on biocompatible polyethyleneglycol methacrylate (PEGMA) [160]. Hybrid polymer–protein (PEGMA–trypsin) conjugates are promising candidates for biomedical applications. The first hybrid (diblock conjugate) and the second hybrid (triblock) demonstrated behaviour depending on their architectures but also their enzymatic activities—hydrolysis of peptide and protein substrates were different for various hybrids. This is an example of polymer–protein conjugates with varied architectures, and it can be used to regulate the properties of the protein polymer hybrids in terms of stability and reactivity.

3.5.2 pH-Responsive Conjugates

A pH-sensitive polymeric carrier for drug release in cancer therapy made of poly(vinylpyrrolidone-co- dimethylmaleic anhydride) (PVD) was conjugated with the drug adriamycin (ADR) [161]. At pH 8.5 no release of the drug from the conjugate was observed. In contrast, at neutral pH (7.0) and slightly acidic pH (6.0), fully active drug in the native form was released.

Also, anticancer polymer [P(*N*-(2-hydroxypropyl)methacrylamide)] drug conjugates, containing doxorubicin (DOX) attached via a pH-responsive hydrolytically labile spacer susceptible to hydrolysis (hydrazone conjugates) showed stability in pH 7.4 buffer but released DOX in response to pH change (from 7.4 to 5.6) [162].

3.5.3 Glutathione-responsive conjugates

N-acetyl-l-cysteine (NAC) is an antioxidant and anti-inflammatory agent with significant potential for applications in the treatment of stroke, neuro-inflammation and cerebral palsy. However, NACs with free sulfhydryl groups display high plasma binding, resulting in low stability and reduced drug efficacy. Conjugates of NAC with thiol-terminated multiarm (6 and 8) poly(ethylene–glycol) (PEG) with disulfide linkages involving sulfhydryls of NAC released the drug at intracellular GSH levels [163]. At physiological extracellular glutathione concentration (2 μM), both conjugates were stable and release of the NAC was not observed. NAC was also conjugated to poly(amidoamine) (PAMAM) dendrimers [164, 165]. PAMAM dendrimers, G4–NH_2 and G3.5–COOH, all with cleavable disulfide linkages, were designed for intracellular delivery. Based on PEG, a dendritic system for intracellular peptide delivery was manufactured via cleavable disulfide bonds [166]. The variable quantity of the disulphide linker allowed the adjustment of the cleavage and release of the drug peptide. Disulphide bonds were also used for the preparation of triazine dendrimer-paclitaxel (PAX) conjugates, as was an ester bond [167]. *N*-(2-hydroxypropyl)methacrylamide (HPMA) copolymer and TNP-470 ([*O*-(chloracetyl-carbomoyl) fumagillol]), an angiogenesis inhibitor, were covalently bound to GFLG (Gly-Phe-Leu-Gly) linker via an enzymatically degradable bond, ethylenediamine [168]. When the concentration of lysosomal cysteine proteases such as cathepsin B increased (this happens in many tumour endothelial cells), cleavage of the linker took place. This conjugate was studied further in vivo and in vitro and went to preclinical trials under the name caplostatin [169, 170].

3.5.4 Dual-Response Conjugates

Dual-response conjugates are also known. A biotin-terminated poly(*N*-isopropylacrylamide)-*b*-poly(acrylic acid) (PNiPAAm)-*b*-(PAA) was conjugated to streptavidin (SA) via the terminal biotin on the PNiPAAm block [171]. Interestingly, the usual aggregation and phase separation of PNiPAAm-SA following the thermally triggered collapse and dehydration of PNIPAAM (the lower critical solution temperature of PNiPAAm is 32°C in water) was prevented by the shielding of the PAA block. In addition, the aggregation properties of the [(PNiPAAm)-*b*-(PAA)]-SA conjugate were pH dependent. By varying temperature and pH, the sizes of these particles differed from 60 nm (pH 7.0, temperatures above the lower critical solution temperature of PNiPAAm) to 218 nm (pH 5.5 and 20°C). This was explained by hydrogen bonding between the –COOH groups of PAA with other –COOH groups and also with the –CONH– groups of PNIPAAM. The aggregation properties of the block copolymer–streptavidin conjugate differ from those of the free block copolymer.

4 APPLICATIONS OF STIMULI-RESPONSIVE POLYMERS IN NANOMEDICINE

The need for accurate and non-invasive diagnostic tools is essential for early intervention to prevent disease progression. In this regard, the development of nanodevices capable of detecting specific and meaningful analytes associated with syndromes, of visualizing the location and distribution of affected cells, and of reporting the activity of a therapeutic agent are highly desirable.

In therapy, the introduction of these agents into the body (regardless of the administration route employed) is confronted by a set of efficient biological barriers, constituting the body's system defenses. Building smart nanoscale systems that are able to circumvent such barriers is seen as a potential way to administer therapeutic agents in a safe, selective, and efficient manner.

As described previously, polymeric systems are available in a variety of forms and structures, from bulk to supramolecular assemblies. In addition, because of their unique properties, stimuli-responsive polymers offer many opportunities to introduce functionalities into nanostructures and allow the fabrication of various smart systems.

The exploitation of polymer responses to stimuli finds wide-ranging application in the biomedical field: smart systems are useful in imaging and sensing (diagnosis), controlled drug delivery and regenerative medicine (therapy), but also in bioseparation, gating valves, or transport and microfluidics [22, 104, 172–180].

In the next sections, we will highlight the most relevant applications of such polymers in several subfields of nanomedicine, and pay particular attention to the advantages and drawbacks associated with those techniques. We focus on systems exploiting the intrinsic properties of stimuli-responsive polymers, i.e. where the functioning of nanostructures is a direct result of polymer chain properties that change upon activation by a given stimulus. Therefore, stimuli such as a magnetic field and ultrasound fall beyond the scope of this review, because they are applied to nanoparticles found within a self-assembled system.

4.1 Diagnosis

Polymer sensors that respond to relevant biomolecules and analytes, as well as pH and temperature, may be very useful in the detection of diseases that are usually accompanied by a significant imbalance in chemicals or variations of physical variables in the environment. Because monitoring these changes and gradients is vital to the diagnosis of certain diseases, great efforts have been made in the field of polymeric biosensors. Another important feature of nanodevices used in biomedical applications is their ability to self-report effective functioning (delivery in a specific location for instance) with the use of imaging techniques.

4.1.1 Sensors

In the field of polymer sensors, the most relevant examples in literature make use of smart surfaces (either composed of self-assembled multilayers or thin polymer films) responding to a change in the conformation of polymer chains, smart polymer probes that respond to chemical modification of polymer chains, and self-immolative dendrimers [181]. In the next sections, these systems are reviewed and classified according to their specific applications.

4.1.1.1 Systems for the Detection of Physical Variables (pH and T)

Several groups exploited the motion of particles, such as gold nanoparticles or quantum dots linked to responsive polymer brushes anchored to a surface, in order to design polymeric nanosensors [182–184]. In such devices, conformational changes of the polymer chains caused by a given stimulus induce a vertical motion to the nanoparticles which can be easily monitored using surface plasmon resonance spectroscopy (SPR). In one example of a pH nanosensor, poly(2-vinylpyridine) (P2VP) polymer brushes reversibly collapsed due to a pH switch from 2 to 5 [183]. Surfaces acting as nano-thermometers were developed using a similar approach with core/shell CdSe/ZnS quantum dots attached to PNiPAAm polymer brushes [185].

Another type of sensor, known as a fluorescent polymeric sensor, presents the advantage of being based solely on the intrinsic properties of polymers. In these systems, a combination of stimuli-sensitive monomers and polymerizable fluorescent dyes compose the segments of the copolymers. Because the dye fluorescence is strongly dependent on its environment, significant changes in the fluorescence signal are observed upon changes in polymer chain hydrophilicity induced by stimuli. Such a copolymer of

PNiPAAm and benzofurazan dye-modified units was reported by Uchiyama et al. [186], and showed a clear and reversible response to temperature cycles, associated with PNiPAAm chain conformational changes and the polarity sensitivity of the benzofurazan moieties (Fig. 6A). The same group reported other polymers based on the same concept using a variety of dyes [187]. It should be mentioned that, in these systems, the temperature is correlated to the fluorescence intensity variations, which may be influenced by local concentration gradients, and that difficulties associated with measurement may occur (signal to noise ratio).

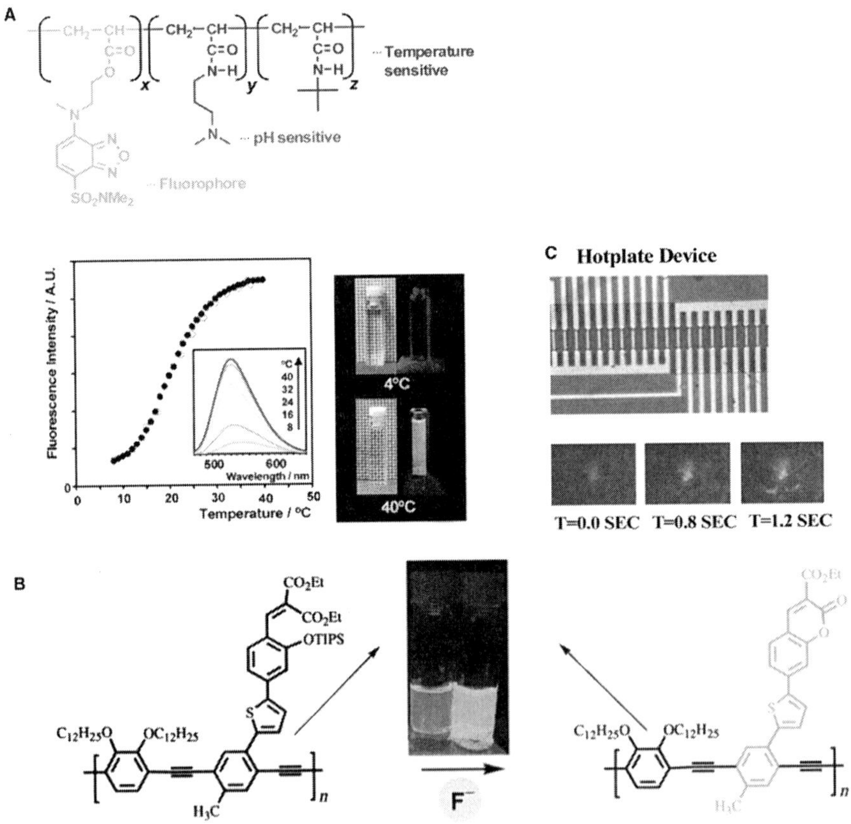

Figure 6. **A** Fluorescent polymer sensor for temperature [186]. **B** Fluorescent polymer sensor for the detection of fluoride ions [194]. **C** Micrograph showing the microfluidic hot plate with gold hot lines and fluorescence microscopy images showing thermally triggered release of fluorescein-labeled myoglobin from the PNiPAAm surfaces [192]

To address this drawback, devices from which the temperature (or other stimuli like pH) may be correlated to emissions at different wavelengths were proposed. In this regard, we describe here some examples using the fluorescence resonance energy transfer technique (FRET).

The transition from coil to globular conformation of responsive polymers was used in combination with FRET to produce pH and temperature sensors. As an example, a diblock copolymer of poly(ethylene glycol) and poly(sulfadimethoxine) (PEG–PSDM) was synthesized, with a FRET donor as a linker between the two chains, and a FRET acceptor as an end-group on the PSDM chain. When pH switches from 7.6 to 6.8 (values framing SDM pKa), the pH-responsive PSDM chains switch from coil to globular conformation. Consequently, the distance between the FRET molecules varies as a function of pH, and the emission wavelength changes accordingly [188].

In recent work, Wu et al. [189] reported the fabrication of silica nanoparticles coated with PNiPAAm temperature-responsive polymer brushes labeled with FRET molecules. 4-(2-acryloyloxyethylamino)-7-nitro-2,1,3-benzoxadiazole (NBDAE), and 10-(2-methacryloxyethyl)-30,30-dimethyl-6-nitro-spiro(2H-1-benzo-pyran-2,20-indoline) (SPMA), were copolymerized with NiPAAm to yield P(NiPAAm-co-NBDAE)-*b*-P(NiPAAm-co-SPMA) copolymer brushes. According to the temperature variations that induce PNiPAAm collapse, specific emissions from FRET moieties were observed.

These two systems represent good examples of fluorescent pH- and thermo-meters.

4.1.1.2 Systems for the Detection of Small Analytes and Biomolecules

Detectors based on SPR spectroscopy have also been used successfully for immunoassay devices based on the enzyme-catalyzed degradation of polymer films. Sumner et al. coated substrates with poly(ester amide) films sensitive to chymotrypsin, and poly(trimethylene) succinate films sensitive to lipase. The decrease in polymer film thickness resulting from the gradual degradation of the polymer chains activated by the enzymes and monitored with SPR was shown to be directly proportional to the enzyme concentration. Therefore, the sensor was proposed as a simplified alternative to ELISA tests [190].

Another array nanodevice based on a microfluidic hot plate grafted with PNiPAAm polymer was reported [191]. It was shown that, depending on the temperature of the hot line, the surface adsorbed and desorbed proteins within seconds (Fig. 6C) [192]. As competitive adsorption/desorption between two proteins occurs interdependent with heating time, the system can be used for selective analysis and separation of proteins.

Another type of detection based on the sensing of analytes via specific chemical reactions changing the properties of polymers has also been reported. An example of fluorescent amplification via enzymatic degradation of a polymer chain was reported recently by Tanaka et al. [193]. A polymer with a phosphate-caged fluorescein main chain was synthesized via polycondensation with diol linkers. Although the polymer obtained was not fluorescent, digestion of the backbone with alkaline phosphatase released highly fluorescent moieties, and the polymer was used to assess the enzymatic activity of a cell lysate.

Chemically induced response was also proposed by several groups to detect potentially toxic elements in drinking water. Although this application may not be core nanomedicine, we mention it in this review because it represents an improvement to prevent future complications and diseases. Kim et al. [194] synthesized a polymer with coumarin derivatives as side groups, able to detect fluoride ions (F^-). The structure of coumarin derivatives can be converted back to coumarin by fluoride ions, thus restoring their fluorescent properties (Fig. 6B). This represents a good example of a fluorescent polymeric sensor for F^-.

The detection of highly toxic mercury using fluorescent polymers was also reported, using a copolymer of poly(ethylene oxide)-*b*-poly(*N*-isopropylacrylamide-*co*-RhBHA) [PEO-*b*-P(NiPAAm-*co*-RhBHA)], where RhBHA is a rhodamine-based Hg^{2+}-sensitive dye [195]. Detection is based on the selective ring-opening of the RhBHA moieties by Hg^{2+} to yield fluorescent acyclic moieties. In this account, authors also investigated the effect of the thermo-induced self-assembly of the amphiphilic block copolymer on the fluorescence intensity and found that, upon formation of micelles, the fluorescent moieties were located inside the hydrophobic core, significantly enhancing the fluorescence. Many other systems exist for the detection of different analytes, such as metalloproteins and transition metals [196, 197].

The group of Sun developed several sensors based on wettability switching (i.e. a reversible transition from superhydrophilicity to superhydrophobicity) of surfaces grafted with PNiPAAm [178]. They synthesized block copolymers comprising PNiPAAm segments and blocks able to recognize different biomolecules. For instance, poly(*N*-isopropylacrylamide)-poly(phenyl boronic acid) (PNiPAAm-PBA) surfaces exhibiting a dramatic change in the presence of glucose, or PNiPAAm comprising oligopeptide units able to bind specific saccharide enantiomers based on chiral recognition, have been reported and used to monitor activity and concentration levels.

A novel class of recently developed molecules called self-immolative dendrimers showed promising use in different applications, including diagnostics and drug delivery. The self-immolative dendrimer molecules comprise a triggerable focal point, which initiates a cascade-like fragmentation of the structure into its building blocks upon activation. It is possible to design the building blocks as active molecules that can be detected once cleaved (these molecules being known as reporters). The release of these subunits can be seen as an amplification of the activation signal (physical, chemical or biological).

Using this approach, Danieli et al. [198] built dendrons with a phenylacetamide group as a point of focus, and two different probes as reporters. As the phenylacetamide group is a substrate of bacterial enzyme penicillin-*G*-amidase (PGA), the dendrimers readily degraded upon enzymatic activation, and subsequent detection of the two reporters allowed the evaluation of enzymatic activity. Because of the limitations of dendrimers, especially the limited number of building blocks due to steric hindrance, the concept was adapted to linear polymers, coined self-immolative polymers [199], to improve the amplification of the signal. One drawback of these self-immolative systems is that the chemistry used in the cascade-like degradation has been, until now, exclusively based on aromatic compounds and the toxicity of such cleaved compounds represents a potential issue in terms of biocompatibility [200].

4.1.2 Imaging

It is interesting to note that the concept of fluorescent polymeric sensors presented previously may be used reversibly, as an imaging technique for the detection of diseased tissues that show slightly elevated temperatures or acidic pH. A good example was reported using polymers comprising dyes sensitive to near infrared (NIR), which is the ideal wavelength range for biomedical applications, since it has superior depth penetration in tissue as opposed to other wavelengths. In this work, Lee et al. [201] made use of Pluronic triblock copolymers [poly(ethylene oxide)-poly(propylene oxide)-poly(ethylene oxide) (PEO–PPO–PEO)] end-capped with a cyanine dye (Cy5.5). Contrary to PNiPAAm polymers showing an intrinsic responsive property, Pluronic block copolymers react to temperature via changes in their supramolecular interactions. Upon heating, the polymer chains evolve from a dissolved state to a micellar aggregation state. According to this work, the transition from dissolved chains to micelles is accompanied by fluorescence quenching of the Cy5.5 terminal dye. In turn, these structures can be used as NIR thermo-probes for imaging.

Another imaging system using stimuli-responsive dendrimers was developed by Criscione et al. [202]. They synthesized PAMAM dendrimers with fluorinated end groups that self-assembled into nano- and micro-particles. The system can deliver drugs under pH-induced disassembly, and the fluorine spins can be used for in vivo imaging using ^{19}F magnetic resonance imaging (^{19}F MRI). Experiments with mice show that the dendrimers can be tracked with non-invasive imaging (Fig. 7A). Interestingly, a shift in relaxation time was observed according to changes in environmental pH, meaning that the system can also be used as a powerful imaging technique for the localization of tumor, with acidic pH.

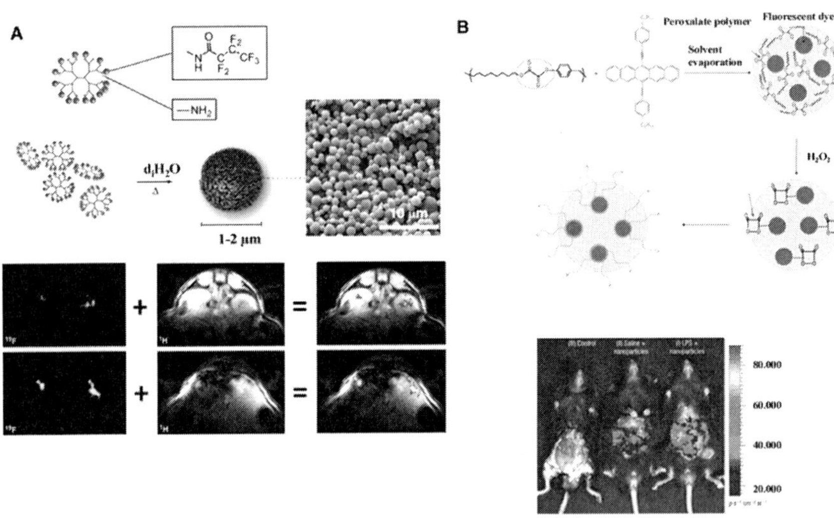

Figure 7. **A** Self-assembly of fluorinated PAMAM dendrimers with fluorine groups for ^{19}F MRI imaging. ^1H and ^{19}F images showing accumulation in vivo after IV injection of the nanoparticles: Overlaid picture of showing localization of the particles in the renal vasculature and localization of the particles in the liver after efficient filtration. [202] **B** Peroxalate–pentacene nanoparticles and H_2O_2-induced reaction yielding fluorescence, and in vivo imaging of hydrogen peroxide production in the inflamed peritoneal cavities of mice [203]

The detection of hydrogen peroxide is very desirable, as it is over-produced in a number of diseases. A smart system capable of imaging H_2O_2 in vivo was proposed by Lee et al. [203]. The nanoparticles were built from peroxalate polymers embedding a fluorescent dye, pentacene. The polymers reacted with hydrogen peroxide to form dixioetanedione intermediates that, in turn, excited the fluorescent dye, leading to light emission in the 460–630 nm wavelength region. Imaging efficiency was investigated in vivo with mice injected with lipopolysaccharide, inducing an inflammatory response. As shown in Fig. 7B, the nanoparticles were capable of imaging H_2O_2 production in the peritoneal cavity of mice.

4.2 Therapy

In this section, the use of stimuli-responsive polymers is classified into two categories. The first deals with devices used as nanocarriers for the transport and delivery of therapeutic agents. As mentioned earlier, the delivery of compounds to a specific location of the body is subject to a variety of obstacles, known as biological barriers, including the reticulo-endothelial system, endothelial/epithelial membranes, complex networks of blood vessels, abnormal flow of blood, and interstitial pressure gradients and the blood–brain barrier [9]. According to the nature of the therapeutic agent, these barriers may simply reduce the efficacy of the treatment, or completely prevent or annihilate its effect. Therefore, one can easily understand the benefit of using a protective vehicle to avoid early screening or biodegradation of a given cargo, with the goal of improving pharmacodynamics and pharmacokinetics, and delivering an intact molecule to a specific target in a controlled manner.

In the second section, we present some interesting works dealing with the use of stimuli-responsive polymers in the field of regenerative medicine. Synthetic polymers have been used to produce scaffolds and supports for cell growth, and the functionalities offered by stimuli-responsive polymers have actually improved those systems a great deal in the direction of biomimetic materials.

4.2.1 Delivery Systems

Delivery applications of smart polymers constitute an overwhelming collection of articles, referring to virtually all polymeric nanostructures described previously. The most trivial structures used for the entrapment and subsequent release of small hydrophobic molecules are micelles. However, the use of classic micellar structures is limited to the encapsulation of hydrophobic drugs in the core, at a time when the demand for carriers able to encapsulate hydrophilic compounds is ever growing. Polymeric vesicles, or polymersomes, have the advantage of encapsulating hydrophobic and hydrophilic therapeutic agents. As reported by Onaca et al. [176], they find applications as nanocarriers for hydrophilic and hydrophobic low molecular weight drugs, proteins, enzymes, and genes. A number of other polymeric nanostructures have shown great potential in drug delivery, including dendrimers, smart surfaces, and in situ forming nanogels, and will be briefly addressed in this review.

Due to the length of the present review, we focused on the most relevant works, with very promising or demonstrated applications in nanomedicine. The examples described below pertain to the triggered delivery of low molecular weight drugs, proteins and enzymes, as well as genes. The stimulus used may be external (i.e. external application of localized light irradiation, ultrasound, or temperature) or internal (i.e. the system responds to local hyperthermia, elevated pH, or over-expression of proteins and enzymes in a tumor environment) [204].

4.2.1.1 Delivery of Low Molecular Weight Drugs

Most of the low molecular weight drugs are hydrophobic molecules, and as such may be limited in their use due to solubility issues. Therefore, their pharmacodynamics and pharmacokinetics are greatly enhanced by solubilization in the hydrophobic domains of micelle cores, or dendrimers, or even the membranes of polymersomes, or by conjugation to polymers. Their release in the body can then be mediated by a number of different stimuli.

Doxorubicin (DOX), an anticancer hydrophobic drug, is perhaps most studied. However, many other small drugs have been used, including paclitaxel (PAX), camptothecin, cisplatine, dexamethasone, indomethacin, *N*-acetyl cysteine, ...

As discussed previously, a number of systems exploit the pH differences found in the body, whether in the vicinity of a tumor, or in endosomes. In those systems, the pH effect may result in the cleavage of pH-sensitive bonds (hydrozone, acetal), as was shown with drug–polymer conjugates releasing doxorubicin, paclitaxel, indomethacin, and camptothecin, which were covalently attached to polymer blocks forming micelles via acid-labile linkages [108, 205–208]. As an example, polymer–DOX conjugates were designed with hydrazone or amide pH-sensitive bonds linking the drug to a poly(ethylene glycol)-poly(caprolactone) (PEG–PCL) diblock copolymer [208]. The pH-triggered release and cellular uptake were evaluated in vitro with MDA-435/LCC6^{WT} and MDA-435/LCC6^{MDR} cells. The therapeutic effect was also investigated in vivo on mice bearing tumors, and tumor regression was shown to be more significant for mice treated with the polymer–DOX micelles (Fig. 8A).

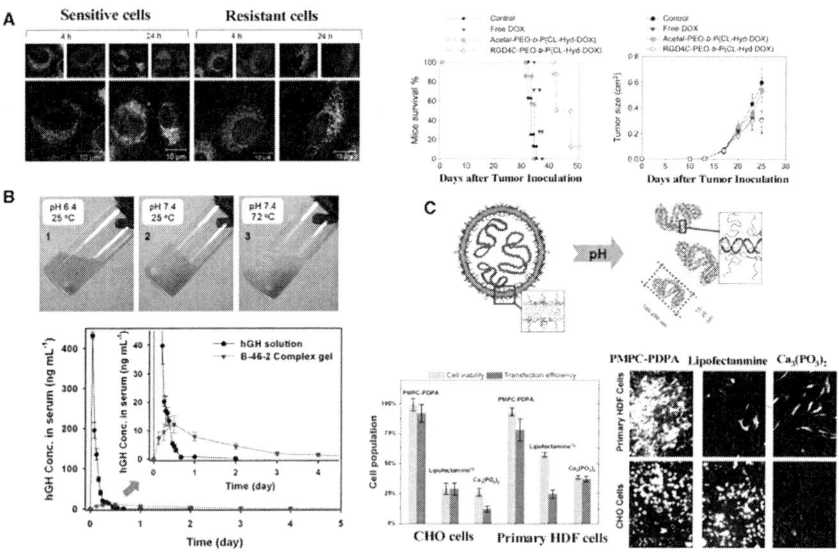

Figure 8. **A** Mitochondrial, endosomal and nuclear distribution of DOX in MDA-435/LCC6^{WT} and MDA-435/LCC6^{MDR} cells after internalization of pH-sensitive DOX–polymer conjugates: *Pink color* shows localization of DOX (*red*) in nucleus (*blue*), while *yellow* color is an indication of localization of DOX (*red*) in mitochondria (*green*) or endo/lysosomes. *Curves* showing mice survival and tumor size evolution for mice treated with DOX–polymer conjugates versus other groups [208]. **B** Photographs of phase transitions of PAEU–PEG–PAEU copolymers with respect to pH or temperature, and hGH concentration in blood of SD rats after injection of hGH solution, and hGH-gel formulation [256]. **C** Scheme depicting concept of pH-responsive PMPC–PDPA vesicles used for gene transfection and the cell viability assay and enhanced GFP expression [234]

The liberation of DOX was also shown using a dendritic polyester with pH-sensitive linkers [209]. Dendrimers as drug delivery systems have advantages over classic polymers, due to their well-defined architecture (low polydispersity, specific morphology, high density of functional groups) [210]. Drugs can be entrapped in dendrimer structures via encapsulation, complexation through electrostatic interactions, or covalent attachment (conjugation) [210]. Drug–polymer conjugates are more attractive than drug–dendrimer complexes, because of their increased stability and higher payloads.

As reported by Ahmed et al., polymersomes have also been used as nanocarriers for smaller drugs. They reported on polymer vesicles capable of encapsulating a cocktail of anticancer drugs, PAX (hydrophobic, entrapped in the membrane) and DOX (DOX–HCl salt, hydrophilic, encapsulated in the inner pool) [211]. The contents can be released from polymersomes via poration in the membrane induced by pH-triggered degradation of the PLA blocks. The system was tested in vivo, and tumors in rats were shown to shrink significantly (by 50% in 5 days). The limitation of the system, and of biodegradable polyesters in general, is due to the rather slow rate of poly(lactic acid) hydrolysis.

The reducing intracellular environment, due to the presence of glutathione, or the action of enzymes (including NADH-oxidase and disulfide isomerase) was also used to trigger the release of smaller drugs via cleavage of reduction-sensitive linkages. As an example *N*-acetyl cysteine (anti-inflammatory agent) was conjugated to polyamidoamine (PAMAM) dendrimers via disulfide linkages, and released in the intracellular domain, in the presence of reducing agents (glutathione, cysteine) [164]. The efficacy of the system was assessed by measuring the reactive oxygen species level in microglial cells. After 72 h, up to a 125% reduction of H_2O_2 was observed for cells treated with the loaded dendrimers. The efficacy of micelles sensitive to a reducing environment was also demonstrated, with a system based on camptothecin–polymer conjugates [212].

Responsiveness to temperature was exploited as well. Most of the temperature-sensitive systems are based on PNiPAAm. Using a thermo-responsive block copolymer, PEO- PNiPAAm, Qin et al. [213]. prepared vesicles which can encapsulate doxorubicin, and sequester a hydrophobic dye in their membranes. Upon cooling to temperatures below PNiPAAm LCST, the membrane is dissolved, and both contents are released upon complete dissociation of the vesicles. Quan et al. [214] designed thermo-responsive micelles from a poly(*N*-acroyloxysuccinimide)-*b*-poly(*N*-isopropylacrylamide)-*b*-poly(caprolactone) (PNAS-*b*-PNiPAAm-*b*-PCL) triblock copolymer for the delivery of DOX to HeLa cells. The micelles are internalized in HeLa cells, and above the LCST of PNIPAAM, i.e. at physiological temperature, 97% of the DOX payload is released.

In targeted drug delivery, it is also of interest to feature sensitivity towards a specific enzyme. A self-immolative dendrimer structure for the release of PAX activated by enzyme was reported [215]. The dendrimer was linked with an enzyme-responsive moiety to a *N*-(2-hydroxypropyl)-methacrylamide (HPMA) polymer for solubilization enhancement. Upon activation with cathepsin B (a lysosomal cystein protease), three PAX molecules were released. Cell growth inhibition assay using TRAMP-2 cells revealed a clear inhibition of cell proliferation when compared to controls. Polymeric micelles sensitive to lysosime were also reported [216].

Ionic interactions were used to deliver drugs, using the concept of PIC micelles, i.e. structures formed via electrostatic interactions between charged macromolecules and oppositely charged polymer chains. While conventional polymer micelles are mainly used for solubilization of hydrophobic drugs, hydrophilic, charged macromolecules (i.e. metal complexes, proteins, nucleic acids, and peptides) can be encapsulated in PIC micelles, and easily released via addition of counterions or pH switches [172, 217]. Cisplatine, a platinum complex-based anticancer drug, was bound to the carboxylic acids of poly(glutamic acid), which acted as ligands for Pt, and the complex was released upon ligand exchange with chloride ions in the body [218]. The micelles accumulated in tumor tissues of mice via EPR effect, leading to complete tumor regression.

Ionic interactions may also mediate the sol–gel transition of polyelectrolytes. Sol–gel polymers undergo a reversible gelation caused by a stimulus. They have application in drug delivery, where they can be formulated as a solution that embeds drugs, transforming into a gel when in contact with the body [179]. The drug is then released by diffusing through the gel, or upon gel degradation in the case of biodegradable polymers. As an example, alginate polymers containing pilocarpine (an alkaloid used in the treatment of glaucoma) undergo a sol–gel transition upon the addition of calcium ions, present in lachrymal fluid. Eye-drops of an alginate solution containing pilocarpine showed a significant decrease of intra-ocular pressure in rabbits over 10 h, due to the diffusion-controlled release of the drug [219]. Thermally induced gel formation was also reported in an ocular drug delivery system, with Pluronic and PNiPAAm based systems, for the delivery of pilocarpine and timolol maleate [220, 221].

4.2.1.2 Protein and Enzyme Delivery

The release of proteins and enzymes is also very challenging. These biomacromolecules are often fragile and present net charges. Therefore, they need to be shielded from potentially harmful species in the body, either via encapsulation in the lumen of polymeric vesicles, or reversible association with polyelectrolytes to form PIC micelles.

It should be noted that, although the encapsulation (and subsequent release) of functional proteins into responsive polymersomes has been demonstrated [116, 222, 223], to the best of our knowledge the triggered release of a therapeutic protein with demonstrated biomedical applications has never been shown [174]. Therefore, although polymersomes represent an attractive nanocarrier for protein delivery, in vivo medical applications are yet to be reported.

As described previously, the dissociation of PIC micelles may be triggered through the use of different stimuli responsive polymers, either via a charge conversion induced by the addition of counterions or pH change, the degradation of chemical bonds via pH or a reducing condition, or via temperature changes [224]. Using such charge conversion, lysozyme was encapsulated in PIC micelles composed of poly(ethylene glycol)-poly[(*N*'-citraconyl-2-aminoethyl)aspartamide] (PEG-pAsp(EDACit)). The PIC micelles degraded in response to the endosomal pH and released lysozyme [225].

Another approach to controlled drug delivery of proteins is to use smart surfaces responsive to temperature, chemical stimuli, or electric stimulus. Polymer films grafted on surfaces are good candidates for drug delivery because they have high storage and high retention capability, and can uptake and release biomacromolecules on demand [22]. As an example, polypyrrole (PPy) offers an opportunity to build electrically responsive systems. Nerve growth factor (NGP) was loaded on a polypyrrole conductive film, and was released upon electrical activation [226]. A similar system was used to release adenosine triphosphate (ATP) [227].

Smart polymer films can also be used as stimuli-activated gates to control release of molecules. The use of thermo-responsive PNiPAAm as an on–off gate was reported by Yavuz et al. [228]. PNiPAAm was covalently attached to gold nanocages via thiolate linkage. Using the photothermal effect of the gold nanocages, PNiPAAm underwent reversible conformational changes resulting in an on–off gating of the pores. The controlled release of DOX and lysozyme was investigated, and in vitro experiments respectively showed significant decreases in cell viability after 5 min of irradiation with IR light, and 80% bioavailabilty of the enzyme.

In situ-forming polymer gels are another class of materials built of stimuli-responsive polymers and having great potential in drug delivery. As an example, poly(ethylene glycol) (PEG) and poly(β-amino ester urethane) (PAEU) copolymer undergo pH- and temperature-induced gelation under physiological conditions [229]. These materials were used to deliver human growth hormone (hGH) to rats. Results show that the hGH concentration in the serum of rats was maintained at a higher level than in the control, due to the controlled release rate obtained with the gel (Fig. 8B).

4.2.1.3 Gene Delivery

The delivery of genes, or gene therapy, was proven very effective in the treatment of several diseases. As with proteins and enzymes, the transport of DNA into a cell is a difficult process, because of the charge and size of such molecules. Therefore, the need for gene carriers that can safely and effectively administer these materials in vivo is growing.

A method of choice is to use PIC micelles. As described above, these structures can help the vectorization of charged macromolecules using polyelectrolytes. Plasmid DNA complexed with a α-lactosyl-poly(ethylene glycol)-poly(2-(dimethylamino)ethyl methacrylate) block copolymer (lactose–PEG–PAMA) was efficiently transfected to HepG2 cells [230]. Another example of PIC micelles was reported by Xiong et al. [231], where siRNA was delivered to metastatic human MDA435/LCC6 cancer cells, and efficient gene silencing was observed.

Recently, an example of a block copolymer for gene delivery bearing a pH-sensitive linkage between hydrophilic and hydrophobic segments was reported. The poly[(2-dimethylamino)ethyl methacrylate] (PDMAEMA) and PEG blocks are connected via an ortho-ester, which can be cleaved upon pH-triggering [232]. Transfection efficiency was proven with the encapsulation of luciferase and EGFP gene expression plasmids, and their pH-triggered release in the endosome of 293T cells.

An example of an instantaneously pH-responsive polymer vesicle was described by Armes and coworkers [233]. They developed a highly biocompatible and pH-sensitive block copolymer, poly[2-(diisopropylamino)ethyl methacrylate]-poly[2-(methacryloyloxy)ethyl phosphorylcholine] (PDPA-PMPC). The PDPA block is deprotonated and insoluble at pH above 7 (pKa around 5.8–6.6). Water-soluble doxorubicin was encapsulated within PDPA–PMPC vesicles, and released upon lowering the solution pH. The system also proved useful for the physical encapsulation and intracellular delivery of GFP-encoding DNA plasmid [234, 235]. As shown in Fig. 8C, superior GFP expression is obtained with the polymer vesicles when compared to Lipofectamine TD and calcium phosphate.

Polypeptide-based block copolymers also show temperature induced conformal changes, from α-helical to β-sheets structures. As an example, polymersomes built of PLL-*b*-PBLG-d7-*b*-PLL have been synthesized, where PLL and PBLG-d7 are poly(l-lysine hydrochloride) and poly(γ-benzyl-d7-lglutamate), respectively [236]. In vitro encapsulation and release of plasmid DNA was shown.

In an example of structures similar to polymer–drug conjugates by Oishi et al. [237], micelles in which the corona-forming block itself is a therapeutic agent have been synthesized. The oligonucleotides, connected to the hydrophobic block with a pH-sensitive spacer, were released upon pH change.

As emphasized in several reviews, dendrimers are also very useful as transfection vectors, for different DNA molecules [210, 238, 239].

4.2.2 Regenerative Medicine

Stimuli-responsive polymers also find application in regenerative medicine. In this regard, they can be classified into polymers for the design of smart surfaces, and polymers that undergo sol–gel transitions for injectable implants. Smart surfaces may be used as supports or scaffolds, with excellent controllability of surfaces properties, that can, in turn, be used for adsorption and desorption of biomacromolecules and cells. It is known that cell behavior and attachment is greatly influenced by the wettability of a surface, and that biomacromolecules have higher affinity for hydrophobic surfaces. Therefore, depending on the application, stimuli-responsive polymers grafted on surfaces provide possibilities to design scaffolds for tissue engineering.

4.2.2.1 Smart surfaces for tissue engineering

Cells in tissues grow in a rather complex fashion, surrounded by an extracellular matrix (ECM) that plays an essential role as a support. In addition, ECM elicits a wide range of biological signals and releases various biological factors, controlling both cell behavior and proliferation. In order to build viable cell sheets for tissue engineering, synthetic materials should mimic functionalities, similar to ECM. Thus, the use of stimuli-responsive polymers to design smart surfaces as ECM biomimetic materials to be used as scaffolds for the growth of new cells and tissue engineering is currently a fast growing research area. In order to advantageously replace other existing materials and allow the growth and proliferation of cell sheets, smart surfaces should display reversible changes in their affinity for biomolecules and their cell adhesion properties, as well as provide sustained release of biomacromolecules.

Although polymer substrates have been used previously in cell culture (with polystyrene, for instance), the use of stimuli-responsive polymers represents a gentler alternative to mechanical or enzymatic digestion (protease) for cell detachment procedures needed in these systems. It guarantees the collection of intact cell sheets using a non-invasive cell recovery method, and these cell sheets can then be implanted in the body for tissue engineering applications.

As an example, thermo-responsive polymer films have been shown to be very useful in the control of cell recognition, adhesion and detachment. In this field, pioneering work was performed by Okano et al. [133, 240, 241] using PNiPAAm as the thermo-responsive polymer. Various cells, including hepatocytes, endothelial cells, fibroblasts, keratinocytes, epithelial cells, macrophages, and microglial cells, adhere and proliferate on such surfaces. When temperature is lowered under the LCST of PNiPAAm, the surface gradually switches from hydrophobic to hydrophilic, leading to cell desorption, without the need to use EDTA or trypsin [242].

In order to improve selective cell adhesion, biologically active moieties have been integrated into smart surfaces. As an example, dynamic surfaces controlling the presentation of recognition and regulatory signals were investigated [243]. In these systems, immobilized RGD sequences promote cell adhesion, and are shielded upon lowering temperature.

As mentioned earlier, the immobilization and programmed release of biologically active agents is desirable in order to promote cell adhesion and direct cell behavior. Such molecules can be hosted on smart surfaces via electrostatic interactions, conjugation, or encapsulation. Release of proteins was shown using ionic strength-sensitive [244] and thermo-responsive systems [245–250].

Even though temperature responsive surfaces based on PNiPAAm have been studied the most, other stimuli have also been investigated, such as light and electrical signals. Nerve regeneration is crucial, because it is very difficult to reconnect severed nerves by surgical means. The use of electro-responsive surfaces based on conductive polypyrol (PPy) was explored, and PC-12 as well as chicken sciatic nerve explants were shown to grow and proliferate preferentially on PPy surfaces submitted to an electric stimulus, when compared to controls [251].

Light was used with spiropyran-based polymers to efficiently detach cells from surfaces in a reversible manner [145]. Platelets and mesenchymal stem cells were shown to adhere to a poly(nitrobenzospiropyran)-poly(methyl methacrylate) copolymer, where the photo-sensitive groups are in a closed, non-polar spiropyran isomer conformation (hydrophobic surface). Upon UV irradiation, the spiropyran is converted to a zwitterionic merocyanine isomer, facilitating cell detachment (hydrophilic surface). Interestingly, light activated systems allow the manipulation of cell sheets, via the selective irradiation of a given region, thus creating patterns (Fig. 9A) [144].

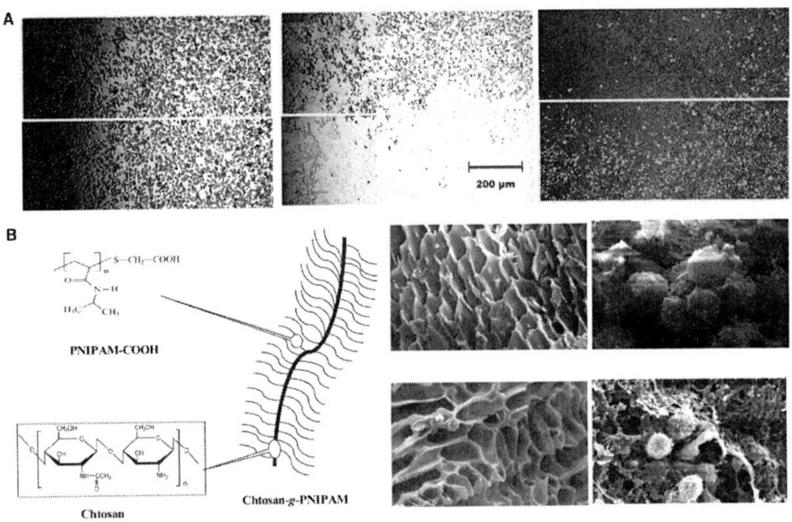

Figure 9 **A** Manipulation of CHO-K1 cell sheets with UV irradiation and temperature: microscopic images of photoresponsive culture surface before (*left*) and after (*middle*) regional UV irradiation followed by the low-temperature washing, and after second regional UV irradiation followed by the low-temperature washing (*right*). *Yellow rectangles* indicate UV-irradiated regions [144]. **B** SEM pictures of injectable nanogels formed by chitosan–PNiPAAm copolymers (*left*): SEM micrographs of chitosan–PNiPAAm hydrogel scaffold and hydrogel scaffold after temperature cycling between 25 and 37°C 100 times (*up*), chondrocytes and meniscus cells cultured in chitosan–PNiPAAm hydrogel scaffolds for 21 days (*bottom*) [254]

Another application of smart surfaces is the controlled fabrication of biomimetic ceramics. Recently, a thermo-responsive surface built of PLA and Bioglass with grafted PNiPAAm showed an interesting application in biomineralization. The production of bonelike apatite is of prime interest for regeneration and tissue engineering, especially for orthopedic applications. In their work, Shi et al. [252] showed that calcification could be controlled by temperature, and yielded apatite material with bone-like structure.

4.2.2.2 Sol–Gel Transition Polymers as Injectable Implants

Most of these systems are used exclusively for in vitro cell cultures, followed by cell desorption: for in vivo use, surgery must be performed to implant the cell sheets. To avoid this, a class of materials known as injectable implants is used. These systems are based on the gelation of a polymer solution upon injection into the body, and can promote cell delivery or other useful therapeutic agents such as growth factor.

The basis for using injectable polymers is that the matrix temporarily replaces damaged tissue, allowing proliferation and growth of cells until a new cell sheet or extracellular matrix is produced on site. Among the physiological stimuli used for gelling, temperature is the most studied and the most advantageous for in vivo application, due to its ease of use. Chitosan–PNiPAAm copolymer-forming gels have been employed as thermo-responsive injectable nanogels as scaffolds for tissue engineering [253, 254]. Mesenchymal stem cells embedded in the copolymer solution were able to differentiate into chondrocytes (cells found in cartilaginous matrix) in vitro (Fig. 9B). The cell–polymer mixture was injected into rabbit bladders, where the formation of new cartilage on the polymer matrix was detected [253]. Another thermo-responsive in situ forming gel based on chitosan and Pluronic polymers was shown to exhibit superior haemostatic properties [255].

5 SUMMARY AND CONCLUSIONS

Progress in medicine today relies on the advent of new systems and approaches that serve to detect pathological events in early stages, permit precise, safe surgery, and treat a specific region efficiently with minimal side effects. In this respect, stimuli-responsive systems are of particular interest. Stimuli-responsiveness represents a key property in medical applications because it serves to allow for controllable response from biological compartments, such as the release of an encapsulated/entrapped active compound, the triggering of a signaling process, or the detection of a specific biomolecule. A variety of systems that are intended to response to stimuli or a combination of stimuli has been developed based on polymers. There are two possible ways to obtain responsiveness: by using an SR polymer or by using a stimuli-responsive compound combined with a non-responsive polymer serving as a template. Stimuli-responsive polymers represent a smart, synthetic way to mimic the behavior of biopolymers, such as proteins, that undergo drastic conformational change at a critical point while remaining stable over a wide range of environmental conditions. Here, we have focused on stimuli-responsive polymers and have indicated both the variety of changes to physical, chemical and biological stimuli, and the possible medical applications. The response of a given polymer is based either on a dramatic alteration of its structure or on a change in its properties, such as charge, solubility, or polarity. An alteration to the polymer structure takes place when the polymer is degraded by breaking chemical bonds in the backbone or at specific positions where cross-linking moieties are inserted in its structure for this purpose. The change in properties is achieved by introducing functional groups that support or even induce changes in chain dimension, secondary structure or supramolecular assembly architecture. Changes in properties are mediated by changes in intermolecular interactions, by undergoing a specific chemical reaction, or by the presence of modified physical conditions.

A large variety of SR polymer-based systems has been developed, both in solution and on solid support, to serve diagnostic and therapeutic purposes. In solution, various architectures have been introduced, ranging from dendrimers to supramolecular assemblies generated by the self-assembly of amphiphilic copolymers, such as micelles and vesicles. On solid support, polymer mono- and multilayers undergo a change in properties as a response to an external stimulus and thus generate smart, active surfaces—especially important in biosensing approaches. However, the multitude of polymer systems and assemblies is dramatically reduced when medical application is intended, due to the complex requirements related to use inside the body. In this respect only SR polymers that are biocompatible and biodegradable can be used without toxicity problems. In addition, size, charge, flexibility, and shape of supramolecular assemblies are properties that should be modulated so as to allow for an optimum administration route and simultaneous high efficacy. Multifunctionality is another key factor that serves to increase the potential of polymer systems in medical applications in terms of developing targeting approaches, or theragnostic strategies. We have presented various medical applications here, in which SR polymer systems represent ideal candidate systems, starting with diagnostic approaches and extending to therapeutic treatment and tissue regeneration. However, using SR polymer systems/assemblies at the nanometer scale is an emerging field that will benefit greatly from more and extended studies on biodisposability, biodistribution, and toxicity in order to provide safe solutions and improve a patient's condition. The modulation of polymer properties for an efficient response to a stimulus represents an important parameter that must be adjusted in medical applications, but must always take into account the overall behavior of the system as it copes with the challenges presented under biological conditions, especially inside the body.

NOTES

E. Cabane, X. Zhang contributed equally to this work.

This article is part of the Topical Collection "In Focus: Nanomedicine".

DECLARATIONS

Acknowledgments

Financial support by the SNSF and the NCCR Nanosciences is gratefully acknowledged. In addition, we thank Mark Inglin from the University of Basel for reading the manuscript. WM thanks in particular Prof. P. Vajkoczy.

Open Access

REFERENCES

1. The Lancet (2003) Lancet 362:673
2. Jeong B, Gutowska A (2002) Trends Biotechnol 20:305–31
3. Galaev IY, Mattiasson B (1999) Trends Biotechnol 17:335–340
4. Hoffman AS, Stayton PS, Bulmus V, Chen G, Chen J, Cheung C, Chilkoti A, Ding Z, Dong L, Fong R, Lackey CA, Long CJ, Miura M, Morris JE, Murthy N, Nabeshima Y, Park TG, Press OW, Shimoboji T, Shoemaker S, Yang HJ, Monji N, Nowinski RC, Cole CA, Priest JH, Harris JM, Nakamae K, Nishino T, Miyata T (2000) J Biomed Mater Res 52:577–586
5. Kikuchi A, Okano T (2002) Prog Polym Sci 27:1165–1193
6. Qiu Y, Park K (2001) Adv Drug Deliv Rev 53:321–339

7. Kumar A, Srivastava A, Galaev IY, Mattiasson B (2007) Prog Polym Sci 32:1205–1237

8. Roy D, Cambre JN, Sumerlin BS (2010) Prog Polym Sci 35:278–301

9. Sanhai WR, Sakamoto JH, Canady R, Ferrari M (2008) Nat Nano 3:242–244

10. Schmaljohann D (2006) Adv Drug Deliv Rev 58:1655–1670

11. Gil ES, Hudson SM (2004) Prog Polym Sci 29:1173–1222

12. Delcea M, Möhwald H, Skirtach AG (2011) Adv Drug Deliv Rev 63:730–747

13. Liechty WB, Kryscio DR, Slaughter BV, Peppas NA (2010) Annu Rev Chem Biomol Eng 1:149–173

14. Delcea M, Möhwald H and Skirtach AG (2011) Adv Drug Deliv Rev. (In Press, Corrected Proof)

15. Zhang L, Xu T, Lin Z (2006) J Membr Sci 281:491–499

16. Liu Y, Meng L, Lu X, Zhang L, He Y (2008) Polym Adv Technol 19:137–143

17. Ohya S, Sonoda H, Nakayama Y, Matsuda T (2005) Biomaterials 26:655–659

18. Suwa K, Morishita K, Kishida A, Akashi M (1997) J Polym Sci Part A Polym Chem 35:3087–3094

19. Na K, Lee KH, Lee DH, Bae YH (2006) Eur J Pharm Sci 27:115–122

20. Sosnik A, Cohn D (2004) Biomaterials 25:2851–2858

21. Anal AK (2007) Recent Pat Endocr Metab Immune Drug Discov 1:83–90

22. Mendes PM (2008) Chem Soc Rev 37:2512–2529

23. Jones DP, Carlson JL, Samiec PS, Sternberg P Jr, Mody V C Jr, Reed RL, Brown LAS (1998) Clinica Chimica Acta 275:175–184

24. Koo AN, Lee HJ, Kim SE, Chang JH, Park C, Kim C, Park JH, Lee SC (2008) Chem Commun 44:6570–6572

25. Gong JP, Nitta T, Osada Y (1994) J Phys Chem 98:9583–9587

26. Tanaka T, Nishio I, Sun S-T, Ueno-Nishio S (1982) Science 218:467–469

27. Roggan A, Friebel M, Dörschel K, Hahn A, Müller G (1999) J Biomed Opt 4:36–46

28. Ichimura K, Oh S-K, Nakagawa M (2000) Science 288:1624–1626

29. Ichimura K, Suzuki Y, Seki T, Hosoki A, Aoki K (1988) Langmuir 4:1214–1216

30. Wang S, Song Y, Jiang L (2007) J Photochem Photobiol C Photochem Rev 8:18–29

31. Yoshida M, Lahann J (2008) ACS Nano 2:1101–1107

32. Jiang X, Lavender CA, Woodcock JW, Zhao B (2008) Macromolecules 41:2632–2643

33. Li Y, Jia X, Gao M, He H, Kuang G, Wei Y (2009) J Polym Sci Part A Polym Chem 48:551–557

34. Czaun M, Hevesi L, Takafuji M, Ihara H (2008) Chem Commun 44:2124–2126

35. Szabó D, Szeghy G, Zrínyi M (1998) Macromolecules 31:6541–6548

36. Korth BD, Keng P, Shim I, Bowles SE, Tang C, Kowalewski T, Nebesny KW, Pyun J (2006) J Am Chem Soc 128:6562–6563

37. Marin A, Muniruzzaman M, Rapoport N (2001) J Controlled Release 71:239–249

38. Norris P, Noble M, Francolini I, Vinogradov AM, Stewart PS, Ratner BD, Costerton JW, Stoodley P (2005) Antimicrob Agents Chemother 49:4272–4279

39. Rapoport NY, Christensen DA, Fain HD, Barrows L, Gao Z (2004) Ultrasonics 42:943–950

40. Dissemond J, Witthoff M, Brauns TC, Haberer D, Goos M (2003) Hautarzt 54:959–965

41. Rofstad EK, Mathiesen B, Kindem K, Galappathi K (2006) Cancer Res 66:6699–6707

42. Vaupel P, Kallinowski F, Okunieff P (1989) Cancer Res 49:6449–6465

43. Gupta P, Virmani K, Garg S (2002) Drug Discov Today 7:569–579

44. Park SY, Bae YH (1999) Macromol Rapid Commun 20:269–273

45. Lee YM, Shim JK (1997) Polymer 38:1227–1232

46. Soppimath KS, Kulkarni AR, Aminabhavi TM (2001) J Controlled Release 75:331–345

47. Abdelaal MY, Abdel-Razik EA, Abdel-Bary EM, El-Sherbiny IM (2007) J Appl Polym Sci 103:2864–2874

48. Park H-Y, Song I-H, Kim J-H, Kim W-S (1998) Int J Pharm 175:231–236

49. Kurisawa M, Yui N (1998) Macromol Chem Phys 199:1547–1554

50. Lee JW, Kim SY, Kim SS, Lee YM, Lee KH, Kim SJ (1999) J Appl Polym Sci 73:113–120

51. Nakamura K, Murray RJ, Joseph JI, Peppas NA, Morishita M, Lowman AM (2004) J Controlled Release 95:589–599

52. Zhang J, Peppas NA (1999) Macromolecules 33:102–107

53. Sideratou Z, Tsiourvas D, Paleos CM (2000) Langmuir 16:1766–1769

54. Burke SE, Barrett CJ (2003) Biomacromolecules 4:1773–1783

55. Toncheva V, Wolfert MA, Dash PR, Oupicky D, Ulbrich K, Seymour LW (1998) Biochim Biophys Acta 1380:354–368

56. Corpart JM, Candau F (1993) Macromolecules 26:1333–1343

57. Kathmann EEL, White LA, McCormick CL (1997) Macromolecules 30:5297–5304

58. Salamone JC, Rodriguez EL, Lin KC, Quach L, Watterson AC, Ahmed I (1985) Polymer 26:1234–1238

59. Morrison ME, Dorfman RC, Clendening WD, Kiserow DJ, Rossky PJ, Webber SE (1994) J Phys Chem 98:5534–5540

60. Szczubiałka K, Jankowska M, Nowakowska M (2003) J Mater Sci Mater Med 14:699–703

61. Leong KW, Brott BC, Langer R (1985) J Biomed Mater Res 19:941–955

62. Mathiowitz E, Jacob JS, Jong YS, Carino GP, Chickering DE, Chaturvedi P, Santos CA, Vijayaraghavan K, Montgomery S, Bassett M, Morrell C (1997) Nature 386:410–414

63. Cohen S, Yoshioka T, Lucarelli M, Hwang LH, Langer R (1991) Pharm Res 8:713–720

64. Shenoy D, Little S, Langer R, Amiji M (2005) Pharm Res 22:2107–2114

65. Cerritelli S, Velluto D, Hubbell JA (2007) Biomacromolecules 8:1966–1972

66. Oh JK, Siegwart DJ, Lee H-I, Sherwood G, Peteanu L, Hollinger JO, Kataoka K, Matyjaszewski K (2007) J Am Chem Soc 129:5939–5945

67. Matsumoto S, Christie RJ, Nishiyama N, Miyata K, Ishii A, Oba M, Koyama H, Yamasaki Y, Kataoka K (2008) Biomacromolecules 10:119–127

68. Yoshida R, Yamaguchi T, Kokufuta E (1999) J Artif Organs 2:135–140

69. Chaterji S, Kwon IK, Park K (2007) Prog Polym Sci 32:1083–1122

70. Chambin O, Dupuis G, Champion D, Voilley A, Pourcelot Y (2006) Int J Pharm 321:86–93

71. Sinha VR, Kumria R (2001) Int J Pharm 224:19–38

72. Vandamme TF, Lenourry A, Charrueau C, Chaumeil JC (2002) Carbohydr Polym 48:219–231

73. Itoh Y, Matsusaki M, Kida T, Akashi M (2006) Biomacromolecules 7:2715–2718

74. Akala EO, Kopečková P, Kopeček J (1998) Biomaterials 19:1037–1047

75. Rao NA (1990) Trans Am Ophthalmol Soc 88:787–850

76. Nobuhiko Y, Jun N, Teruo O, Yasuhisa S (1993) J Controlled Release 25:133–143

77. Miyata K, Kakizawa Y, Nishiyama N, Harada A, Yamasaki Y, Koyama H, Kataoka K (2004) J Am Chem Soc 126:2355–2361

78. Boas U, Heegaard PMH (2004) Chem Soc Rev 33:43

79. Cheng Y, Xu Z, Ma M, Xu T (2007) J Pharm Sci 97:123–143

80. Patri AK, Majoros IJ, Baker J R Jr (2002) Curr Opin Chem Biol 6:466–471

81. Zhou J, Wang L, Ma J, Wang J, Yu H, Xiao A (2010) Eur Polym J 46:1288–1298

82. Yang Z, Zhang W, Zou J, Shi W (2007) Polymer 48:931–938

83. Haba Y, Harada A, Takagishi T, Kono K (2004) J Am Chem Soc 126:12760–12761

84. Haba Y, Kojima C, Harada A, Kono K (2006) Macromolecules 39:7451–7453

85. Koyama T, Hatano K, Matsuoka K, Esumi Y, Terunuma D (2009) Molecules 14:2226–2234

86. Burakowska E, Zimmerman SC, Haag R (2009) Small 5:2199–2204

87. Chen W, Tomalia DA, Thomas JL (2000) Macromolecules 33:9169–9172

88. Liu Y, Bryantsev VS, Diallo MS, Goddard W A III (2009) J Am Chem Soc 131:2798–2799

89. Yang K, Weng L, Cheng Y, Zhang H, Zhang J, Wu Q, Xu T (2011) J Phys Chem B 115:2185–2195

90. Kohman RE, Zimmerman SC (2009) Chem Commun 45:794–796

91. Cho SY, Allcock HR (2007) Macromolecules 40:3115–3121

92. Kostiainen MA, Rosilo H (2009) Chem Eur J 15:5656–5660

93. Ong W, McCarley RL (2006) Macromolecules 39:7295–7301

94. Azagarsamy MA, Sokkalingam P, Thayumanavan S (2009) J Am Chem Soc 131:14184–14185

95. Azagarsamy MA, Yesilyurt V, Thayumanavan S (2010) J Am Chem Soc 132:4550–4551

96. Jin Q, Liu G, Ji J (2010) J Polym Sci Part A Polym Chem 48:2855–2861

97. Kotharangannagari VK, Sánchez-Ferrer A, Ruokolainen J, Mezzenga R (2011) Macromolecules 44:4569–4573

98. Chan Y, Wong T, Byrne F, Kavallaris M, Bulmus V (2008) Biomacromolecules 9:1826–1836

99. Katayama Y, Sonoda T, Maeda M (2001) Macromolecules 34:8569–8573

100. González DC, Savariar EN, Thayumanavan S (2009) J Am Chem Soc 131:7708–7716

101. Savariar EN, Ghosh S, Thayumanavan S (2008) J Am Chem Soc 130:5416–5417

102. Discher DE, Ahmed F (2006) Annu Rev Biomed Eng 8:323–341

103. Meng F, Zhong Z (2011) J Phys Chem Lett 2:1533–1539

104. Meng F, Zhong Z, Feijen J (2009) Biomacromolecules 10:197–209

105. Chécot F, Lecommandoux S, Klok HA, Gnanou Y (2003) Eur Phys J E Soft Matter 10:25–35

106. Checot F, Rodriguez-Hernandez J, Gnanou Y, Lecommandoux S (2007) Biomol Eng 24:81–85

107. Shi Z, Zhou Y, Yan D (2008) Macromol Rapid Commun 29:412–418

108. Giacomelli C, Schmidt V, Borsali R (2007) Macromolecules 40:2148–2157

109. Holowka EP, Sun VZ, Kamei DT, Deming TJ (2006) Nat Mater 6:52–57

110. Blanazs A, Massignani M, Battaglia G, Armes SP, Ryan AJ (2009) Adv Funct Mater 19:2906–2914

111. Chen X, Ding X, Zheng Z, Peng Y (2006) New J Chem 30:577–582

112. Zhou Y, Yan D, Dong W, Tian Y (2007) J Phys Chem B 111:1262–1270

113. Li Y, Lokitz BS, McCormick CL (2006) Angew Chem Int Ed 45:5792–5795

114. Sanson C, Le Meins JF, Schatz C, Soum A, Lecommandoux S (2010) Soft Matter 6:1722

115. Smith AE, Xu X, Kirkland-York SE, Savin DA, McCormick CL (2010) Macromolecules 43:1210–1217

116. Napoli A, Boerakker MJ, Tirelli N, Nolte RJM, Sommerdijk NAJM, Hubbell JA (2004) Langmuir 20:3487–3491

117. Kim KT, Cornelissen JJLM, Nolte RJM, van Hest JCM (2009) Adv Mater 21:2787–2791

118. Xu H, Meng F, Zhong Z (2009) J Mater Chem 19:4183

119. Napoli A, Valentini M, Tirelli N, Muller M, Hubbell JA (2004) Nat Mater 3:183–189

120. Tong X, Wang G, Soldera A, Zhao Y (2005) J Phys Chem B 109:20281–20287

121. Lin L, Yan Z, Gu J, Zhang Y, Feng Z, Yu Y (2009) Macromol Rapid Commun 30:1089–1093

122. Mabrouk E, Cuvelier D, Brochard-Wyart FO, Nassoy P, Li M-H (2009) Proc Nat Acad Sci 106:7294–7298

123. Milner ST (1991) Science 251:905–914

124. Jordan R (2006) Surface-initiated polymerization I. Springer-Verlag, New York

125. Sakellariou G, Park M, Advincula R, Mays JW, Hadjichristidis N (2006) J Polym Sci Part A Polym Chem 44:769–782

126. Mittal KL, Lee K-W (1997) Polymer surfaces and interfaces: characterization, modification and applications. VSP, Utrecht

127. Norrman K, Ghanbari-Siahkali A, Larsen NB (2005) Annu Rep Sect C (Physical Chemistry) 101:174–201

128. Wise DL, Wnek GE, Trantolo DJ, Cooper TM, Gresser JD (1998) Electrical and optical polymer systems. Marcel Dekker, New York

129. Pelton R (2000) Adv Colloid Interface Sci 85:1–33

130. Chung JE, Yokoyama M, Okano T (2000) J Controlled Release 65:93–103

131. Dimitrov I, Trzebicka B, Müller AHE, Dworak A, Tsvetanov CB (2007) Prog Polym Sci 32:1275–1343

132. Dreher MR, Simnick AJ, Fischer K, Smith RJ, Patel A, Schmidt M, Chilkoti A (2007) J Am Chem Soc 130:687–694

133. Okano T, Yamada N, Sakai H, Sakurai Y (1993) J Biomed Mater Res 27:1243–1251

134. Takezawa T, Mori Y, Yoshizato K (1990) Bio Technol 8:854–856

135. Yamada N, Okano T, Sakai H, Karikusa F, Sawasaki Y, Sakurai Y (1990) Die Makromolekulare Chemie Rapid Commun 11:571–576

136. Sun T, Wang G, Feng L, Liu B, Ma Y, Jiang L, Zhu D (2004) Angew Chem Int Ed 43:357–360

137. Jonas AM, Hu Z, Glinel K, Huck WTS (2008) Nano Lett 8:3819–3824

138. Lee C-S, Baker SE, Marcus MS, Yang W, Eriksson MA, Hamers RJ (2004) Nano Lett 4:1713–1716

139. Yang W, Baker SE, Butler JE, Lee C-S, Russell JN, Shang L, Sun B, Hamers RJ (2005) Chem Mater 17:938–940

140. Mendes PM, Christman KL, Parthasarathy P, Schopf E, Ouyang J, Yang Y, Preece JA, Maynard HD, Chen Y, Stoddart JF (2007) Bioconj Chem 18:1919–1923

141. Yeo W-S, Yousaf MN, Mrksich M (2003) J Am Chem Soc 125:14994–14995

142. Miller LL, Zhou XQ (1987) Macromolecules 20:1594–1597

143. Wong JY, Langer R, Ingber DE (1994) Proc Nat Acad Sci 91:3201–3204

144. Edahiro J-I, Sumaru K, Tada Y, Ohi K, Takagi T, Kameda M, Shinbo T, Kanamori T, Yoshimi Y (2005) Biomacromolecules 6:970–974

145. Higuchi A, Hamamura A, Shindo Y, Kitamura H, Yoon BO, Mori T, Uyama T, Umezawa A (2004) Biomacromolecules 5:1770–1774

146. Auernheimer JR, Dahmen C, Hersel U, Bausch A, Kessler H (2005) J Am Chem Soc 127:16107–16110

147. Hayashi G, Hagihara M, Dohno C, Nakatani K (2007) J Am Chem Soc 129:8678–8679

148. Pearson D, Downard AJ, Muscroft-Taylor A, Abell AD (2007) J Am Chem Soc 129:14862–14863

149. Garcia A, Marquez M, Cai T, Rosario R, Hu Z, Gust D, Hayes M, Vail SA, Park C-D (2006) Langmuir 23:224–229

150. Nakanishi J, Kikuchi Y, Inoue S, Yamaguchi K, Takarada T, Maeda M (2007) J Am Chem Soc 129:6694–6695

151. Nakanishi J, Kikuchi Y, Takarada T, Nakayama H, Yamaguchi K, Maeda M (2004) J Am Chem Soc 126:16314–16315

152. Minko S (2006) J Macromol Sci Part C Polym Rev 46:397–420

153. Netz RR, Andelman D (2003) Phys Rep 380:1–95

154. Tam TK, Ornatska M, Pita M, Minko S, Katz E (2008) J Phys Chem C 112:8438–8445

155. Motornov M, Tam TK, Pita M, Tokarev I, Katz E, Minko S (2009) Nanotechnology 20:434006

156. Zhou J, Wang G, Hu J, Lu X, Li J (2006) Chem Commun 42:4820–4822

157. Sun H, Liu S, Ge B, Xing L, Chen H (2007) J Membr Sci 295:2–10

158. Kontturi K, Mafé S, Manzanares JA, Svarfvar BL, Viinikka P (1996) Macromolecules 29:5740–5746

159. Li M, De P, Gondi SR, Sumerlin BS (2008) Macromol Rapid Commun 29:1172–1176

160. Yaşayan G, Saeed AO, Fernández-Trillo F, Allen S, Davies MC, Jangher A, Paul A, Thurecht KJ, King SM, Schweins R, Griffiths PC, Magnusson JP, Alexander C (2011) Polym Chem 2:1567

161. Kamada H (2004) Clin Cancer Res 10:2545–2550

162. Ulbrich K, Etrych T, Chytil P, Jelínková M, Říhová B (2004) J Drug Target 12:477–489

163. Navath RS, Wang B, Kannan S, Romero R, Kannan RM (2010) J Controlled Release 142:447–456

164. Kurtoglu YE, Navath RS, Wang B, Kannan S, Romero R, Kannan RM (2009) Biomaterials 30:2112–2121

165. Navath RS, Kurtoglu YE, Wang B, Kannan S, Romero R, Kannan RM (2008) Bioconj Chem 19:2446–2455

166. Hamilton SK, Harth E (2009) ACS Nano 3:402–410

167. Lim J, Chouai A, Lo S-T, Liu W, Sun X, Simanek EE (2009) Bioconj Chem 20:2154–2161

168. Satchi-Fainaro R, Puder M, Davies JW, Tran HT, Sampson DA, Greene AK, Corfas G, Folkman J (2004) Nat Med 10:255–261

169. Chesler L, Goldenberg DD, Seales IT, Satchi-Fainaro R, Grimmer M, Collins R, Struett C, Nguyen KN, Kim G, Tihan T, Bao Y, Brekken RA, Bergers G, Folkman J, Weiss WA (2007) Cancer Res 67:9435–9442

170. Satchi-Fainaro R, Mamluk R, Wang L, Short SM, Nagy JA, Feng D, Dvorak AM, Dvorak HF, Puder M, Mukhopadhyay D, Folkman J (2005) Cancer Cell 7:251–261

171. Kulkarni S, Schilli C, Grin B, Müller AHE, Hoffman AS, Stayton PS (2006) Biomacromolecules 7:2736–2741

172. Cohen Stuart MA, Huck WTS, Genzer J, Muller M, Ober C, Stamm M, Sukhorukov GB, Szleifer I, Tsukruk VV, Urban M, Winnik F, Zauscher S, Luzinov I, Minko S (2010) Nat Mater 9:101–113

173. Mano JF (2008) Adv Eng Mater 10:515–527

174. Christian DA, Cai S, Bowen DM, Kim Y, Pajerowski JD, Discher DE (2009) Eur J Pharm Biopharm 71:463–474

175. Li M-H, Keller P (2009) Soft Matter 5:927–937

176. Onaca O, Enea R, Hughes DW, Meier W (2009) Macromol Biosci 9:129–139

177. Shaikh R, Pillay V, Choonara Y, du Toit L, Ndesendo V, Bawa P, Cooppan S (2010) AAPS Pharm Sci Tech 11:441–459

178. Sun T, Qing G (2011) Adv Mater 23:H57–H77

179. Van Tomme SR, Storm G, Hennink WE (2008) Int J Pharm 355:1–18

180. Xiong X-B, Falamarzian A, Garg SM and Lavasanifar A (2011) J Controlled Release. (In Press, Corrected Proof)

181. Hu J, Liu S (2010) Macromolecules 43:8315–8330

182. Mitsuishi M, Koishikawa Y, Tanaka H, Sato E, Mikayama T, Matsui J, Miyashita T (2007) Langmuir 23:7472–7474

183. Tokareva I, Minko S, Fendler JH, Hutter E (2004) J Am Chem Soc 126:15950–15951

184. Westenhoff S, Kotov NA (2002) J Am Chem Soc 124:2448–2449

185. Tagit O, Tomczak N, Benetti EM, Cesa Y, Blum C, Subramaniam V, Herek JL, Vancso GJ (2009) Nanotechnology 20:185501

186. Uchiyama S, Kawai N, de Silva AP, Iwai K (2004) J Am Chem Soc 126:3032–3033

187. Uchiyama S, Matsumura Y, de Silva AP, Iwai K (2003) Anal Chem 75:5926–5935

188. Hong SW, Ahn C-H, Huh J, Jo WH (2006) Macromolecules 39:7694–7700

189. Wu T, Zou G, Hu J, Liu S (2009) Chem Mater 21:3788–3798

190. Sumner C, Krause S, Sabot A, Turner K, McNeil CJ (2001) Biosens Bioelectron 16:709–714

191. Huber DL, Manginell RP, Samara MA, Kim B-I, Bunker BC (2003) Science 301:352–354

192. Bunker BC (2008) Mater Sci Eng R Rep 62:157–173

193. Tanaka K, Kitamura N, Chujo Y (2010) Macromolecules 43:6180–6184

194. Kim T-H, Swager TM (2003) Angew Chem Int Ed 42:4803–4806

195. Hu J, Li C, Liu S (2009) Langmuir 26:724–729

196. Grabchev I, Qian X, Xiao Y, Zhang R (2002) New J Chem 26:920–925

197. Sandanaraj BS, Demont R, Aathimanikandan SV, Savariar EN, Thayumanavan S (2006) J Am Chem Soc 128:10686–10687

198. Danieli E, Shabat D (2007) Bioorg Med Chem 15:7318–7324

199. Sagi A, Weinstain R, Karton N, Shabat D (2008) J Am Chem Soc 130:5434–5435

200. Lee CC, MacKay JA, Frechet JMJ, Szoka FC (2005) Nat Biotech 23:1517–1526

201. Lee S-Y, Lee S, Youn I-C, Yi DK, Lim YT, Chung BH, Leary JF, Kwon IC, Kim K, Choi K (2009) Chem Eur J 15:6103–6106

202. Criscione JM, Le BL, Stern E, Brennan M, Rahner C, Papademetris X, Fahmy TM (2009) Biomaterials 30:3946–3955

203. Lee D, Khaja S, Velasquez-Castano JC, Dasari M, Sun C, Petros J, Taylor WR, Murthy N (2007) Nat Mater 6:765–769

204. MacEwan SR, Callahan DJ, Chilkoti A (2010) Nanomedicine 5:793–806

205. Bae Y, Fukushima S, Harada A, Kataoka K (2003) Angew Chem Int Ed 42:4640–4643

206. Ponta A, Bae Y (2010) Pharm Res 27:2330–2342

207. Alani AWG, Bae Y, Rao DA, Kwon GS (2010) Biomaterials 31:1765–1772

208. Xiong X-B, Ma Z, Lai R, Lavasanifar A (2010) Biomaterials 31:757–768

209. Lee CC, Gillies ER, Fox ME, Guillaudeu SJ, Fréchet JMJ, Dy EE, Szoka FC (2006) Proc Nat Acad Sci 103:16649–16654

210. Astruc D, Boisselier E, Ornelas C (2010) Chem Rev 110:1857–1959

211. Ahmed F, Pakunlu RI, Brannan A, Bates F, Minko T, Discher DE (2006) J Controlled Release 116:150–158

212. Cabral H, Nakanishi M, Kumagai M, Jang W-D, Nishiyama N, Kataoka K (2009) Pharm Res 26:82–92

213. Qin S, Geng Y, Discher DE, Yang S (2006) Adv Mater 18:2905–2909

214. Quan C-Y, Wu D-Q, Chang C, Zhang G-B, Cheng S-X, Zhang X-Z, Zhuo R-X (2009) J Phys Chem C 113:11262–11267

215. Erez R, Segal E, Miller K, Satchi-Fainaro R, Shabat D (2009) Bioorg Med Chem 17:4327–4335

216. Malugin A, Kopeckova P, Kopecek J (2006) Mol Pharm 3:351–361

217. Miyata K, Christie RJ, Kataoka K (2011) React Funct Polym 71:227–234

218. Nishiyama N, Okazaki S, Cabral H, Miyamoto M, Kato Y, Sugiyama Y, Nishio K, Matsumura Y, Kataoka K (2003) Cancer Res 63:8977–8983

219. Cohen S, Lobel E, Trevgoda A, Peled Y (1997) J Controlled Release 44:201–208

220. Nanjawade BK, Manvi FV, Manjappa AS (2007) J Controlled Release 122:119–134

221. Cao Y, Zhang C, Shen W, Cheng Z, Yu L, Ping Q (2007) J Controlled Release 120:186–194

222. Cabane E, Malinova V, Menon S, Palivan CG, Meier W (2011) Soft Matter 7:9167–9176

223. Lee JCM, Bermudez H, Discher BM, Sheehan MA, Won Y-Y, Bates FS, Discher DE (2001) Biotechnol Bioeng 73:135–145

224. Lee Y, Kataoka K (2009) Soft Matter 5:3810–3817

225. Lee Y, Fukushima S, Bae Y, Hiki S, Ishii T, Kataoka K (2007) J Am Chem Soc 129:5362–5363

226. George PM, LaVan DA, Burdick JA, Chen CY, Liang E, Langer R (2006) Adv Mat 18:577–581

227. Pernaut J-M, Reynolds JR (2000) J Phys Chem B 104:4080–4090

228. Yavuz MS, Cheng Y, Chen J, Cobley CM, Zhang Q, Rycenga M, Xie J, Kim C, Song KH, Schwartz AG, Wang LV, Xia Y (2009) Nat Mater 8:935–939

229. Huynh CT, Kang SW, Li Y, Kim BS, Lee DS (2011) Soft Matter 7:8984–8990

230. Wakebayashi D, Nishiyama N, Yamasaki Y, Itaka K, Kanayama N, Harada A, Nagasaki Y, Kataoka K (2004) J Controlled Release 95:653–664

231. Xiong X-B, Uludağ H, Lavasanifar A (2010) Biomaterials 31:5886–5893

232. Lin S, Du F, Wang Y, Ji S, Liang D, Yu L, Li Z (2007) Biomacromolecules 9:109–115

233. Du J, Tang Y, Lewis AL, Armes SP (2005) J Am Chem Soc 127:17982–17983

234. Lomas H, Canton I, MacNeil S, Du J, Armes SP, Ryan AJ, Lewis AL, Battaglia G (2007) Adv Mater 19:4238–4243

235. Lomas H, Massignani M, Abdullah KA, Canton I, Lo Presti C, MacNeil S, Du J, Blanazs A, Madsen J, Armes SP, Lewis AL, Battaglia G (2008) Faraday Discuss 139:143–159

236. Iatrou H, Frielinghaus H, Hanski S, Ferderigos N, Ruokolainen J, Ikkala O, Richter D, Mays J, Hadjichristidis N (2007) Biomacromolecules 8:2173–2181

237. Oishi M, Sasaki S, Nagasaki Y, Kataoka K (2003) Biomacromolecules 4:1426–1432

238. Jain NK, Asthana A (2007) Expert Opin Drug Deliv 4:495–512

239. Shcharbin D, Klajnert B, Bryszewska M (2009) Biochemistry (Moscow) 74:1070–1079

240. Kushida A, Yamato M, Konno C, Kikuchi A, Sakurai Y, Okano T (1999) J Biomed Mater Res 45:355–362

241. Hirose M, Kwon OH, Yamato M, Kikuchi A, Okano T (2000) Biomacromolecules 1:377–381

242. Yamato M, Konno C, Utsumi M, Kikuchi A, Okano T (2002) Biomaterials 23:561–567

243. Ebara M, Yamato M, Aoyagi T, Kikuchi A, Sakai K, Okano T (2004) Biomacromolecules 5:505–510

244. Anikin K, Röcker C, Wittemann A, Wiedenmann J, Ballauff M, Nienhaus GU (2005) J Phys Chem B 109:5418–5420

245. Nath N, Chilkoti A (2003) Anal Chem 75:709–715

246. Comolli N, Neuhuber B, Fischer I, Lowman A (2009) Acta Biomater 5:1046–1055

247. Kim SY, Lee SC (2009) J Appl Polym Sci 113:3460–3469

248. Frey W, Meyer DE, Chilkoti A (2003) Langmuir 19:1641–1653

249. Hyun J, Lee W-K, Nath N, Chilkoti A, Zauscher S (2004) J Am Chem Soc 126:7330–7335

250. Cunliffe D, de las Heras Alarcón C, Peters V, Smith JR, Alexander C (2003) Langmuir 19:2888–2899

251. Schmidt CE, Shastri VR, Vacanti JP, Langer R (1997) Proc Nat Acad Sci 94:8948–8953

252. Shi J, Alves NM, Mano JF (2007) Adv Funct Mater 17:3312–3318

253. Cho JH, Kim S-H, Park KD, Jung MC, Yang WI, Han SW, Noh JY, Lee JWJW (2004) Biomaterials 25:5743–5751

254. Chen J-P, Cheng T-H (2006) Macromol Biosci 6:1026–1039

255. Ryu JH, Lee Y, Kong WH, Kim TG, Park TG, Lee H (2011) Biomacromolecules 12:2653–2659

256. Huynh CT, Kang SW, Li Y, Kim BS, Lee DS (2008) Soft Matter 7:8984–8990

5

CHAPTER

ELECTROCHEMICAL DNA HYBRIDIZATION SENSORS BASED ON CONDUCTING POLYMERS

Md. Mahbubur Rahman [1], Xiao-Bo Li [1], Nasrin Siraj Lopa [1], Sang Jung Ahn [2] and Jae-Joon Lee [1,*]

[1] *Nanotechnology Research Center and Department of Applied Life Science, College of Biomedical and Health Science, Konkuk University, Chungju 380-701, Korea*

[2] *Center for Advanced Instrumentation, Korea Research Institute of Standards and Science (KRISS), Daejeon 305-340, Korea*

ABSTRACT

Conducting polymers (CPs) are a group of polymeric materials that have attracted considerable attention because of their unique electronic, chemical, and biochemical properties. This is reflected in their use in a wide range of potential applications, including light-emitting diodes, anti-static coating, electrochromic materials, solar cells, chemical sensors, biosensors, and drug-release systems. Electrochemical DNA sensors based on CPs can be used in numerous areas related to human health. This review summarizes the recent progress made in the development and use of CP-based electrochemical DNA hybridization sensors. We discuss the distinct properties of CPs with respect to their use in the immobilization of probe DNA on electrode surfaces, and we describe the immobilization techniques used for developing DNA hybridization sensors together with the various transduction methods employed. In the concluding part of this review, we present some of the challenges faced in the use of CP-based DNA hybridization sensors, as well as a future perspective.

Keywords: electrochemical DNA sensors; conducting polymers; immobilization techniques; transduction mechanism

1. INTRODUCTION

Deoxyribonucleic acid (DNA) detection plays a prominent role in myriad disciplines related to human health, including the diagnosis of infectious diseases, identification of genetic mutations, drug discovery, forensics, and food technology [1–3]. For analyzing specific DNA sequences, reliable techniques based on either direct sequencing or DNA hybridization have been developed [2,3]. DNA sequencing is the process of determining the precise order of four nucleotides bases (adenine, guanine, cytosine, and thiamine) in a strand of DNA. DNA sequencing technology based on 2-D thin-layer chromatography was invented by Maxam and Gilbert and Sanger et al. in the 1970s [4,5]. The Maxam–Gilbert sequencing method rapidly became popular owing to the short sequence-reading times involved and because purified DNA could be directly used in this method. However, the requirement of using large amounts of purified DNA and complicated purification steps, combined with a shortage of available sequencers, limited the use of this method. Other major challenges associated with the Maxam–Gilbert method have been the extensive use of hazardous chemicals and difficulties with sample scale-up. By contrast, the chain-termination method developed by Sanger and coworkers made DNA sequencing comparatively more practical, because it required lesser amounts of purified DNA than the Maxam–Gilbert method did and it also provided multiple options for labeling the sequencing template. Of the two methods, the Sanger method is more efficient, uses fewer toxic chemicals, and requires the use of lower amounts of radioactivity. Furthermore, radioactive phosphorus labeling or the use of a primer labeled at the 5′ end with a fluorescent dye allows an optical set-up to be employed in the sequencing performed using the Sanger method; this facilitates easy analysis and the use of inexpensive automation. In order to enhance the sensitivity of this method, dye-terminator sequencing chemistry has been introduced [6–8]. However, dye-terminator sequencing has limited practical utility owing to the "dye effect" that arises from the difference in the incorporation of the dye-labeled chain terminators into the DNA fragment, which generates unequal peak heights and shapes in the DNA sequencing chromatogram. DNA sequencing by hybridization onto a solid support (e.g., nitrocellulose, nylon membrane, or lysine-coated glass slide) performed using fluorescently or radioactively tagged DNA became a common method for DNA analysis in the early 1990s [9,10]. This detection method appeared to be a promising tool for the real-time analysis of multiple DNA sequences, and it depended on the anchoring of multiple DNA-specific probes onto solid surfaces [11–13]. Such an array system might be useful in genome-wide genetic mapping, physical mapping, proteomics, and gene expression studies. However, the main challenges involved in using solid supports are the lack of commonly used DNA probes in "user-friendly" assays and an immobilization method that is fully compatible with the hybridization process, and low sensitivity and reproducibility [14]. To enable rapid, sensitive, and label-free DNA detection, numerous approaches have been suggested based on optical [15–17], acoustic [18], and electrochemical techniques [19–21].

Electrochemical methods are typically inexpensive and rapid methods that allow distinct analytes to be detected in a highly sensitive and selective manner [22–25]. Although electrochemical DNA sensors exploit a range of distinct chemistries, they all take advantage of the nanoscale interactions among the target present in solution, the recognition layer, and the solid electrode surface. This has led to the development of simple signal transducers for the electrochemical detection of DNA hybridization by using an inexpensive analyzer. DNA hybridization can be detected electrochemically by using various strategies that exploit the electrochemistry of the redox reaction of reporters [26] and enzymes immobilized onto an electrode surface [27], direct or catalytic oxidation of DNA bases [28–31], electrochemistry of nanoparticles [32–35], conducting polymers (CPs) [35–37], and quantum dots [38].

CPs are organic conjugated polymers that feature an extended π-orbital system through which electrons can move from one end of the polymer to the other. In 2000, H. Shirakawa, A. MacDiarmid, and A. Heeger were awarded the Nobel Prize in chemistry for their revolutionary research on the conductive behavior of polymers and provocative research based on CPs. Unlike saturated polymers, CPs exhibit several distinctive properties such as excellent electrical conductivity, low ionization potentials, and high electron affinity. The electrical conductivity of CPs is responsible for the excitation of polarons, bipolarons, and solitons during the doping processes. The ground state p-bonds (p – p*) of CPs are partially localized as a result of Peierls distortion [39]. However, depending on the doping concentration, the formation of polarons, bipolarons, and solitons creates distinct band gaps between the self-localized excitations and the localized electronic sates. CPs also exhibit very high flexibility, which can be modulated together with their electrical properties by using appropriate chemical modeling and synthesis [40–42]. These distinctive properties of CPs have broadened their application in various technological fields, such as in the design of light-emitting diodes [43], anti-static coating [44], electrochromic devices [45], solar cells [46], anti-corrosion coatings [47], chemical sensors and biosensors [48], and drug-release systems [49,50]. To date, diverse CPs have been developed and used in sensing applications, such as poly(acetylene), polypyrrole (PPy), polythiophene (PTh), poly(terthiophene), polyaniline (PANI), poly(fluorine), poly(3-alkylthiophene), poly(tetrathiafulvalene), poly(naphthalene), poly(p-phenylene sulfide), poly(para-phenylene vinylene), and poly(thionine) (PTH); these CPs are reviewed elsewhere [51]. Previous studies have also examined the growth and stabilities of PANI [52,53], PPy [54,55], poly(azulene) [56], and PTh [57,58]. Apart from these commonly used CPs, various reversibly doped and undoped CPs, which exhibit considerable changes in conductivity, have been studied using electrochemical methods [59]. The electrical conductivity of CPs depends substantially on the pH and the applied potentials, which can vary over several orders of magnitude [60]. Moreover, the grafting of organometallic compounds can aid in the tuning of the physical properties of CPs [61–64]. The electronic structure, chain conformation, and orientation of CPs can cause extremely sensitive changes in the polymeric chain environment of CPs. For example, a change in the delocalized electronic structure of CPs during DNA hybridization alters their optical and electrical properties [65]. These advantages offered by CPs make them suitable for developing sensitive electrochemical DNA hybridization sensors. In this article, we comprehensively review recent works on various CPs, as well as their application and implementation for electrochemical DNA sensing. Furthermore, we discuss the most commonly used methods of immobilizing DNA probes for developing DNA biosensors along with the transduction mechanism employed.

2. IMMOBILIZATION TECHNIQUES USED FOR DEVELOPING DNA BIOSENSORS

To enhance the sensitivity of DNA biosensors, the DNA probe used must be sufficiently immobilized [66]. Nonspecific adsorption and stabilization of the immobilized DNA probe are crucial for achieving high sensitivity and specificity. Moreover, minimizing nonspecific adsorption is essential for ensuring the high reactivity, accessibility, orientation, and stability of surface-confined DNA probes [66]. DNA probes immobilized on sensor surfaces can be used in a manner similar to enzyme-based biosensors; these probes are immobilized by means of adsorption, covalent immobilization, or avidin (or streptavidin)-biotin interaction [67,68]. Figure 1 shows the unique design of a CP-based DNA sensor. In this sensor, single-stranded DNA (ss-DNA) probes are immobilized on or within a CP layer, and target DNA base-pairing to the probe generates a recognition signal that can be recorded using an electrode (e.g., an electrode made of gold, platinum, or glassy carbon). Identification occurs at the CP/electrolyte edge, and the generated recognition signal reaches the transducer through the CP layer. In this section, we discuss the various methods used for immobilizing a DNA probe in order to develop a DNA hybridization sensor.

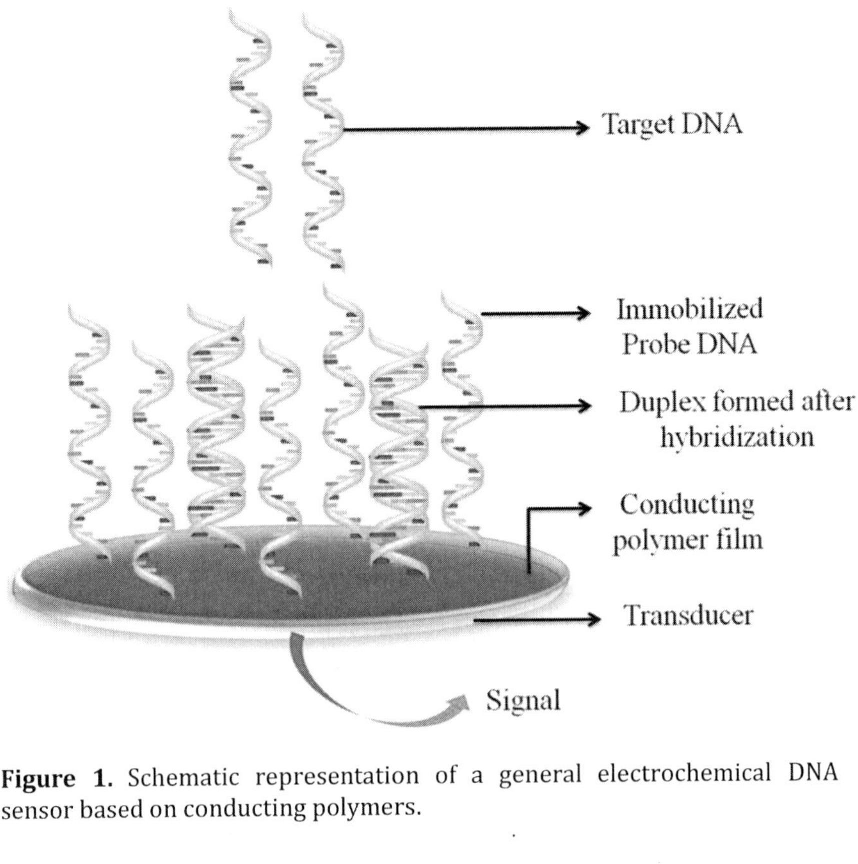

Figure 1. Schematic representation of a general electrochemical DNA sensor based on conducting polymers.

2.1. Adsorption

Adsorption is the simplest immobilization method in which a DNA probe can be immobilized without any modification of the probe [69]. Hirayama et al. developed an enhanced and simplified dry-adsorption protocol for DNA probe immobilization that increased hybridization sensitivity [70]. The efficiencies of DNA adsorption and retention were increased 1.4–6.5 and 4.2–19.6 times, respectively, compared with the efficiencies achieved using conventional methods such as incubation and decantation. Moreover, the use of this simple protocol reduces the consumption of DNA and increases the hybridization efficiency substantially. Another method involves the use of the amino group of the natural cationic chitosan polymer that can readily form a strong complex with the negatively charged phosphate backbone of DNA [71]. Xu et al. successfully immobilized a DNA probe labeled with aminoferrocene (AFC) on a chitosan-modified glassy carbon electrode (GCE) by means of adsorption [72]. The AFC-labeled DNA probe formed a duplex only with the complementary target DNA, and the detection limit was 2.0 nM. DNA that is either physically or chemically adsorbed onto a solid electrode surface can also be used for studying the electrochemical behavior of DNA and its interaction with other molecules [73–76]. For example, Azek et al. developed a disposable DNA sensor by physically adsorbing amplified human cytomegalovirus DNA onto a screen-printed electrode (SPE) [77]. The extent of hybridization of the target DNA was determined using horseradish peroxidase-conjugated streptavidin, and the detection limit was measured to be 6×10^{-7} nM.

DNA probes and DNA composites can also be immobilized on electrodes by applying an electric potential [78,79]. Wang et al. electrochemically adsorbed a DNA probe on an electrochemically pretreated carbon-paste electrode (CPE) at an applied potential of −0.5 V (vs. Ag/AgCl) [80]. This DNA sensor requires only nanogram quantities of DNA owing to the low background response of the potentiometric-stripping mode. Recently, Wu et al. electrodeposited silver nanoparticle-DNA composites at a controlled dimension on a GCE by reducing silver in the presence of DNA [81]. The inclusion of DNA with the silver nanoparticles prevents nanoparticle aggregation and enhances the catalytic activity. Lahiji et al. electrochemically deposited a uniform DNA-carbon nanotube (CNT) composite on an Au substrate by maintaining a +0.5 V potential [82]. This immobilization technique does not require prior DNA or substrate functionalization, and this is combined with a new method of generating a modified DNA electrode that offers the advantage of the high electron-transporting capacity of CNTs for sensing.

2.2. Covalent Immobilization

DNA probes have frequently been immobilized through covalent binding to various solids [83–88]. Ligaj et al. developed a stearic acid-modified conventional CPE in order to covalently immobilize DNA probes through an ethylenediamine linker [86]. This process does not alter the structural flexibility of the DNA probe, and it enhances DNA hybridization. Raymond et al. developed a simple and specific method that does not require the labeling of the target before hybridization [87]. The amino-linker of the probe DNA allows it to be covalently attached to a functionalized glass surface. Functionalized CP films, such as poly(3-pyrrolylacrylic acid) (Py-co-PAA), poly(5-(3-pyrrolyl) 2,4-pentadienoic acid) (Py-co-PPDA), and poly(3-pyrrolylpentanoic acid) (PPA), were previously used for covalently immobilizing a DNA probe [88]. Based on the principle of chemisorption, thiol-metal interactions have also been frequently employed to covalently bind thiol-functionalized DNA probes onto gold surfaces [89–92].

The distinctive electrical, thermal, chemical, mechanical, and 3-D spatial characteristics of CNTs suggest that it is possible to construct DNA biosensors with high sensitivity, selectivity, and multiplexing by exploiting Watson–Crick base-pairing [93]. Rodriguez et al. covalently attached single-walled CNTs (SWNTs) to a gold surface modified with 11-amino-1-undecanethiol (AUT) and subsequently immobilized a DNA probe on the Au/AUT/SWNTs through covalent linkage [94]. The interaction and covalent immobilization of DNA probes on functionalized CNTs and CNT/CP-composite-modified electrodes have been widely studied [95,96]. Yang et al. developed a sensitive electrochemical DNA sensor based on the synergistic effects of a PANI nanofiber and a multiwall CNT (MWNT) composite on a chitosan film [97]. The covalent immobilization of the DNA probe on the PANI-MWNT-composite film enhanced DNA hybridization, which was highly reproducible and stable.

2.3. Avidin/Streptavidin-Biotin Interaction

Avidin and streptavidin are large tetrameric proteins containing four identical biotin-binding sites that can be used for forming tetravalent avidin/streptavidin-biotin bonds in order to develop DNA-coated electrodes; in these electrodes, surface-confined avidin/streptavidin reacts with biotinylated DNA [98]. Because of the innate aqueous immobility of the avidin-biotin complex, this system is easy to use. Pan et al. generated a mixture of self-assembled monolayers (SAMs) by using 2-mercaptoethanol (ME) and 11-mercaptoundecanoic acid (MUA) on an Au electrode and attached a DNA probe to the activated MUA through streptavidin-biotin chemistry, as shown in Figure 2 [99]. Caruso et al. employed the quartz crystal microbalance (QCM) technique in order to immobilize biotin-DNA on an avidin-modified QCM electrode [100]. Several strategies have been developed for attaching biotin to modified CP electrodes, including the biotin-sandwich technique for immobilizing DNA [101–106]. Guillerez et al. designed a DNA sensor by using an electropolymerized biotinylated-PPy film [101]. Biotinylated DNA probes were immobilized onto PPy-biotin films through an intercalated layer of avidin (PPy-biotin/avidin/DNA probe). Furthermore, to detect DNA hybridization, fluorescently labeled or non-labeled avidin and biotinylated DNA probes have also been immobilized on a biotin derivative (photobiotin)-modified poly(dimethylsiloxane) (PDMS) chip by using biotin/avidin/biotin chemistry [106].

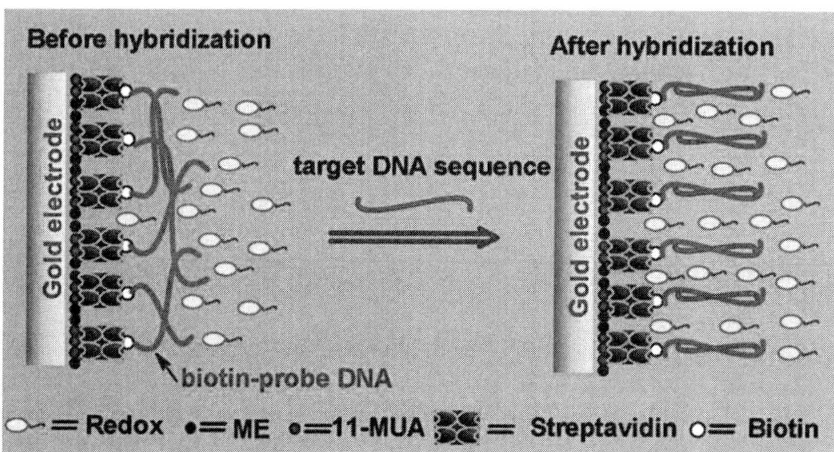

Figure 2. Schematic illustration of DNA probe immobilization through avidin-biotin chemistry and the hybridization of target DNA on a self-assembled monolayer (SAM)-modified Au electrode. (Reproduced with permission from [99]. Copyright 2014, American Chemical Society.)

3. TRANSDUCTION METHODS

Sensing and/or transduction must occur in order to convert a recognition event into a readable signal. Depending on the type of measurement, transduction can occur by electrochemical, optical, mass-based, or thermal means [107–109]. Among these, electrochemical transduction has been shown to be appropriate for DNA sensing, wherein a biorecognition event directly gives rise to an electrical signal, and this allows the sensing system to be miniaturized [110]. Recently, various convenient systems for analysis and on-site monitoring have been developed using distinct solid contact materials such as gold, ferrocene-labeled PNAs (peptide nucleic acids), CNTs, and CPs [111–113]. CPs have been widely used because they can create a high redox capacitance that makes the recorded signal highly stable [114–116]. The transduction mechanism of poly(3,4-ethylenedioxythiophene) (PEDOT) was comprehensively studied using various electrochemical techniques, including electrochemical impedance spectroscopy (EIS) [117,118]. Although CPs clearly offer advantages over other transducing materials, a few drawbacks are also associated with CPs depending on their type. For example, PPy exhibits slight chemical instability in the presence of certain components of ambient media, such as oxygen, acids, bases, redox reactants, nonreactive ions, and surfactants [119,120]. The possible formation of a water layer at the boundary between CPs and polymeric ion-selective membranes is also a hurdle that must be addressed [121]. In the case of CP-based electrochemical DNA sensors, the polymers not only serve as an immobilization template, but also actively participate in signal transduction. When using electrochemical methodologies, reversible doping and dedoping of CPs markedly alter the electrochemical responses. Doping and dedoping modulate the interaction of the probe-target complex and can regulate the sensitivity and stability of DNA detection [122,123]. The change in the signal after probe-DNA immobilization and target-DNA hybridization can also be quantified using diverse approaches, such as by measuring the change in current as a function of the applied potential (voltammetry), the change in current at a fixed applied potential (amperometry), or the change in conductivity (conductometry), impedance (impedimetry), or potential (potentiometry). Among these transduction methods, voltammetric methods (e.g., cyclic voltammetry (CV), differential pulse voltammetry (DPV), and square wave voltammetry (SWV)) are most widely used for the detection of DNA hybridization. Voltammetric measurement requires the use of two- or three-electrode electrochemical cell systems together with a potentiostat, which allows the application of the potential and the measurement of the resultant current. Voltammetry such as CV, DPV, and SWV depend on the pattern of the applied potential, which also potently controls the sensitivity of the current response.

4. DEVELOPMENT OF DNA SENSORS BASED ON DISTINCT CPS

4.1. DNA Sensors Based on Polypyrrole and Its Derivatives

PPy is formed from a number of connected pyrrole ring systems and is highly biocompatible. PPy synthesized at neutral pH is extensively used as a versatile immobilization matrix in the design of biosensors, such as catalytic biosensors, immunosensors, and DNA sensors and in molecular-imprinting technologies [124]. When deposited on the electrode surface, PPy provides an effective DNA-sensing platform, in which PPy itself acts as an interface for the attachment of the DNA probe. Notably, PPy-modified electrodes have also facilitated the development of indicator- and label-free detection of DNA [125]. Livache et al. developed a novel method for electrosynthesizing a PPy-DNA composite through co-polymerization [126]. To synthesize a PPy film harboring covalently linked DNA, a mixture of pyrrole and a pyrrole bearing a specific DNA probe was electrooxidized. This was the first study to detect DNA hybridization on the surface of modified macroelectrodes. Soon after this study, the same group also generated a DNA chip constructed of three components: silicon chips bearing a matrix of 50-μm or 4-μm microelectrodes for genotyping hepatitis C virus (HCV), a QCM, and a non-patterned gold/glass slide featuring 500-μm spots [127]. Wang et al. described the incorporation of DNA dopants into a PPy network by using an electrochemical QCM (EQCM) and showed that it exhibited a strong affinity for target DNA [128,129]. Youssoufi et al. developed a new type of electrochemical DNA hybridization sensor based on DNA-functionalized PPy [130]. The prepared PPy precursor contained a loosely bound ester group that was directly substituted with an amino-labeled DNA probe of various sequence lengths. The electrochemical response of this sensor was analyzed in aqueous media containing distinct target DNA sequences. The voltammetric signals obtained for DNA-PPy remained unchanged in the presence of a noncomplementary target DNA sequence; however, the signal changed considerably when a complementary target DNA was added. This was quantified using amperometry, and the detection limit of the biosensor was determined to be approximately 1×10^{-2} nmol in the absence of any signal processing. Other researchers also developed a similar type of DNA hybridization sensor by functionalizing PPy; these researchers introduced PPy nanotubes in which PPy was functionalized with, for example, poly[pyrrole-co-4-(3-pyrrolyl) butanoic acid] or carboxylic acid [131–133]. The acid-functionalized PPy is a favorable alternative for fabricating label-free sensors because it enables versatile immobilization of DNA, proteins, and enzymes by using various pendant groups: –SH, –NH_2, and –COOH. Peng et al. prepared a poly[pyrrole-co-4-(3-pyrrolyl) butanoic acid]-modified platinum electrode for DNA hybridization, which exhibited high electroactivity in aqueous medium [131]. An NH_2-substituted DNA probe was covalently grafted onto the surfaces of this polymer in a one-step procedure. Komarova et al. developed a prototype amperometric sensor for the detection of a biowarfare pathogen, the virus Variola major, based on DNA-doped ultrathin PPy films [134]; the investigators determined that thinner films harboring smaller or more highly concentrated dopant ions produced stronger amperometric signals than did thicker films bearing larger or less concentrated dopant ions. After the film surface was blocked with fragmented calf-thymus DNA, the nonspecific signal disappeared completely when ultrathin (Langmuir–Blodgett) films were tested; however, the specific signal from the complementary DNA remained unaffected. Under optimal conditions, the detection limit for the target DNA was 16×10^{-3} nM. Ease of use and rapid detection are the primary advantages of this sensor; however, steric hindrance and poor accessibility of the probe to the analyte in the film can reduce hybridization efficiency and substantially limit sensitivity and selectivity [68].

Several groups have developed DNA sensors based on PPy and PPy derivatives by using a modified fluorine-doped tin oxide (FTO) electrode [135–137]. Eguiluz et al. developed a PPy/FTO electrode and used Ag/Au-nanoparticle labels to detect Alicyclobacillus acidoterrestris in pure cultures by means of reverse-transcription polymerase chain reaction (RT-PCR) [136]. The sensor sensitivity could also be enhanced by performing asymmetric nested RT-PCR of the amplicon and using Ag/Au-based electrochemical detection, which was able to detect 2 colony-forming units/mL of spores. In this electrochemical bioassay, the detection and quantification limits for the target A. acidoterrestris were 7.07 and 23.6 nM, respectively. Riccardi et al. developed a new type of label-free PPy-based DNA sensor for identifying HCV [137]; in this approach, HCV is detected through the electrostatic modulation of the ion-exchange kinetics of PPy films. Here, the PPy layer was electropolymerized in order to immobilize a synthetic, single-stranded, 18-mer HCV genotype-1-specific probe DNA on a 2,5-bis (2-thienyl)-N-(3-phosphoryl-n-alkyl)pyrrole film. HCV DNA sequences (244-mer) obtained through RT-PCR amplification of the original viral RNA were examined based on the disruption of the ion-exchange properties of the PPy film. However, with this sensor, the selectivity, sensitivity, and reproducibility of DNA detection were poor.

In order to overcome the poor selectivity, sensitivity, and detection limit of DNA sensors, researchers have introduced PPy composites, such as PPy-CNTs, PPy-nanoparticles, PPy-nanoengineered materials, and pyrrole-derivative bilayers [138–140]. Xu et al. developed an impedimetric DNA biosensor based on a GCE modified with a PPy-MWNT composite [138]. COOH-MWNTs and PPy were electrodeposited on the GCE to facilitate the immobilization of the NH_2-DNA probe. The hybridization reaction of this DNA/PPy/MWNT-COOH/GCE results in a decrease in impedance, which is attributed to the electron-transfer resistance through double-stranded DNA (ds-DNA) being lower than that through ss-DNA. The PPy/MWNT-COOH-modified electrode exhibited high electron-transport capacity and also featured an increased specific surface area. Consequently, the sensitivity and selectivity of DNA hybridization were increased and the detection limit was 5.0×10^{-3} nM. Table 1 summarizes the characteristics of some of the reported electrochemical DNA sensors based on PPy and PPy derivatives, together with their immobilization methods, detection method, detection limit, and sensitivity.

Table 1. DNA sensors based on Polypyrrole (PPy), PPy derivatives, and PPy composites and their performance in the detection of DNA hybridization.

Matrix	Immobilization Method	Detection Method	Detection Limit	Sensitivity	Ref.
Copolymerizations of 5′ pyrrole-labeled DNA and pyrrole	DNA entrapment	Fluorescence microscopy	-	$>10^{-11}$ M	[127]
Copolymerizations of DNA probe within PPy	DNA entrapment	CV amperometry	>6 μg	1.08 nA μg	[129]
Poly[3-acetic acid pyrrole 3-*N*-hydroxyphthalimide pyrrole]]	Covalent	CV	1×10^{-7} nmol	-	[130]
Poly[pyrrole-*co*-4-(3-pyrrolyl) butanoic acid]	Covalent	CV EIS	-	10.5, 3.0 and 1.7 $\mu A\ cm^{2}\ nM$ of complementary DNA for 23, 57 and 114 nm film thicknesses, respectively	[131]
Carboxylic acid-functionalized PPy nanotubes (CPPy NTs)	Covalent	Photoluminescence	-	High sensitivity ($\Delta R/R_0 = 1.7$) even at low concentration (1 nmol) of target DNA	[132]
Poly [3-acetic acid pyrrole, 3-N-hydroxyphthalimide pyrrole]	Covalent	EIS	1×10^{-3} nmol	21.6 $\Omega\ cm^{-2}\ \mu M$	[133]
PPy doped with an DNA	DNA entrapment	Chronoamperometry	16×10^{-3} nM	-	[134]
poly(Py-co-PAA)	Covalent	QCM EIS	0.98 nM	-	[135]
PPy-DNA	DNA entrapment	CV Linear Sweep Voltammetry (LSV)	-	-	[136]
2,5-bis(2-thienyl)-*N*-(3-phosphorylpropyl)pyrrole	DNA entrapment, Mg^{2+} ion serve as a linker	CV LSV	1.82×10^{-11} nM	-	[137]
PPy MWNTs	Carbodiimide cross linking between amine and carboxyl group	CV EIS	5.0×10^{-3} nM	-	[138]
PPy–polyaniline–Au	HS-DNA band on Au via Au-thiol chemistry	EIS	1.0×10^{-4} nM	-	[141]
PPy–poly(3,4-ethylenedioxythiophene)–Ag	HS-DNA band on Ag via Ag-thiol chemistry	EIS	5.4×10^{-6} nM	-	[142]
Copolymer of PPy and 3-pyrrolylacrylic acid (PAA)	Covalent	EIS	-	-	[143]

4.2. DNA Sensors Based on Polythiophene and Its Derivatives

PTh and functionalized PTh demonstrate a variety of remarkable solid-state properties and hold tremendous potential for use in molecular electronic devices, solid-state batteries, and sensors [144,145]. Carboxylic acid- and ester group-functionalized PTh polymers (e.g., 3′-carboxyl-5,2′,5′,2″-terthiophene, poly(thiophen-3-yl-acetic acid 1,3-dioxo-1,3-dihydro-isoindol-2-yl ester) (PTAE), and 3-((2′:2″,5 ″:2‴-terthiophene)-3″-yl) acrylic acid (TAA)) have been widely used for developing electrochemical DNA sensors [146–149]. Lee et al. used a poly(3′-carboxyl-5,2′,5′,2″-terthiophene)-modified GCE and reported that only a short hybridization time (1 h) was required [146]. The amine group linked to the 5′ end of the DNA probe (a 19-mer) was covalently attached with the carboxyl (-COOH) group-terminated polymer, which corroborated the hybridization of the target DNA (Figure 3I). The hybridization of fully complementary target DNA induced a significant decrease of the impedance values (Figure 3II). The difference in impedance values before and after hybridization of target DNA can be ascribed by the change in conductivity and capacitive current. This method is more advantageous than other methods because of its selectivity, short response time, and minimal use of intercalators and fluorescent tags. Cha et al. described a synthetic route for thiophen-3-yl-acetic acid 1,3-dioxo-1,3-dihydro-isoindol-2-yl ester (TAE), which can be readily electropolymerized on a Pt chip electrode and allow for the direct substitution of its exiting group with a prosthetic group that contains a terminal amino group on the DNA probe [149]. The sensitivity of this sensor was 0.62 μA/nmol and the detection limit was 1 nmol.

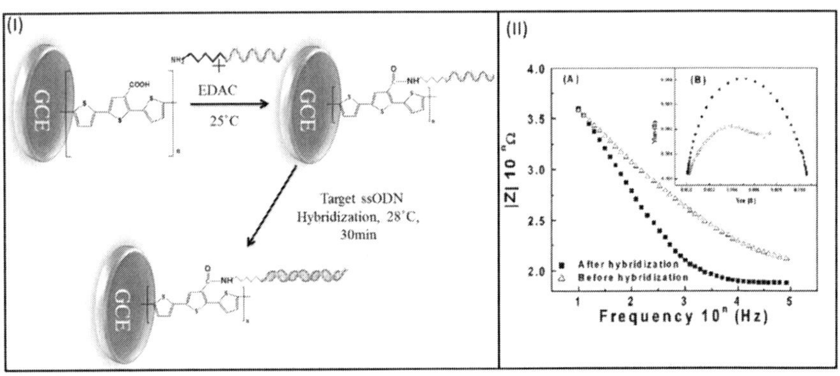

Figure 3. Schematic representation of the immobilization of probe DNA and the hybridization of a target sequence (**I**), and plots of (A) impedance and (B) admittance before and after hybridization in a phosphate buffer solution (**II**). (Redrawn and reproduced with permission from [146]. Copyright 2014, American Chemical Society.)

Peng et al. fabricated a poly(TAA)-modified DNA sensor that exhibited a qualitatively unique response when compared to functionalized PPy sensors [148,149]. The applicability of these two polymers as active substrates for DNA sensors was confirmed by covalently attaching NH_2-DNA probes to the –COOH group of both polymers. Here, the hybridization of complementary DNA can be detected by an increase in the admittance without the requirement for an indicator or any sample modification. Experimental results suggested that PPy functionalized with long unsaturated carbon side chains exhibited more favorable DNA-sensing properties and a larger difference in the impedance signal. This difference was due to the disparities in the movement of the dominant ions ($CF_3SO_3^-$ and ClO_4^-) at the interface of the polymer film and the electrolyte, which was confirmed using an EQCM.

Considerable emphasis has been placed on producing portable and inexpensive devices for DNA detection because of their importance in forensics, medical diagnostics, and evolutionary studies [150–153]. Shiddiky et al. developed an ultrasensitive technique for detecting DNA and proteins based on poly-5,2′:5′,2″-terthiophene-3′-carboxylic acid (pTTCA) (Figure 4) [154]. Dendrimer (DEN) and hydrazine were covalently linked to the pTTCA film and the signal was amplified by the pTTCA/DEN assembly loaded with Au nanoparticles (AuNPs). The target DNA- or protein-linked hydrazine labels (avidin-hydrazine) adsorbed onto the pTTCA/DEN film, and DPV measurements revealed a linear dynamic range for the electrocatalytic detection of DNA and protein. The simplicity, low detection limit, and reproducibility (RSD < 4.3% for n = 10) of the sensor make this a promising tool that can be developed in the future for practical applications.

Functionalized PTh was also used for developing a DNA sensor. Fang et al. developed a novel methodology for detecting DNA by using ferrocene-functionalized PTh deposited on a nanogold-modified electrode [155]. Nanogold-modified electrodes substantially increase the quantity of immobilized PNA probes and thus cause an increase in the electrical signal. Positively charged ferrocene-functionalized PTh does not bind electrostatically with the PNA probes because of the absence of anionic phosphate groups. This limitation can be resolved by performing an initial DNA–PNA hybridization. Adsorption of cationic PTh onto the DNA backbone results in the generation of a detectable hybridization-recognition signal in DPV. Thus, PNA could be used as a highly sensitive, selective, and reversible coupling substrate for DNA immobilization. Table 2 summarizes the characteristics of electrochemical DNA sensors based on PTh and PTh derivatives, together with their immobilization techniques, detection method, detection limit, and sensitivity.

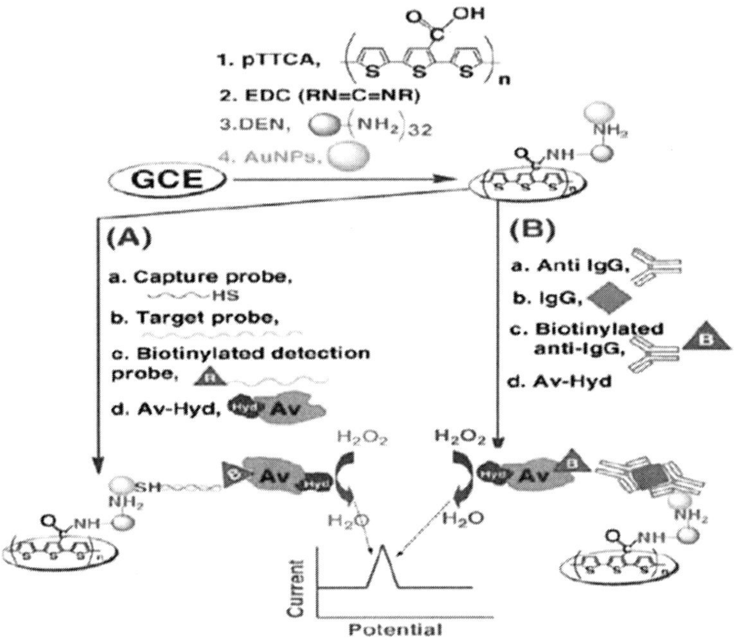

Figure 4. Schematic illustration of the poly-5,2′:5′,2″-terthiophene-3′-carboxylic acid (pTTCA)/ dendrimer (DEN)/ Au nanoparticles (AuNP)/biomolecule-linked avidin-hydrazine assembly developed for (**A**) DNA and (**B**) protein sensors, based on the electrocatalytic activity of hydrazine. (Reproduced with permission from [154]. Copyright 2014, American Chemical Society.)

Table 2. Polythiophene (PTh)- and PTh-derivative-based DNA sensors and their performance in the detection of DNA hybridization.

Matrix	Immobilization Method	Detection Method	Detection Limit	Sensitivity	Ref.
PTh	Covalent	Fluorescence	-	-	[144]
Poly(3′-carboxyl-5,2′,5′,2″-terthiophene)	Covalent	EIS	-	5.608 (ng/cm^2)/Hz	[146]
Poly(thiophen-3-yl-acetic acid 1,3-dioxo-1,3-dihydro-isoindol-2-yl ester	Covalent	CV	1 nmol	0.62 μA/nmol	[149]
Poly (3-[(S)-5-amino-5-carboxyl-3-oxapentyl]-2,5-thiophenylene hydrochloride)	Hydrogen bond	Fluorometric	1×10^{-2} nmol	-	[151]
Cationinc PTh	Covalent	Fluorometric	3.6×10^{-13} nM	-	[152]
Poly(5,2′:5,2″-terthiophene-3′-carboxylic Acid)	Covalent	Electrophoresis	1.14×10^{-4} nM	0.20 nA (fg/μL)$^{-1}$	[153]
PTh functionalized- methylene blue	HS-DNA bind on Au via Au-thiol chemistry	DPV	-	-	[156]

4.3. DNA Sensors Based on Polyaniline and Its Derivatives

The environmental stability of PANI and its easy synthesis (either chemically or electrochemically) have broadened the application of PANI as a chemical sensor [157,158] and biosensor [159–161]. PANI is the best-recognized semi-flexible rod-like CP with chemically and structurally flexible $-NH_2$ linkage on its surrounding molecules, thus making it suitable for binding biomolecules [159]. PANI undergoes two redox reactions and can be functionalized, and this makes it a favorable material for DNA sensing. The electrical conductivity of PANI has been established to strongly depend on pH, and most previous studies on PANI were performed at a pH below 4.0. However, using a neutral pH solution is critical for developing biosensors because most biocatalytic and immunological reactions occur at neutral pH [162]. Therefore, a challenge is to incorporate biological molecules in the conventional pH-dependent PANI. Research has shown that N-substituted aniline does not exhibit pH sensitivity because an alkyl chain is covalently bound to the nitrogen atom in order to prevent the formation of an emeraldine base (the deprotonated form). Moreover, self-doped PANI, commonly referred to as sulfonated PANI, exhibits redox activity even at neutral pH [163].

Wu et al. synthesized PANI-intercalated graphite-oxide nanocomposites (PANI/GO) enveloped in CPE [164]. This PANI/GO-modified CPE displayed electrochemical activity and two sharp peaks at 668 and 207 mV in SWV measurements. These results indicated that ss- and ds-DNA transformed the redox characteristics of PANI/GO, and this could be used to monitor probe immobilization and the hybridization of complementary DNA, with the hybridization peak occurring at −270 mV. DNA detection performed using the PANI/GO-modified CPE was highly stability and reproducible. Gu et al. developed an impedimetric DNA hybridization sensor based on a PANI/polyacrylate (PANI/PAA)-modified boron-doped diamond (BDD) electrode [165]. An ultrathin film of the PANI-PAA copolymer was electropolymerized onto BDD surfaces in order to enhance the availability of the −COOH for binding the DNA probe. The hybridization event was sensed based on the direct oxidation of guanine and adenine in the DNA double helix.

Immobilization of probe DNA on a polymer matrix revealed the limitations in the selectivity and specificity of hybridization [166]. In this context, PNA is recognized to provide enhanced stability and specificity in the detection of targets containing a single mismatch [167]. Gao et al. developed a novel signal-amplification method for ultrasensitive detection of DNA, which involved enzymatically catalyzed PANI formation and template-guided deposition for enhancing DNA hybridization [168]. The hybridization was

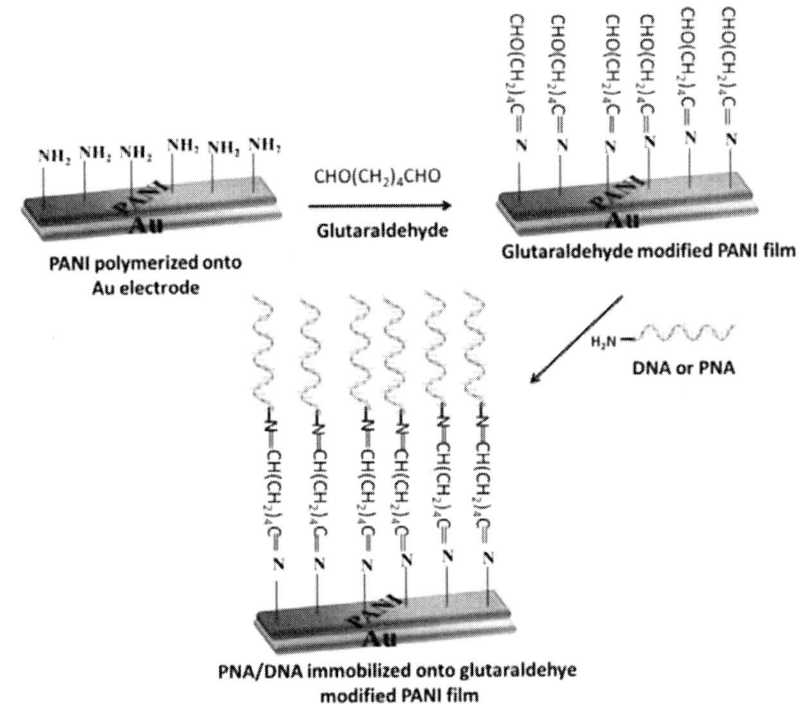

Figure 5. NH2-DNA and peptide nucleic acid (PNA) immobilization on polyaniline-coated gold film. (Redrawn with permission from [174]. Copyright 2014, American Chemical Society.)

Table 3. Polyaniline (PANI)- and PANI-derivative-based DNA sensors and their performance in the detection of DNA hybridization.

Matrix	Immobilization Method	Detection Method	Detection Limit	Sensitivity	Ref.
PANI-intercalated graphite oxide nanocomposite	Covalent	SWV	-	0.77 μA/μg/mL	[164]
PANI/PAA-modified borondoped diamond (BDD)	Covalent	EIS	20 nM	-	[165]
Avidin modified-PANI	Avidin interaction	DPV	5×10^{-10} nmole	-	[166]
Polyaniline nanowire	Covalent immobilization	DPV	1×10^{-3} nM	-	[169]
PANI nanotube	Covalent	CV/DPV	1×10^{-6} nM	1 pM	[170]
Graphene/PANI	Non-covalent binding	DPV	3.2×10^{-5} nM	-	[175]
PANI-Au	HS-DNA bind on Au via Au-thiol chemistry	DPV	0.1 nM	-	[176]

quantified by examining the electroactivity of the deposited PANI by using SWV. This DNA sensor was extremely sensitive—it had a femtomolar detection limit—and it was highly selective for sequences mismatched by one, two, and three bases. The biosensor was used for detecting His4, RCA1, and GAPDH, and the results obtained were similar to those obtained from northern-blotting analysis of the same samples.

An ultrasensitive technique for detecting DNA hybridization has also been developed by using PANI nanowires or nanotubes or methylene blue (MB) as an indicator [169,170]. MB was used to distinguish between ss- and ds-DNA by using various electrochemical techniques [171,172]. MB specifically binds to unpaired bases of DNA/PNA. The redox reaction of MB was used for monitoring the native and denatured states of DNA [173]. Prabhakar et al. covalently immobilized 20-base-long NH_2-DNA and PNA probes on a PANI/Au electrode in order to detect DNA hybridization and used MB as an indicator (Figure 5) [174]. PNA-PANI/Au and DNA-PANI/Au electrodes were used for detecting the presence of complementary target Mycobacterium tuberculosis by using SWV, and the response time was short (30 s). The study revealed that the PNA electrode exhibits a higher affinity for complementary DNA sequence and an improved detection limit and higher specificity than do electrodes used in other methods. Table 3 lists the characteristics of electrochemical DNA sensors based on PANI and PANI derivatives, together with their immobilization techniques, detection method, detection limit, and sensitivity.

4.4. DNA Sensors Based on Quinone and Its Derivatives

Quinone is an electronically conductive redox polymer that has attracted substantial interest as a chemical sensor [177] and biosensor [178,179] because of its capacity to transport charges inside films and the nature of the ionic flux at the interface of the polymer and the solution (anionic and cationic, respectively). The characteristics of quinone-based polymers differ from those of conventional p-doped electronically conducting polymers (ECPs). The sensitivity of ECP-based biosensors depends on the amplification of the interaction between the electrochemical transducer and biomolecules [180].

Piro et al. constructed a new electroactive film, poly(JUG-co-JUGA), by co-electrooxidizing 5-hydroxy-1,4-naphthoquinone and 5-hydroxy-3-thioacetic acid-1,4-naphthoquinone [180]. This poly(JUG-co-JUGA) copolymer presents both electroactive and chemically reactive groups for sensing DNA and L-lactate [181]. An NH_2-DNA probe was covalently immobilized on poly(JUG-co-JUGA) and the electroactivity of the quinone group was used for detecting hybridization. The main feature of this DNA hybridization sensor is the transformation between ss- and ds-DNA at the solution/polymer interface. Therefore, the rate of charge (ion) diffusion from the solution to the polymer/solution interface is primarily affected by hybridization. After hybridization, the current increases in SWV measurements because of electrostatic and/or steric effects. Piro and coworkers also examined PNA probe-based DNA hybridization sensors (Figure 6) [182]. The PNA probe was covalently attached to poly(JUG-co-JUGA), and upon hybridization with the target complementary DNA, the flexibility of the PNA probe was altered, which caused changes in the electrochemical signal at the polymer/solution interface. This simple and reagent-free DNA hybridization sensor can discriminate single mismatches and can be regenerated after a simple dehybridization step. The main drawback of this sensor is its detection limit, 10 nM, which does not constitute the theoretical limit. In order to improve the detection limit, Acevedo et al. fabricated a sequential multilayer CNT that featured an increased area for the oxidation of soluble redox couples [183]. The electropolymerization of quinine and quinone derivatives onto the MWNT-modified electrode produced an interpenetrated CP/CNT network that is electroactive in both aqueous and nonaqueous media. The effective current response was enhanced up to 19 times, which increased the sensitivity and lowered the detection limit.

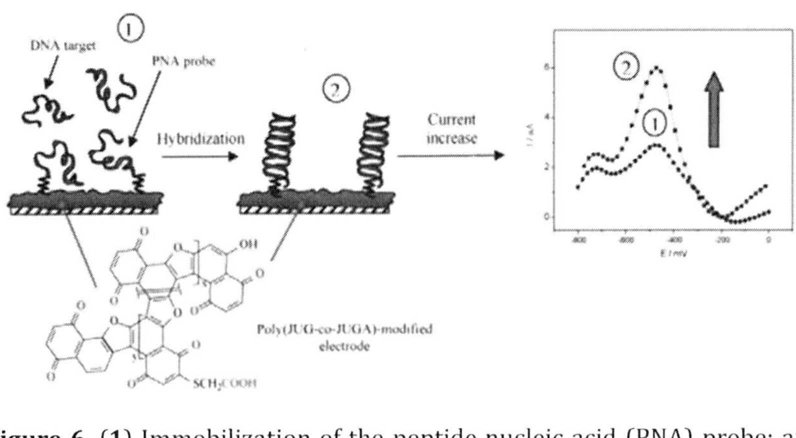

Figure 6. (**1**) Immobilization of the peptide nucleic acid (PNA) probe; and (**2**) hybridization of the target DNA onto poly(JUG-co-JUGA)-modified glassy carbon electrode (GCE) together with the corresponding square wave voltammetry (SWV) signals for (**1**) PNA probe immobilization, and (**2**) target DNA hybridization. (Reproduced with permission from [182]. Copyright 2014, Elsevier.)

4.5. DNA Sensors Based on Miscellaneous Conducting Polymers and Their Derivatives

Poly(triamine) (PTyr) contains one primary aliphatic amine per triamine moiety, which represents an extremely high concentration of surface reactive sites for biomolecule immobilization [184]. Tran et al. first prepared a PTyr film by electrooxidizing 4-hydroxiphenylethylamine in perchloric acid, which left one reactive amine group per moiety [185]. DNA was immobilized on the polymer film through a phosphoramidate covalent bond [186], and this yielded a high surface concentration of DNA (~500 pmol/cm^2). DNA can also readily bind to PTyr through nonspecific adsorption, and thus differentiating between adsorbed and covalently bonded DNA could be challenging. However, DNA probes that were weakly adsorbed were removed by washing with SL salmon-testis DNA. The remaining covalently bonded DNA probes were used for the hybridization of GEM DNA, a complementary DNA sequence derived from HIV gag protein. For detecting DNA hybridization, Li et al. prepared a new type of conjugated CP, poly(indole-5-carboxylic acid) (PICA), on GCE by means of anodic oxidation (Figure 7) [187]. PICA exhibited optimal electrochemical behavior and thermal stability, with a conductivity of 10^{-2} S/cm and high redox activity compatible with the concept of molecular-wire transduction. The PICA-modified sensor showed comparable sensitivity and its detection limit was 1.0 nM, which can be further improved by increasing the side chain length because longer side chains permit greater freedom of movement and more enhanced hybridization [35].

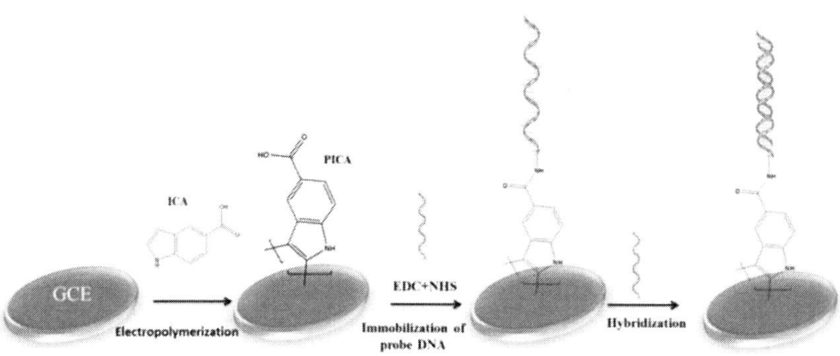

Figure 7. Schematic representation of the preparation of an electrochemical DNA sensor based on a poly(indole-5-carboxylic acid) (PICA) conducting polymer. (Redrawn from [187]. Copyright 2014, Elsevier.)

Certain other CPs have been synthesized and employed in DNA sensing applications in which fluorescence techniques are used. These CPs include poly({2,5-bis[3-(N,N-diethylamino)-1-oxapropyl]-para-phenylenevinylene}-alt-para-phenylenevinylene)dibromide [188], poly(9,9-bis(6′-N,N,N-trimethylammonium)-hexyl)-fluorene phenylene) [189], and poly(fluorene-co-phenylene) [190]. The potential application of these CPs for the development of an electrochemical DNA sensor is in high demand.

5. CONCLUSIONS AND OUTLOOK

This review has summarized the diverse strategies used for developing DNA biosensors by using electroactive CPs. CPs have been extensively used for developing electrochemical label-free methods of DNA detection. This is because CPs exhibit highly favorable electrical conductivity or charge-transport properties. Over the last decade, researchers have developed numerous types of CPs and their derivatives for highly sensitive and selective electrochemical detection of DNA. Adsorption, covalent immobilization, and avidin-biotin interactions have been used for developing DNA biosensors by modifying the electrode surface with CPs and their derivatives. To develop CP-based DNA sensors, regeneration of a surface-immobilized probe and the reuse of DNA biosensors must be addressed. The key factors that must be considered for probe immobilization are the following: the immobilization chemistry must be stable, the probes must retain their functionality after attachment, and immobilized biomolecules must maintain proper orientation and configuration [191]. CP films must be deposited on inert substrates because hydrophobic interactions and the consequential electrochemical oxidation and reduction accompanied by the movement of ions in and out of the CP film can result in the delamination of CP films. With regard to the immobilization of DNA probes, most detection methods that are used for developing DNA biosensors and microarrays are open to criticism [192]. CP-based DNA sensors are expected to be highly sensitive, selective, and reproducible and to enable multi-analyte determination.

Although we have focused on CPs and various CP-CNT composites, other materials could also be used for developing electrochemical DNA hybridization sensors. For example, graphene, a new allotrope of carbon composed of sp^2-hybridized carbon atoms arranged in a honeycomb lattice, is an ideal 2-D material for developing DNA sensors. Graphene can be readily functionalized and doped with various functional groups (e.g., –COOH, –OH) and atoms, which might facilitate the immobilization of DNA probes in a biocompatible manner.

ACKNOWLEDGMENTS

This work was supported by Konkuk University.

CONFLICTS OF INTEREST

The authors declare no conflict of interest.

REFERENCES

1. Barhoumi, A.; Halas, N.J. Label-Free detection of DNA hybridization using surface enhanced raman spectroscopy. J. Am. Chem. Soc. 2010, 132, 12792–12793.

2. Cai, B.; Wang, S.; Huang, L.; Ning, Y.; Zhang, Z.; Zhang, G.-J. Ultrasensitive label-free detection of PNA-DNA hybridization by reduced graphene oxide field-effect transistor biosensor. ACS Nano. 2014, 8, 2632–2638.

3. Ahmed, M.U.; Nahar, S.; Safavieh, M.; Zourob, M. Real-time electrochemical detection of pathogen DNA using electrostatic interaction of a redox probe. Analyst 2013, 138, 907–915.

4. Maxam, A.M.; Gilbert, W. A new method for sequencing DNA. Proc. Natl. Acad. Sci. USA 1997, 74, 560–564.

5. Sanger, F.; Nicklen, S.; Coulson, A.R. DNA sequencing with chain-terminating inhibitors. Proc. Natl. Acad. Sci. USA 1977, 74, 5463–5467.

6. Marciacq, F.; Sauvaigo, S.; Issartelte, J.P.; Mouret, J.F.; Molko, D. Synthesis and enzymatic incorporation of Morpholino thymidine-S-triphosphate in DNA fragments. Tetrahedron Lett. 1999, 40, 4673–4676.

7. Reeve, M.A.; Fuller, C.W. A novel thermostable polymerase for DNA sequencing. Nature 1995, 376, 796–797.

8. Lee, L.G.; Spurgeon, S.L.; Heiner, C.R.; Benson, S.C.; Rosenblum, B.B.; Menchen, S.M.; Graham, R.J.; Constantinescu, A.; Upadhya, K.G.; Cassel, J.M. New energy transfer dyes for DNA sequencing. Nucleic Acids. Res. 1977, 25, 2816–2822.

9. Wallace, R.B.; Shaffer, J.; Murphy, R.F.; Bonner, J.; Hirose, T.; Itakura, K. Hybridization of synthetic oligodeoxyribonucleotides to phi chi 174 DNA: The effect of single base pair mismatch. Nucleic Acids Res. 1979, 6, 3543–3557.

10. Wallace, R.B.; Johnson, M.J.; Hirose, T.; Miyake, T.; Kawashima, E.H.; Itakura, K. The use of synthetic oligonucleotides as hybridization probes. II. Hybridization of oligonucleotides of mixed sequence to rabbit beta-globin DNA. Nucleic Acids Res. 1981, 9, 879–894.

11. Pease, A.C.; Solas, D.; Sullivan, E.J.; Cronin, M.T.; Holmes, C.P.; Fodor, S.P. Light-generated oligonucleotide arrays for rapid DNA sequence analysis. Proc. Natl. Acad. Sci. USA 1994, 91, 5022–5026.

12. Fodor, S.P.; Rava, R.P.; Huang, X.C.; Pease, A.C.; Holmes, C.P.; Adams, C.L. Multiplexed biochemical assays with biological chips. Nature 1993, 364, 555–556.

13. Shalon, D.; Smith, S.J.; Brown, P.O. A DNA microarray system for analyzing complex DNA samples using two-color fluorescent probe hybridization. Genome Res. 1996, 6, 639–645.

14. Adessi, C.; Matton, G.; Ayala, G.; Turcatti, G.; Mermod, J.J.; Mayer, P.; Kawashima, E. Solid Phase DNA amplification: Characterisation of primer attachment and amplification mechanism. Nucleic Acids Res. 2000, 28.

15. Liu, J.; Tian, S.; Tiefenauer, L.; Nielsen, P.E.; Knoll, W. Simultaneously amplified electrochemical and surface plasmon optical detection of DNA hybridization based on ferrocene-streptavidin conjugates. Anal. Chem. 2005, 77, 2756–2761.

16. Peng, H.; Zhang, L.; Kjaellman, T.H.M.; Soeller, C.; Travas-Sejdic, J. DNA hybridization detection with blue luminescent quantum dots

and dye-labeled single stranded DNA. J. Am. Chem. Soc. 2007, 129, 3048–3049.

17. Yin, M.-J.; Wu, C.; Shao, L.-Y.; Chan, W.K.E.; Zhang, A.P.; Lu, C.; Tam, H.-Y. Label-free, disposable fiber-optic biosensors for DNA hybridization detection. Analyst. 2013, 138, 1988–1994.

18. Papadakis, G.; Tsortos, A.; Bender, F.; Ferapontova, E.E.; Gizeli, E. Direct detection of DNA conformation in hybridization processes. Anal. Chem. 2012, 84, 1854–1861.

19. Drummond, T.G.; Hill, M.G.; Barton, J.K. Electrochemical DNA sensor. Nat. Biotechnol. 2003, 21, 1192–1199.

20. Alfonta, L.; Singh, A.K.; Willner, I. Liposomes labeled with biotin and horseradish peroxidase: A probe for the enhanced amplification of antigen-antibody or oligonucleotide-DNA sensing processes by the precipitation of an insoluble product on electrodes. Anal. Chem. 2011, 73, 91–102.

21. Ji, H.; Yan, F.; Lei, J.; Ju, H. Ultrasensitive electrochemical detection of nucleic acids by template enhanced hybridization followed with rolling circle amplification. Anal. Chem. 2012, 84, 7166–7171.

22. Rahman, M.M.; Ahammad, A.J.S.; Jin, J.H.; Ahn, S.J.; Lee, J.J. A comprehensive review of glucose biosensors based on nanostructured metal-oxides. Sensors 2010, 10, 4855–4886.

23. Kim, Y.J.; Rahman, M.M.; Lee, J.J. Ultrasensitive and label-free detection of annexin A3 based on quartz crystal microbalance. Sens. Actuators B Chem. 2013, 177, 172–177.

24. Rahman, M.M.; Li, X.B.; Kim, J.C.; Lim, B.O.; Ahammad, A.J.S.; Lee, J.J. A cholesterol biosensor based on a bi-enzyme immobilized on conducting poly(thionine) film. Sens. Actuators B Chem. 2014, 202, 536–542.

25. Rahman, M.M.; Li, X.B.; Lopa, N.S.; Lee, J.J. Electrodeposition of Gold on Fluorine-doped tin oxide: Characterization and application for

catalytic oxidation of nitrite. Bull. Korean Chem. Soc. 2014, 35, 2072–2076.

26. Bonanni, A.; Chua, C.K.; Zhao, G.; Sofer, Z.; Pumera, M. Inherently electroactive graphene oxide nanoplatelets as labels for single nucleotide polymorphism detection. ACS Nano. 2012, 6, 8546–8551.

27. Walter, A.; Surkus, A.E.; Flechsig, G.U. Hybridization detection of enzyme-labeled DNA at electrically heated electrodes. Anal. Bioanal. Chem. 2013, 405, 3907–3911.

28. Hsieh, K.; Xiao, Y.; Soh, H.T. Electrochemical DNA detection via exonuclease and target-catalyzed transformation of surface-bound probes. Langmuir 2010, 26, 10392–10396.

29. Kerman, K.; Morita, Y.; Takamura, Y.; Tamiya, E. Label-free electrochemicaldetection of DNA hybridization on gold electrode. Electrochem. Commun. 2003, 5, 887–891.

30. Kang, B.; Yeo, U.; Yoo, K.H. Anodized aluminum oxide-based capacitance sensors for the direct detection of DNA hybridization. Biosens. Bioelectron. 2010, 25, 1592–1596.

31. Mousavi-Sani, S.Z.; Raoof, J.B.; Ojani, R.; Hamadi-Asl, E. Nano-Gold modified geneosenor for direct detection of DNA hybridization. J. Chin. Chem. Soc. 2013, 60, 650–656.

32. Cai, H.; Wang, Y.; He, P.; Fang, Y. Electrochemical detection of DNA hybridization based on silver-enhanced gold nanoparticle label. Anal. Chim. Acta. 2002, 469, 165–172.

33. Castañeda, M.T.; Alegret, S.; Merkoçi, A. Electrochemical detection of DNA hybridization using micro and nanoparticles. Methods Mol. Biol. 2009, 504, 127–143.

34. Wang, J.; Liu, G.; Polsky, R.; Merkoci, A. Electrochemical stripping detection of DNA hybridization based on cadmium sulfide nanoparticle tags. Electrochem. Commun. 2002, 4, 722–726.

35. Bangar, M.A.; Shirale, D.J.; Purohit, H.J.; Chen, W.; Myung, N.V.; Mulchandani, A. Single conducting polymer nanowire based sequence-specific, sase-pair-length dependant label-free DNA sensor. Electroanalysis 2011, 23, 371–379.

36. Garnier, F.; Korri-Youssoufi, H.; Srivastava, P.; Mandrand, B.; Delair, T. Toward intelligent polymers: DNA sensors based on oligonucleotide-functionalized polypyrroles. Synth. Met. 1999, 100, 89–94.

37. Ma, H. Effect of electron-electron interactions on the charge carrier transitions in trans-polyacetylene. J. Phys. Chem. A. 2010, 144, 5439–5444.

38. Patel, M.K.; Singh, J.; Singh, M.K.; Agrawal, V.V.; Ansari, S.G.; Malhotra, B.D. Tin oxide quantum dot based DNA sensor for pathogen detection. J. Nanosci. Nanotechnol. 2013, 13, 1671–1678.

39. Heeger, A.J. Nobel lecture: Semiconducting and metallic polymers: The fourth generation of polymeric materials. Rev. Mod. Phys. 2001, 73, 681–700.

40. Pernaut, J.M.; Reynolds, J.R. Use of conducting electroactive polymers for drug delivery and sensing of bioactive molecules. A redox chemistry approach. J. Phys. Chem. B 2000, 104, 4080–4090.

41. Park, S.M. Electrochemistry of π-conjugated polymers. In Handbook of Organic Conductive Molecules and Polymers; Nalwa, H.S., Ed.; Wiley: Chichester, UK, 1997; Volume 3, pp. 429–469.

42. Guiseppi-Elie, A.; Wallace, G.G.; Matsue, T. Handbook of Conducting Polymers, 2nd ed.; Skotheim, T.A., Elsenbaumer, R., Reynolds, J.R., Eds.; Dekker M.: New York, NY, USA, 1998; Volume 963.

43. Kim, Y.H.; Lee, J.; Hofmann, S.; Gather, M.C.; Müller-Meskamp, L.; Leo, K. Achieving high efficiency and improved stability in ITO free transparent organic light-emitting diodes with conductive polymer electrodes. Adv. Funct. Mater. 2013, 23, 3763–3769.

44. Defieuw, G.; Samijn, R.; Hoogmartens, I.; Vanderzande, D.; Gelan, J. Antistatic polymer layers based on poly(isothianaphthene) applied from aqueous compositions. Synth. Met. 1993, 57, 3702–3706.

45. Shen, K.Y.; Hu, C.W.; Chang, L.C.; Ho, K.C. A complementary electrochromic device based on carbon nanotubes/conducting polymers. Sol. Energy Mater. Sol. Cells. 2012, 98, 294–299.

46. Mengistie, D.A.; Ibrahem, M.A.; Wang, P.C.; Chu, C.W. Highly conductive PEDOT: PSS treated with formic acid for ITO-free polymer solar cells. ACS Appl. Mater. Interfaces 2014, 6, 2292–2299.

47. Baldissera, A.F.; Freitas, D.B.; Ferreira, C.A. Electrochemical impedance spectroscopy investigation of chlorinated rubber-based coatings containing polyaniline as anticorrosion agent. Mater. Corros. 2010, 61, 790–801.

48. Rahman, M.M.; Li, X.B.; Jeon, Y.D.; Lee, H.J.; Lee, S.J.; Lee, J.J. Simultaneous determination of ranitidine and metronidazole at poly(thionine) modified anodized glassy carbon electrode. J. Electrochem. Sci. Technol. 2012, 2, 90–94.

49. Svirskis, D.; Travas-Sejdic, J.; Rodgers, A.; Garg, S. Electrochemically controlled drug delivery based on intrinsically conducting polymers. J. Control. Release 2010, 146, 6–15.

50. Leprince, L.; Dogimont, A.; Magnin, D.; Champagne, S.D. Dexamethasone electrically controlled release from polypyrrole-coated nanostructured electrodes. J. Mater. Sci. Mater. Med. 2010, 21, 925–930.

51. Rahman, M.A.; Kumar, P.; Park, D.S.; Shim, Y.B. Electrochemical sensors based on organic conjugated polymers. Sensors 2008, 8, 118–141.

52. Ali, S.R.; Parajuli, R.R.; Balogun, Y.; Ma, Y.; He, H. A nonoxidative electrochemical sensor based on a self-doped polyaniline/carbon nanotube composite for sensitive and selective detection of the

neurotransmitter dopamine: A review. Sensors 2008, 8, 8423–8452.

53. Dhaoui, W.; Bouzitoun, M.; Zarrouk, H.; Ouada, H.B.; Pron, A. Electrochemical sensor for nitrite determination based on thin films of sulfamic acid doped polyaniline deposited on Si/SiO2 structures in electrolyte/insulator/semiconductor (E.I.S.) configuration. Synth. Met. 2008, 158, 722–726.

54. Ruangchuaya, L.; Sirivat, A. Selective conductivity response of polypyrrole-based sensoron flammable chemicals. React. Funct. Polym. 2004, 61, 11–22.

55. Guedon, P.; Livache, T.; Martin, F.; Lesbre, F.; Roget, A.; Bidan, G.; Levy, Y. Characterization and optimization of a real-time, parallel, label-free, polypyrrole-based DNA sensor by surface plasmon resonance imaging. Anal. Chem. 2000, 72, 6003–6009.

56. Shim, Y.B.; Park, S.M. Electrochemistry of conductive polymers XXII, Electrochemical and spectroelectrochemical studies of polyazulene growth and its characterization. J. Electrochem. Soc. 1997, 144, 3027–3033.

57. Uygun, A. DNA hybridization electrochemical biosensor using a functionalized polythiophene. Talanta 2009, 79, 194–198.

58. Kim, H.J.; Piao, M.H.; Choi, S.H.; Shin, C.H.; Lee, Y.T. Development of amperometric hydrogen peroxide sensor based on horseradish peroxidase-immobilized poly(thiophene-co-epoxythiophene). Sensors 2008, 8, 4110–4118.

59. Lee, J.W.; Park, D.S.; Shim, Y.B.; Park, S.M. Electrochemical characterization of poly (1,8- diaminonaphthalene): A functionalized polymer. J. Electrochem. Soc. 1992, 139, 3507–3514.

60. Paul, E.W.; Ricco, A.J.; Wrighton, M.S. Resistance of polyanillne films as a function of electrochemical potential and the fabrication of poiyanlline-based microelectronic devices. J. Phys. Chem. 1985, 89, 1441–1447.

61. Mishra, A.; Ma, C.Q.; Bäuerle, P. Functional oligothiophenes: Molecular design for multidimensional nanoarchitectures and their applications. Chem. Rev. 2009, 109, 1141–1276.

62. Bhat, D.K.; Kumar, M.S. N and P doped poly(3,4-ethylenedioxythiophene) electrode materials for symmetric redox supercapacitors. J. Mater. Sci. 2007, 42, 8158–8162.

63. Turner, M.L. Inorganic and organometallic polymers. Annu. Rep. Prog. Chem. Sect. A Inorg. Chem. 1999, 95, 453–465.

64. Teles, F.R.R.; Fonseca, L.P. Applications of polymers for biomolecule immobilization in electrochemical biosensors. Mater. Sci. Eng. C. 2008, 28, 1530–1543.

65. McQuade, D.T.; Pullen, A.E.; Swager, T.M. Conjugated polymer-based chemical sensors. Chem. Rev. 2000, 100, 2537–2574.

66. Sassolas, A.; Leca-Bouvier, B.D.; Blum, L.J. DNA biosensors and microarrays. Chem. Rev. 2008, 108, 109–139.

67. Fuentes, M.; Mateo, C.; García, L.; Tercero, J.C.; Guisán, J.M.; Fernández-Lafuente, R. Directed covalent immobilization of aminated DNA probes on aminated plates. Biomacromolecules 2004, 5, 883–888.

68. Peng, H.; Zhang, L.; Soeller, C.; Travas-Sejdic, J. Conducting polymers for electrochemical DNA sensing. Biomaterials 2009, 30, 2132–2148.

69. Pividori, M.I.; Merkoci, A.; Alegret, S. Electrochemical genosensor design: Immobilisation of oligonucleotides onto transducer surfaces and detection methods. Biosens. Bioelectron 2000, 15, 291–303.

70. Hirayama, H.; Tamaoka, J.; Horikoshi, K. Improved immobilization of DNA to microwell plates for DNA–DNA hybridization. Nucleic Acids Res. 1996, 24, 4098–4099.

71. Kara, P.; Kerman, K.; Ozkan, D.; Meric, B.; Erdem, A.; Nielsen, P.E.; Ozsoz, M. Label-Free and label based electrochemical detection of hybridization by using methylene blue and peptide nucleic acid probes at chitosan modified carbon paste electrodes. Electroanalysis 2002, 14, 1685–1690.

72. Xu, C.; Cai, H.; He, P.; Fang, Y. Electrochemical detection of sequence-specific DNA using a DNA probe labeled with aminoferrocene and chitosan modified electrode immobilized with ssDNA. Analyst. 2001, 126, 62–65.

73. Wong, E.L.S.; Erohkin, O.; Gooding, J.J. A comparison of cationic and anionic intercalators for the electrochemical transduction of DNA hybridization via long range electron transfer. Electrochem. Commun. 2004, 6, 648–654.

74. Zhao, Y.D.; Pang, D.W.; Zhang, M.; Cheng, J.K. DNA-modified electrodes. Part 6. A new method for the determination of binding constants and binding site sizes. Fresenius J. Anal. Chem. 1999, 363, 708–709.

75. Zhao, Y.D.; Pang, D.W.; Wang, Z.L.; Cheng, J.K.; Qi, Y.P. DNA modified electrodes. Part 2. Electrochemical characterization of gold electrode modified with DNA. J. Electroanal. Chem. 1997, 431, 203–209.

76. Pang, D.W.; Abruña, H.D. Micromethod for the investigation of the interactions between DNA and redox-active molecules. Anal. Chem. 1998, 70, 3162–3169.

77. Azek, F.; Grossiord, C.; Joannes, M.; Limoges, B.; Brossier, P. Hybridization Assay at a Disposable Electrochemical Biosensor for the attomole Detection of Amplified Human Cytomegalovirus DNA. Anal. Biochem. 2000, 284, 107–113.

78. Wang, J.; Cai, X.; Rivas, G.; Shiraishi, H. Stripping potentiometric transduction of DNA hybridization processes. Anal. Chim. Acta. 1996, 326, 141–147.

79. Marrazza, G.; Chianella, L.; Mascini, M. Disposable DNA electrochemical sensor for hybridization detection. Biosens. Bioelectron 1999, 14, 43–51.

80. Wang, J.; Cai, X.; Jonsson, C.; Balakrishnan, M. Adsorptive stripping potentiometry of DNA at electrochemically pretreated carbon paste electrodes. Electroanalysis 1996, 8, 20–24.

81. Wu, S.; Zhao, H.; Ju, H.; Shi, C.; Zhao, J. Electrodeposition of silver-DNA hybrid nanoparticles for electrochemical sensing of hydrogen peroxide and glucose. Electrochem. Commun. 2006, 8, 1197–1203.

82. Lahiji, R.R.; Dolash, B.D.; McDonald, J.; Zemlyanov, D.; Bergstrom, D.E.; Reifenberger, R. Electrodeposition study of ODN:SWCNT hybrids on gold substrates. Phys. Stat. Sol. A 2008, 205, 1408–1411.

83. Tam, P.D.; Tuan, M.A.; Chien, N.D. DNA covalent attachment on conductometric biosensor for modified genetic soybean detection. Commun. Phys. 2007, 17, 234–240.

84. Peng, H.; Soeller, C.; Travas-Sejdic, J. Novel conducting polymers for DNA sensing. Macromolecules 2007, 40, 909–914.

85. Rasmussen, S.R.; Larsen, M.R.; Rasmussen, S.E. Covalent immobilization of DNA onto polystyrene microwells: The molecules are only bound at the 5′ end. Anal. Biochem. 1991, 198, 138–142.

86. Ligaj, M.; Jasnowska, J.; Musiał, W.G.; Filipiak, M. Covalent attachment of single-stranded DNA to carbon paste electrode modified by activated carboxyl groups. Electrochim. Acta. 2006, 51, 5193–5198.

87. Raymond, F.R.; Ho, H.A.; Peytavi1, R.; Bissonnette, L.; Boissinot, M.; Picard, F.J.; Leclerc, M.; Bergeron, M.G. Detection of target DNA using fluorescent cationic polymer and peptide nucleic acid probes on solid support. BMC Biotechnol. 2005, 5.

88. Niveleau, A.; Sage, D.; Reynaud, C.; Bruno, C.; Legastelois, S.; Thomas, V.; Dante, R. Covalent linking of haptens, proteins and

nucleic acids to a modified polystyrene support. J. Immunol. Methods 1993, 159, 177–187.

89. Smith, E.A.; Wanat, M.J.; Cheng, Y.; Barreira, S.V.P.; Frutos, A.G.; Corn, R.M. Formation Spectroscopic characterization, and application of sulfhydryl-terminated alkanethiol monolayers for the chemical attachment of DNA onto gold surfaces. Langmuir 2001, 17, 2502–2507.

90. Ajore, R.; Kumar, R.; Kaur, I.; Sobti, R.C.; Bharadwaj, L.M. DNA immobilization chemical interference due to aggregates study by dip and drop approach. J. Biochem. Biophys. Methods 2007, 70, 779–785.

91. Li, F.; Chen, W.; Zhang, S. Development of DNA electrochemical biosensor based on covalent immobilization of probe DNA by direct coupling of sol–gel and self-assembly technologies. Biosens. Bioelectron. 2008, 24, 781–786.

92. Liu, T.; Tang, J.; Jiang, L. The enhancement effect of gold nanoparticles as a surface modifier on DNA sensor sensitivity. Biochem. Biophys. Res. Commun. 2004, 313, 3–7.

93. Wang, J.; Lin, Y. Functionalized carbon nanotubes and nanofibers for biosensing applications. Trends Anal. Chem. 2008, 27, 619–626.

94. Santiago-Rodríguez, L.; Sánchez-Pomales, G.; Cabrera, C.R. Single-walled carbon nanotubes modified gold electrodes as an impedimetric DNA sensor. Electroanalysis 2010, 22, 399–405.

95. Gangopadhyay, R.; De, A. Conducting polymer nanocomposites: A brief overview. Chem. Mater. 2000, 12, 608–622.

96. Wahab, R.; Ansari, S.G.; Kim, Y.S.; Mohanty, T.R.; Hwang, I.H.; Shin, H.S. Immobilization of DNA on nano-hydroxyapatite and their interaction with carbon nanotubes. Synth. Met. 2009, 159, 238–245.

97. Yang, T.; Zhou, N.; Zhang, Y.; Zhang, W.; Jiao, K.; Li, G. Synergistically improved sensitivity for the detection of specific DNA sequences

using polyaniline nanofibers and multi-walled carbon nanotubes composites. Biosens. Bioelectron. 2009, 24, 2165–2170.

98. Tichoniuk, M.; Ligaj, M.; Filipiak, M. Application of DNA hybridization biosensor as a screening method for the detection of genetically modified food components. Sensors 2008, 8, 2118–2135.

99. Pan, S.L.; Rothberg, L. Chemical control of electrode functionalization for detection of DNA hybridization by electrochemical impedance spectroscopy. Langmuir 2005, 21, 1022–1027.

100. Caruso, F.; Rodda, E.; Furlong, D.N.; Niikura, K.; Okahata, Y. Quartz crystal microbalance study of DNA immobilization and hybridization for nucleic acid sensor development. Anal. Chem. 1997, 69, 2043–2049.

101. Dupont-Filliard, A.; Billon, M.; Livache, T.; Guillerez, S. Biotin/avidin system for the generation of fully renewable DNA sensor based on biotinylated polypyrrole film. Anal. Chim. Acta. 2004, 515, 271–277.

102. Calvo-Muñoz, M.L.; Dupont-Filliard, A.; Billon, M.; Guillerezb, S.; Bidan, G.; Marquette, C.; Blum, L. Detection of DNA hybridization by ABEI electrochemiluminescence in DNA-chip compatible assembly. Bioelectrochemistry 2005, 66, 139–143.

103. Baur, J.; Gondran, C.; Holzinger, M.; Defrancq, E.; Perrot, H.; Cosnier, S. Label-Free Femtomolar Detection of Target DNA by Impedimetric DNA sensor based on poly(pyrrole-nitrilotriacetic acid) film. Anal. Chem. 2010, 52, 1066–1072.

104. Bonanni, A.; Pividori, M.I.; Valle, M.D. Application of the avidin-biotin interaction to immobilize DNA in the development of electrochemical impedance genosensors. Anal. Bioanal. Chem. 2007, 389, 851–861.

105. Cosnier, S.; Galland, B.; Gondran, C.; le Pellec, A. Electrogeneration of Biotinylated functionalized polypyrroles for the simple immobilization of enzymes. Electroanalysis 1998, 10, 808–813.

106. Shamansky, L.M.; Davis, C.B.; Stuart, J.K.; Kuhr, W.G. Immobilization and detection of DNA on microfluidic chips. Talanta 2001, 55, 909–918.

107. Piunno, P.A.E.; Krull, U.J.; Hudson, R.H.E.; Damha, M.J.; Cohen, H. Fiberloptic DNA sensor for fluorometric nucleic acid determination. Anal. Chem. 1995, 67, 2635–2643.

108. Liu, G.; Wan, Y.; Gau, V.; Zhang, J.; Wang, L.; Song, S.; Fan, C. An enzyme-based E-DNA sensor for sequence-specific detection of femtomolar DNA targets. J. Am. Chem. Soc. 2008, 130, 6820–6825.

109. Ferguson, B.S.; Buchsbaum, S.F.; Swensen, J.S.; Hsieh, K.; Lou, X.; Soh, H.T. Integrated micro fluidic electrochemical DNA sensor. Anal. Chem. 2009, 81, 6503–6508.

110. Wang, J. Portable electrochemical systems. Trends Anal. Chem. 2002, 21, 226–232.

111. Jin, Y. Label-free monitoring of site-specific DNA cleavage by EcoRI endonuclease using cyclic voltammetry and electrochemical impedance. Anal. Chim. Acta. 2009, 634, 44–48.

112. Luo, X.; Hsing, I.M. Electrochemistry and adsorptive stripping voltammetric determination of amoxicillin on a multiwalled carbon nanotubes modified glassy carbon electrode. Electroanalysis 2009, 21, 1577–1586.

113. Karadeniz, H.; Erdem, A.; Caliskan, A. Electrochemical monitoring of DNA hybridization by multiwalled carbon nanotube based screen printed electrodes. Electroanalysis 2008, 20, 1932–1938.

114. Cadogan, A.; Gao, Z.; Lewenstam, A.; Ivaska, A.; Diamond, D. All-solid-state sodium-selective electrode based on a calixarene ionophore in a poly(viny1 chloride) membrane with a polypyrrole solid contact. Anal. Chem. 1992, 64, 2496–2501.

115. Michalska, A. Optimizing the analytical performance and construction of ion-selective electrodes with conducting polymer-based ion-to-electron transducers. Anal. Bioanal. Chem. 2006, 384, 391–406.

116. Bobacka, J. Conducting polymer-based solid-state ion-selective electrodes. Electroanalysis 2006, 18, 7–18.

117. Bobacka, J. Potential stability of all-solid-state ion-selective electrodes using conducting polymers as ion-to-electron transducers. Anal. Chem. 1999, 71, 4932–4937.

118. Bobacka, J.; Lewenstam, A.; Ivaska, A. Electrochemical impedance spectroscopy of oxidized poly(3,4-ethylenedioxythiophene) film electrodes in aqueous solutions. J. Electroanal. Chem. 2000, 489, 17–27.

119. Hatchett, D.W.; Josowicz, M. Composites of intrinsically conducting polymers as sensing nanomaterials. Chem. Rev. 2008, 108, 746–769.

120. Maksymiuk, K. Chemical reactivity of polypyrrole and its relevance to polypyrrole based electrochemical sensors. Electroanalysis 2006, 18, 1537–1551.

121. Sutter, J.; Pretsch, E. Response behavior of poly(vinyl chloride)- and polyurethane-based Ca2+-selective membrane electrodes with polypyrrole- and poly(3-octylthiophene)-mediated internal solid contact. Electroanalysis 2006, 18, 19–25.

122. Shen, Z.; Burrows, P.E.; Bulovic, V.; Forrest, S.R.; Thompson, M.E. Three-color, tunable, organic light-emitting devices. Science 1997, 276, 2009–2011.

123. Lidzey, D.G.; Bradley, D.C.; Alvarado, S.F.; Seidler, P.F. Electroluminescence in polymer films. Nature 1997, 386, 135–135.

124. Ramanavičius, A.; Ramanavičienė, A.; Malinauskas, A. Electrochemical sensors based on conducting polymer-polypyrrole. Electrochim. Acta. 2006, 51, 6025–6037.

125. Bidan, G.; Billon, M.; Galasso, K.; Livache, T.; Mathis, G.; Roget, A.; Torres-Rodriguez, L.M.; Vieil, E. Electropolymerization as a versatile route for immobilizing biological species onto surfaces: Application to DNA biochips. Appl. Biochem. Biotechnol. 2000, 89, 183–193.

126. Livache, T.; Roget, A.; Dejean, E.; Barthet, C.; Bidan, G.; Teoule, R. Preparation of a DNA matrix via an electrochemically directed copolymerization of pyrrole and oligonucleotides bearing a pyrrole group. Nucleic Acids Res. 1994, 22, 2915–2921.

127. Livache, T.; Fouque, B.; Roget, A.; Marchand, J.; Bidan, G.; Téoule, R.; Mathis, G. Polypyrrole DNA chip on a silicon device: Example of hepatitis C virus genotyping. Anal. Biochem. 1998, 255, 188–194.

128. Wang, J.; Jiang, M. Toward genolelectronics: Nucleic acid doped conducting polymers. Langmuir 2000, 16, 2269–2274.

129. Wang, J.; Jiang, M.; Fortes, A.; Mukherjee, B. New label-free DNA recognition based on doping nucleic-acid probes within conducting polymer films. Anal.Chim. Acta. 1999, 402, 7–12.

130. Korri-Youssoufi, H.; Yassar, A. Electrochemical probing of DNA based on oligonucleotide functionalized polypyrrole. Biomacromolecules 2001, 2, 58–64.

131. Peng, H.; Soeller, C.; Vigar, N.; Kilmartin, P.A.; Cannell, M.B.; Bowmaker, G.A.; Cooney, R.P.; Travas-Sejdic, J. Label-free electrochemical DNA sensor based on functionalized conducting copolymer. Biosens. Bioelectron. 2005, 20, 1821–1828.

132. Ko, S.; Jang, J. Label-free target DNA recognition using oligonucleotide-functionalized polypyrrole nanotubes. Ultramicroscopy 2008, 108, 1328–1333.

133. Tlili, C.; Jaffrezic-Renault, N.J.; Martelet, C.; Korri-Youssoufi, H. Direct electrochemical probing of DNA hybridization on oligonucleotide-functionalized polypyrrole. Mater. Sci. Eng. C. 2008, 28, 848–854.

134. Komarova, E.; Aldissi, M.; Bogomolova, A. Direct electrochemical sensor for fast reagent-free DNA detection. Biosens. Bioelectron. 2005, 21, 182–189.

135. Peng, H.; Soeller, C.; Vigar, N.A.; Caprio, V.; Travas-Sejdic, J. Label-free detection of DNA hybridization based on a novel functionalized conducting polymer. Biosens. Bioelectron. 2007, 22, 1868–1873.

136. Eguiluz, K.I.V.; Salazar-Banda, G.R.; Elizabeth, M.; Huacca, F.; Alberice, J.V.; Carrilho, E.; Machado, S.A.S.; Avaca, L.A. Sequence-specific electrochemical detection of Alicyclobacillus acidoterrestris DNA using electroconductive polymer-modified fluorine tin oxide electrodes. Analyst. 2009, 134, 314–319.

137. Riccardi, C.S.; Kranz, C.; Kowalik, J.; Yamanaka, H.; Mizaikoff, B.; Josowicz, M. Label-Free DNA detection of hepatitis C virus based on modified conducting polypyrrole films at microelectrodes and atomic force microscopy tip-integrated electrodes. Anal. Chem. 2008, 80, 237–245.

138. Xu, Y.; Ye, X.; Yang, L.; He, P.; Fang, Y. Impedance DNA biosensor using electropolymerized polypyrrole/multiwalled carbon nanotubes modified electrode. Electroanalysis 2006, 18, 1471–1478.

139. Xu, Y.; Jiang, Y.; Cai, H.; He, P.G.; Fang, Y.Z. Electrochemical impedance detection of DNA hybridization based on the formation of M-DNA on polypyrrole/carbon nanotube modified electrode. Anal. Chim. Acta. 2004, 516, 19–27.

140. Qi, H.; Li, X.; Chen, P.; Zhang, C. Electrochemical detection of DNA hybridization based on polypyrrole/ss-DNA/multi-wall carbon nanotubes paste electrode. Talanta 2007, 72, 1030–1035.

141. Wilson, J.; Radhakrishnan, S.; Sumathi, C.; Dharuman, V. Polypyrrole-polyaniline-Au (PPy-PANi-Au) nano composite films for label-free electrochemical DNA sensing. Sens. Actuators B Chem. 2012, 171, 216–222.

142. Radhakrishnan, S.; Sumathi, C.; Umar, A.; Kim, S.J.; Wilson, J.; Dharuman, V. Polypyrrole-poly(3,4-ethylenedioxythiophene)-Ag (PPy-PEDOT-Ag) nanocomposite films for label-free electrochemical DNA sensing. Biosens. Bioelectron. 2013, 47, 133–140.

143. Booth, M.A.; Harbison, S.A.; Travas-Sejdic, J. Effects of redox couple on the response of polypyrrole-based electrochemical DNA sensors. Electroanalysis 2012, 24, 1311–1317.

144. Ho, H.A.; Doré, K.; Boissinot, M.; Bergeron, M.G.; Tanguay, R.M.; Boudreau, D.; Leclerc, M. Direct molecular detection of nucleic acids by fluorescence signal amplification. J. Am. Chem. Soc. 2005, 127, 12673–12676.

145. Floch, F.L.; Ho, H.A.; Leclerc, M. Label-Free electrochemical detection of protein based on a ferrocene-bearing cationic polythiophene and aptamer. Anal. Chem. 2006, 78, 4727–4731.

146. Lee, T.Y.; Shim, Y.B. Direct DNA Hybridization Detection based on the oligonucleotide-functionalized conductive polymer. Anal. Chem. 2001, 73, 5629–5632.

147. Cheng, G.; Zhao, J.; Tu, Y.; He, P.; Fang, Y. A sensitive DNA electrochemical biosensor based on magnetite with a glassy carbon electrode modified by multi-walled carbon nanotubes in polypyrrole. Anal. Chim. Acta. 2005, 533, 11–16.

148. Zhang, Z.; Liang, P.; Zheng, X.; Peng, D.; Yan, F.; Zhao, R.; Feng, C.L. DNA Immobilization/hybridization on plasma-polymerized pyrrole. Biomacromolecules 2008, 9, 1613–1617.

149. Cha, J.; Han, J.I.; Choi, Y.; Yoon, D.S.; Oh, K.W.; Lim, G. DNA hybridization electrochemical sensor using conducting polymer. Biosens. Bioelectron. 2003, 18, 1241–1247.

150. Peng, H.; Soeller, C.; Travas-Sejdic, J. DNA sensor based on conducting polymers functionalized with conjugated side chain. Proceedings of the IEEE International Nanoelectronics Conference

on Emerging Technologies, 28–31 October 2007, Atlanta, GA, USA; pp. 1124–1127.

151. Nilsson, K.P.R.; Inganäs, O. Chip and solution detection of DNA hybridization using a luminescent zwitterionic polythiophene derivative. Nat. Mater. 2003, 2, 419–424.

152. Ho, H.A.; Najari, A.; Leclerc, M. Optical detection of DNA and proteins with cationic polythiophenes. Accounts Chem. Res. 2008, 2, 168–178.

153. Shiddiky, A.J.M.; Shim, Y.B. Trace Analysis of DNA: Preconcentration, separation, and electrochemical detection in microchip electrophoresis using Au nanoparticles. Anal. Chem. 2007, 79, 3724–3733.

154. Shiddiky, A.J.M.; Rahman, M.A.; Shim, Y.B. Hydrazine-catalyzed ultrasensitive detection of DNA and proteins. Anal. Chem. 2007, 79, 6886–6890.

155. Fang, B.; Jiao, S.; Li, M.; Qu, Y.; Jiang, X. Label-free electrochemical detection of DNA using ferrocene-containing cationic polythiophene and PNA probes on nanogold modified electrodes. Biosens. Bioelectron. 2008, 23, 1175–1179.

156. Liu, M.; Luo, C.; Peng, H. Electrochemical DNA sensor based on methylene blue functionalized polythiophene as a hybridization indicator. Talanta 2012, 88, 216–221.

157. Huang, J.; Virji, S.; Weiller, B.H.; Kaner, R.B. Polyaniline nanofibers: Facile synthesis and chemical Sensors. J. Am. Chem. Soc. 2003, 125, 314–315.

158. Janata, J.; Josowicz, M. Conducting polymers in electronic chemical sensors. Nat. Mater. 2003, 2, 19–24.

159. Tahir, Z.M.; Alocilja, E.C.; Grooms, D.L. Indium tin oxide-polyaniline biosensor: Fabrication and characterization. Sensors 2007, 7, 1123–1140.

160. Lukachova, L.V.; Karyakin, A.A.; Karyakina, E.E.; Gorton, L. The improvement of polyaniline glucose biosensor stability using enzyme immobilization from water–organic mixtures with a high content of organic solvent. Sens. Actuators B Chem. 1997, 44, 356–360.

161. Sangodkar, H.; Sukeerthi, S.; Srinivasa, R.S.; Lal, R.; Contractor, A.Q. A biosensor array based on polyaniline. Anal. Chem. 1996, 68, 779–783.

162. Wei, D.; Ivaska, A. Electrochemical biosensors based on polyaniline. Chem. Anal. (Wars.) 2006, 51, 839–852.

163. Lindfors, T.; Ivaska, A. Potentiometric and UV-vis characterization of N-substituted polyanilines. J. Electroanal. Chem. 2002, 535, 65–74.

164. Wu, J.; Zou, Y.; Li, X.; Liu, H.; Shen, G.; Yu, R. A biosensor monitoring DNA hybridization based on polyaniline intercalated graphite oxide nanocomposite. Sens. Actuators B Chem. 2005, 104, 43–49.

165. Gu, H.; Su, X.D.; Loh, K.P. Electrochemical impedance sensing of DNA hybridization on conducting polymer film-modified diamond. J. Phys. Chem. B. 2005, 109, 13611–13618.

166. Arora, K.; Prabhakar, N.; Chand, S.; Malhotra, B.D. Ultrasensitive DNA hybridization biosensor based on polyaniline. Biosens. Bioelectron. 2007, 23, 613–620.

167. Arora, K.; Prabhakar, N.; Chand, S.; Malhotra, B.D. Escherichia coli genosensor based on polyaniline. Anal. Chem. 2007, 79, 6152–6158.

168. Gao, Z.; Rafea, S.; Lim, L.H. Detection of nucleic acids using enzyme-catalyzed template-guided deposition of polyaniline. Adv. Mater. 2007, 19, 602–606.

169. Zhu, N.; Chang, Z.; He, P.; Fang, Y. Electrochemically fabricated polyaniline nanowire-modified electrode for voltammetric

detection of DNA hybridization. Electrochim. Acta. 2006, 51, 3758–3762.

170. Chang, H.; Yuan, Y.; Shi, N.; Guan, Y. Electrochemical DNA biosensor based on conducting polyaniline nanotube Array. Anal. Chem. 2007, 79, 5111–5115.

171. Erdem, A.; Kerman, K.; Meric, B.; Akarca, U.S.; Osos, M. Novel hybridization indicator methylene blue for the electrochemical detection of short DNA sequences related to the hepatitis B virus. Anal. Chim. Acta. 2000, 422, 139–149.

172. Kerman, K.; Ozkan, D.; Kara, P.; Meric, B.; Gooding, J.J.; Ozsoz, M. Voltammetric determination of DNA hybridization using methylene blue and self-assembled alkanethiol monolayer on gold electrodes. Anal. Chim. Acta. 2002, 462, 39–47.

173. Pänke, O.; Kirbs, A.; Lisdat, F. Voltammetric detection of single base-pair mismatches and quantification of label-free target ssDNA using a competitive binding assay. Biosens. Bioelectron. 2007, 22, 2656–2662.

174. Prabhakar, N.; Arora, K.; Singh, H.; Malhotra, B.D. Polyaniline based nucleic acid Sensor. J. Phys. Chem. B 2008, 112, 4808–4816.

175. Du, M.; Yang, T.; Li, X.; Jiao, K. Fabrication of DNA/graphene/polyaniline nanocomplex for label-free voltammetric detection of DNA hybridization. Talanta 2012, 88, 439–444.

176. Saberi, R.S.; Shahrokhian, S.; Marrazza, G. Amplified electrochemical DNA sensor based on polyaniline film and gold nanoparticles. Electroanalysis 2013, 25, 1373–1380.

177. Ge, Y.; Smith, D.K. Development of chemical sensors based on redox-dependent receptors. Preparation and characterization of phenanthrenequinone-modified electrodes. Anal. Chem. 2000, 72, 1860–1865.

178. Geng, L.; Boguslavsky, L.I.; Kovalev, I.P.; Sahni, S.K.; Kalash, H.; Skotheim, T.A. Amperometric biosensors based on dehydrogenase/NAD and heterocyclic qulnones. Biosens. Bioelectron. 1996, 11, 1267–1275.

179. Bucur, B.; Mallat, E.; Gurban, A.M.; Gocheva, Y.; Velasco, C.; Marty, J.L.; Noguer, T. Strategies to develop malic acid biosensors based on malate quinone oxidoreductase (MQO). Biosens. Bioelectron. 2006, 21, 2290–2297.

180. Piro, B.; Reisberg, S.; Noel, V.; Pham, M.C. Investigations of the steric effect on electrochemical transduction in a quinone-based DNA sensor. Biosens. Bioelectron. 2007, 22, 3126–3131.

181. Haccoun, J.; Piro, B.; Tran, L.D.; Dang, L.A.; Pham, M.C. Reagentless amperometric detection of l-lactate on an enzyme-modified conducting copolymer poly(5-hydroxy-1,4-naphthoquinone-co-5-hydroxy-3-thioacetic acid-1,4-naphthoquinone). Biosens. Bioelectron. 2004, 19, 1325–1329.

182. Reisberg, S.; Dang, L.A.; Nguyen, Q.A.; Piro, B.; Noel, V.; Nielsen, P.E.; Le, L.A.; Pham, M.C. Label-free DNA electrochemical sensor based on a PNA-functionalized conductive polymer. Talanta 2008, 76, 206–210.

183. Acevedo, D.F.; Reisberg, S.; Piro, B.; Peralta, D.O.; Miras, M.C.; Pham, M.C.; Barbero, C.A. Fabrication of an interpenetrated network of carbon nanotubes and electroactive polymers to be used in oligonucletide biosensing. Electrochim. Acta. 2008, 53, 4001–4006.

184. Pham, M.C.; Lacaze, P.C.; Dubois, J.E. Voltammetric and XPS Analysis of Metal-Complexed Polytyramine Films: Geometry-Dependent Electron Transfer Therein. J. Electrochem. Soc. 1984, 131, 777–784.

185. Tran, L.D.; Prio, B.; Pham, M.C.; Doan, T.L.; Dao, L.H. Electropolymerized polytriamine films: Covalent binding of oligoneucleotide and hybridization. Synth. Met. 2003, 137, 1439–1440.

186. Robinson, V.L.; Stock, A.M. High energy exchange: Proteins that make or break phosphoramidate bonds. Structure 1999, 7, R47–R53.

187. Li, X.; Xia, J.; Zhang, S. Label-free detection of DNA hybridization based on poly(indole-5-carboxylic acid) conducting polymer. Anal. Chim. Acta. 2008, 622, 104–110.

188. Peng, H.; Soeller, C.; Travas-Sejdic, J. A novel cationic conjugated polymer for homogeneous fluorescence based DNA detection. Chem. Commun. 2006, 35, 3735–3737.

189. Gaylord, B.S.; Heeger, A.J.; Bazan, G.C. DNA Hybridization Detection with Water-Soluble Conjugated Polymers and Chromophore-Labeled Single-Stranded DNA. J. Am. Chem. Soc. 2003, 125, 896–900.

190. Liu, B.; Bazan, G.C. Homogeneous Fluorescence-Based DNA Detection with Water-Soluble Conjugated Polymers. Chem. Mater. 2004, 16, 4467–4476.

191. Campas, M.; Katakis, I. DNA biochip arraying, detection and amplification strategies. Trends Anal. Chem. 2004, 23, 49–62.

192. Khanna, V.K. Existing and emerging detection technologies for DNA (deoxyribonucleic acid) finger printing, sequencing, bio- and analytical chips: A multidisciplinary development unifying molecular biology, chemical and electronics engineering. Biotechnol. Adv. 2007, 25, 85–98.

6
CHAPTER

EMERGING TRANSPARENT CONDUCTING ELECTRODES FOR ORGANIC LIGHT EMITTING DIODES

Tze-Bin Song [1,2] **and Ning Li** [2,*]

[1] *Department of Materials Science & Engineering, University of California, Los Angeles, CA 90095, USA*

[2] *IBM T. J. Watson Research Center, 1101 Kitchawan Road, Yorktown Heights, NY 10598, USA*

ABSTRACT

Organic light emitting diodes (OLEDs) have attracted much attention in recent years as next generation lighting and displays, due to their many advantages, including superb performance, mechanical flexibility, ease of fabrication, chemical versatility, etc. In order to fully realize the highly flexible features, reduce the cost and further improve the performance of OLED devices, replacing the conventional indium tin oxide with better alternative transparent conducting electrodes (TCEs) is a crucial step. In this review, we focus on the emerging alternative TCE materials for OLED applications, including carbon nanotubes (CNTs), metallic nanowires, conductive polymers and graphene. These materials are selected, because they have been applied as transparent electrodes for OLED devices and achieved reasonably good performance or even higher device performance than that of indium tin oxide (ITO) glass. Various electrode modification techniques and their effects on the device performance are presented. The effects of new TCEs on light extraction, device performance and reliability are discussed. Highly flexible, stretchable and efficient OLED devices are achieved based on these alternative TCEs. These results are summarized for each material. The advantages and current challenges of these TCE materials are also identified.

Keywords: transparent electrode; organic light emitting diode; carbon nanotube; metallic nanowire; graphene; conductive polymer

1. INTRODUCTION

Organic light emitting diode (OLED) has emerged as a potential candidate for next generation flexible, large-area, high resolution display and solid state lighting panels, because of its high color quality, attractive appearance, ease of fabrication, low manufacturing and materials cost, etc [1]. With great efforts from both academia and industry, the OLED has been developed based on small molecule and polymer materials and also fabricated with both vacuum deposition and solution processes [2,3,4,5]. In the past few decades, researchers have been focusing on improving the device efficiency and lowering the manufacturing cost. Today, OLED displays are becoming dominant in the high-end market. OLED lighting is also on the verge of becoming widely commercially available, and its performance is competitive with that of its inorganic counterparts.

The basic OLED structure is composed of a stack of several layers: anode/hole transport layer (HTL)/emission layer (EL)/electron transport layer (ETL)/cathode, as shown in Figure 1a. [6] The first OLED was developed by Tang and VanSlyke in 1987 with the structure of an indium tin oxide (ITO)/aromatic diamine/8-hydroxyquinoline aluminum (Alq_3)/Mg-Al metal electrode [7]. Since then, ITO glass has been commonly used as the anode for OLEDs, because ITO simultaneously provides good transparency and conductivity [8]. Moreover, the work function of ITO is around 4.7 eV, which forms a low barrier for hole injection into the emission layer made of commonly used organic materials (Figure 1b) [9]. Despite these advantages, ITO is far from being a perfect candidate for OLED applications for the following reasons. First, it is not ideal for highly flexible electronics, due to its brittleness. Under mechanical bending or stretching, crack generation in the ITO film leads to much deteriorated electrical properties [10]. Second, the sputtering deposition of high quality ITO is a low throughput process and requires elevated temperature. Solution processed ITO also requires high temperature annealing to achieve a good conductivity [11]. It is vital therefore to only use substrates that are stable at high temperatures, which means an increased substrate cost and much reduced performance on plastic substrates. Furthermore, due to the widespread application of ITO as the transparent conducting electrode (TCE) for various optical devices and the limited global reserve of indium, the price of ITO will rise dramatically and further raise the cost of OLEDs. In addition, ITO does not offer ideal performance for OLEDs. It has significant light reflection and also traps the light in the waveguide mode. Its conductivity needs to be further improved, as well, for large area devices.

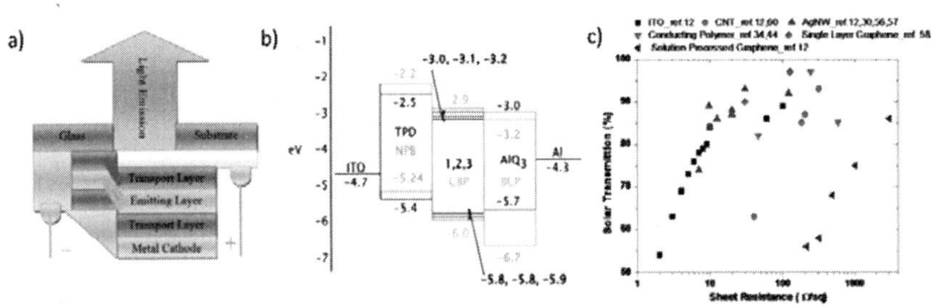

Figure 1. (**a**) Schematic diagram of the organic light emitting diode (OLED) structure; (**b**) engery level diagram of a simple OLED device consisting of N,N′-Bis(3-methylphenyl)-N,N′-diphenylbenzidine (TPD), N,N′-Di(1-naphthyl)-N,N′-diphenyl-(1,1'-biphenyl)-4,4'-diamine (NPB), 4,4'-Bis(N-carbazolyl)-1,1'-biphenyl (CBP), Bathocuproine (BCP), and Tris-(8-hydroxyquinoline)aluminum (Alq3); (**c**) sheet resistance and transmission chart for various types of transparent conducting electrode (TCE) materials including carbon nanotube (CNT), silver nanowire (AgNW), conductive polymers, and graphene.

Considering all these factors, there has been increasing interest and an urgent need to look for alternative TCE materials to replace the conventional ITO. These TCE materials should be highly conductive and optically transparent; meanwhile, they should also be low cost and enable new attractive features. Here, we review the recent progress on the promising next generation TCEs. We focus our attention on the following materials: carbon nanotubes (CNTs), metallic nanowires, conducting polymers and graphene. These materials have shown the potential to fulfill standard requirements on the sheet resistance and transmission values of

TCE and can be formed by low-cost processes, such as spin coating, spray coating and even roll-to-roll processes [12]. The sheet resistance and transmittance of these materials is summarized in Figure 1c using the solar spectrum as a reference for transmission evaluation. Moreover, OLED devices demonstrated with these techniques show great potential for future highly flexible, foldable and wearable opto-electronics. We summarize the progress of each of these TCE materials with the device performance achieved and give comparisons between these techniques.

2. CARBON NANOTUBES

The carbon nanotube (CNT) network is the first nanostructured TCE investigated for OLEDs, leading to a boom of interest in this decade [13]. CNTs exhibit a unique electrical property in that they can be both metallic and semiconducting [14]. Because of this, they are widely applied as high-performance flexible transparent transistors, optical modulators, flexible emitters, as well as TCEs [15,16,17]. Metallic CNTs have a suitable work function (4.7–5.2 eV) for the application as anodes in OLEDs [18,19]. In addition, the high stability, flexibility and mobility of CNTs make the CNT network a potential candidate to replace the rigid ITO substrate, while avoiding the contamination of the organic layers from the oxygen atoms in ITO.

Zhang et al. first developed large area CNT sheets (meter long, 5 cm-wide) as the electrode for OLED, and this nanotube sheet was reported to be as strong as the steel (Figure 2a) [13]. This report demonstrated the huge potential of CNT networks' application in optical electronics and opened a new direction for the nanostructured TCE. Zhang et al. tested various CNTs from different growth methods [20]. The arc discharge nanotubes showed better performance compared to high-pressure CO conversion (HiPCO) nanotubes in surface roughness, sheet resistance and transparency. A sheet resistance of ~160 ohm/sq at 87% transparency can be achieved when the CNTs network is treated with $SOCl_2$, as shown in Figure 2b,c. Li et al. demonstrated that a poly(3,4-ethylenedioxythiophene) poly(styrenesulfonate) (PEDOT:PSS) layer could play an important role in planarizing the roughness from CNT networks (Figure 2d) and decreasing the hole injection barrier from the CNTs to a polymer blend hole transporting layer poly(9,9-dioctylfluorene-co-N-(4-butylphenyl) diphenylamine) (TFB)+ 4,4'-bis[(p-trichlorosilylpropylphenyl) phenylamino]biphenyl ($TPD-Si_2$), which could reduce the leakage current [21]. With a PEDOT:PSS layer modified with methanol, a surface roughness

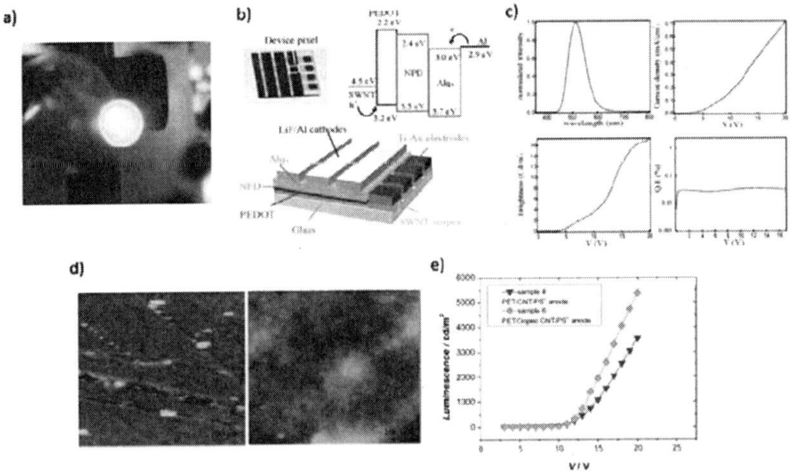

Figure 2. (**a**) Photograph of an OLED that uses an multi-wall nanotube (MWNT) sheet as the anode and poly(3,4-ethylenedioxythiophene) poly(styrenesulfonate) (PEDOT:PSS)/ Poly[2-methoxy-5-(2-ethylhexyloxy)-1,4-phenylenevinylene] (MEH-PPV) /Ca/Al [13]; (**b**) patterned multilayer single wall CNT/PEDOT:PSS/NPB/Alq_3/LiF/Al [20]; (**c**) device performance: photoluminescence spectrum, current density vs. voltage bias curve, brightness vs. voltage bias and quantum efficiency as a function of current density [20]; (**d**) surface roughness of PEDOT:PSS ~4 nm and methanol-modified PEDOT:PSS ~0.96 nm on a CNT network [21]; (**e**) the luminescence vs. the voltage of an OLED with PEDOT:PSS composite (PS^C)-modified CNT on Polyethylene terephthalate (PET) substrate [22].

less than 1 nm can be achieved, due to better PEDOT:PSS wetting onto the CNT network.

Ou et al. further modified the surface of the carbon nanotube network with PEDOT:PSS composite (PS^C) coating, which contained polyethylene glycol (PEG) additive in Baytron P500 and used HNO_3 acid treatment to improve the conductivity of the CNT network and the band alignment for hole injection [22]. Outstanding performance with a maximum luminance of ~9000 cd/m^2 and a luminance efficiency (LE) of ~10 cd/A at 1000 cd/m^2 was achieved, which was comparable to devices on ITO substrates. Yu et al. explored the capability of CNT transparent electrodes as the cathode and anode for flexible and transparent organic light emitting diodes by a lamination method [23]. Furthermore, stretchable OLEDs based on a CNT network as the TCE was built. The electroluminescent efficiency of the devices can be sustained under a 45% strain, which cannot be achieved for traditional ITO substrates [24]. The device stability with the CNT electrode exhibited comparable lifetime with that of the ITO electrode, and the acid resistivity of the CNT electrode to PEDOT:PSS is better than that of ITO electrode for long-term operation.

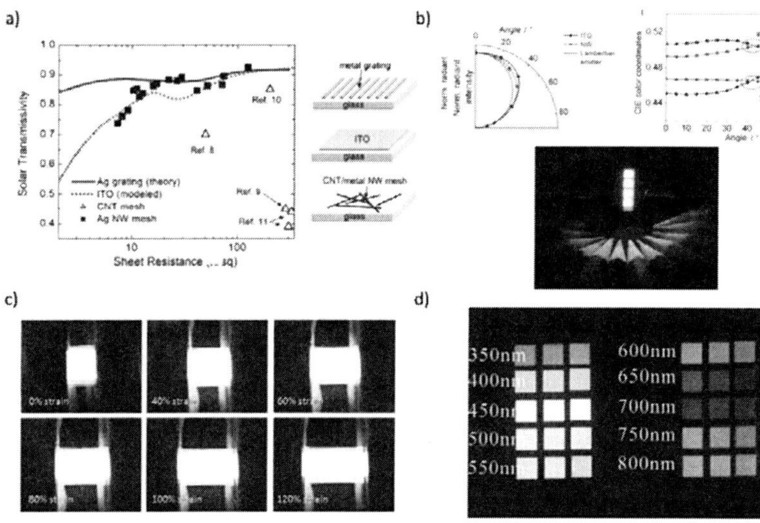

Figure 3. (**a**) Solar photon flux-weighted transmissivity vs. sheet resistance for Ag gratings (blue line), ITO (red dotted line), CNT meshes (Δ) and Ag nanowire meshes (■) deposited on a glass substrate. The Ag line width is a 40 nm and a 400 nm grating period [26]. (**b**) Normalized radiant intensity, color coordinates vs. viewing angle and photographic image of four operating nanowire (NW)-OLEDs [28]. (**c**) Photographs of a polymer light-emitting electrochemical cell (PLEC) (original emission area, 5.0 × 4.5 mm^2) biased at 14 V at specified strains [29]. (**d**) Colors from blue to red can be selected by different period nanowire arrays [32].

3. METALLIC NANOWIRES

Recent studies on the metallic nanowire's application in optical electronics have attracted a lot of attention. Similar to CNTs, high conductivity from the metal material and high transmittance from the open space between nanowires make the metallic nanowire a potential candidate as the TCE.

Compared with CNTs, the metallic nanowire network shows better sheet resistance and transmission values, because the wire to wire contact resistance can be reduced by thermal treatment [25]. The low contact resistance between nanowires can significantly reduce the power loss on the electrodes. However, the metallic nanowire network requires PEDOT:PSS or other hole transport materials to ensure efficient hole injection as an anode, which slightly restricts the fabrication process. Lee et al. demonstrated the potential of a silver nanowire network TCE on a glass substrate with a sheet resistance of 16 ohm/sq and an average transmittance of 86% between the wavelengths of 400 and 800 nm, which is comparable to commercial ITO substrates (Figure 3a) [26]. Yu et al. first demonstrated a composite electrode in which the silver nanowires were embedded in cross-linkable polyacrylate substrate, which could successfully replace the traditional rigid glass substrate [27]. This result opened up the possibility of realizing the high flexibility and high performance OLEDs by incorporating a solution processed metallic nanowire network.

Gaynor et al. investigated the angular dependence of white OLEDs using silver nanowires embedded in poly(methyl methacrylate) (PMMA) as the electrode [28]. The scattering of the silver nanowire network kept a stabilized viewing angle characteristic with reduced color shift and better Lambertian emission for the OLED. By further incorporation of light outcoupling techniques, a power efficiency of 54 lm/W was achieved, as shown in Figure 3b. Liang et al. reported an elastomeric polymer (polyurethane acrylate (PUA)) -based silver nanowire substrate with yellow light-emitting polymers consisting of ethoxylated trimethylolpropane triacrylate (ETPTA), polyethylene oxide (PEO) and lithium trifluoromethanesulphonate (LiTf), and the efficiency was kept at 2.5 cd/A under 120% strain (Figure 3c) [29]. The concern for metallic nanowire electrode is the instability, due to Rayleigh instability and contact ripening, resulting in the loss of the conductive path during operation. These might be the challenges for having long lifetime OLED devices [30,31].

The improvement of the silver nanowire TCE provided a platform for OLEDs to reach wider applications on display and lighting. Furthermore, the dimension of the metallic nanowire could affect the light scattering, light coupling and sheet resistance to transmission values of the TCE, providing us with an additional degree of freedom in improving the device performance. Aligned metal nanowire fabricated by a vacuum process was reported to have improved light outcoupling of the OLEDs (Figure 3d) [32]. The optical effect of the metallic nanowire on OLED and the alignment control of the nanowire through fabrication are still under investigation.

4. CONDUCTIVE POLYMERS

Among various types of conductive polymers, PEDOT:PSS and polyaniline (PANI) are currently the most popular materials to replace the conventional ITO electrode. These two materials are well-studied, conjugated polymers with excellent mechanical stability, flexibility and, more importantly, they can achieve a high conductivity and transparency.

It was shown that PANI has the potential as a solution-processable TCE by Cao et al. [33]. They discovered that the camphor-sulfonic acid (CSA) doped PANI (PANI:CSA), which is soluble in m-cresol or chloroform, is conductive (<10 S/cm) and optically transparent. Gustafsson et al. successfully fabricated the PANI:CSA film on Polyethylene terephthalate (PET) substrate as the anode and demonstrated the first flexible polymer light emitting diode (PLED) device [5]. However, PANI:CSA film needs to be thicker than 250 nm to achieve 160 ohm/sq, while the transparency is about 70% under such a thickness (Figure 4a) [34]. Fehse et al. reported a new dispersion of PANI (D1033), which has a conductivity of 200 S/cm and a lower absorbance at 750 nm in the visible region than that of PANI:CSA (Figure 4b) [35]. Even though the improvement of the PANI properties resulted in better device performance, it is still lower than that of the commercial ITO substrate. PEDOT:PSS was invented in 1991 by Bayer [36] as an antistatic coating material. Cao et al. first introduced PEDOT:PSS into the PLED as a TCE with 500 ohm/sq and 75% transmission within the visible light region [37]. After that, PEDOT:PSS became commonly used as the hole injection layer for OLEDs, as well. In order to improve the properties of the PEDOT:PSS, many treatments have been applied to improve the conductivity without losing the transparency. Kim et al. investigated different organic solvents, including dimethyl sulfoxide (DMSO), N,N-dimethyl formamide (DMF) and tetrahydrofuran (THF), and the addition of the high boiling polar solvent DMSO produced the highest conductivity among the three [38]. Ouyang et al. reported using ethylene glycol (EG) as a polar additive and acquired PEDOT:PSS thin films with a conductivity up to 160 S/cm, demonstrating a comparable performance to ITO/PEDOT:PSS on Poly[2-methoxy-5-(2-ethylhexyloxy)-1,4-phenylenevinylene] (MEH-PPV) PLED devices [39]. Fehse et al. used a commercially available high conductivity PEDOT:PSS Baytron PH500 (conductivity ~500 S/cm) as the TCE in small molecular OLED devices and achieved comparable or even superior performance (on green and blue emitting devices) than ITO anode devices [40]. Various treatments, like using zwitterionic surfactants, co-solvent system or exposing the films to dichloroacetic acid, were reported to achieve better conductivity in recent years [41,42,43]. Kim et al. demonstrated that with the addition of EG and a solvent post-treatment method, they could remarkably increase the conductivity up to 1418 S/cm, around 65 ohm/sq at 80% transmission [44]. Vosgueritchian et al. reported using a combination of DMSO and the fluorosurfactant Zonyl-FS300 (Zonyl) to achieve 46 ohm/sq at 82% transmission [45]. Recently, Xia et al. reported a H_2SO_4 treatment to reach 2400 S/cm and 3065 S/cm from multiple treatments [46]. These values are close to that of commercially available ITO substrates.

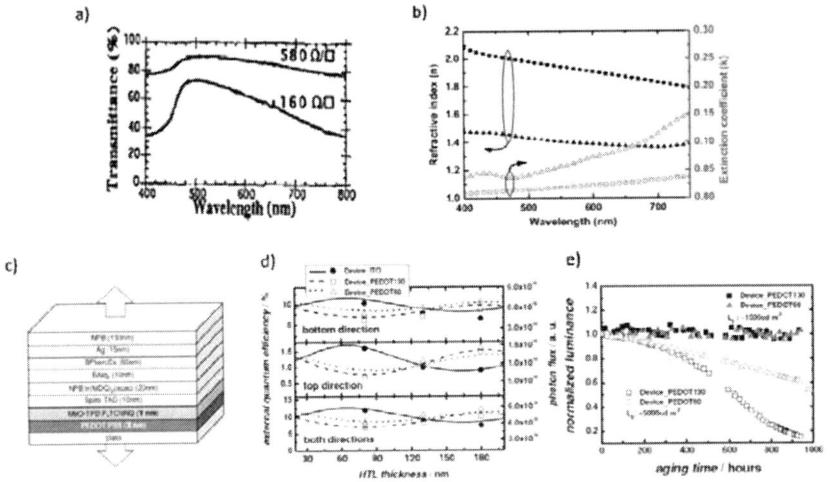

Figure 4. (**a**) Transmittance and sheet resistance of different polyaniline (PANI) thicknesses [34]; (**b**) the wavelength dependence of the refractive index and the extinction coefficient of ITO (squares) and PANI (triangles) films [35]; (**c**) the device structure of ITO-free transparent OLEDs based on a PEDOT:PSS TCE; (**d**) the comparison of external quantum efficiencies (EQEs) and corresponding photon fluxes for bottom and top emission and the sum of both (experiment (symbols) and optical simulation (lines) results); (**e**) aging characteristics of PEDOT:PSS-based OLEDs with a different PEDOT:PSS thickness. The encapsulated devices are aged over around 900 hours with an initial luminance of around 1500 cd/m^2. Constant currents are applied for each sample according to the corresponding luminance [47]. HTL, hole transport layer.

Kim et al. demonstrated long lifetime (over 900 h under 1500 cd/m^2) and high performance (over 12% external quantum efficiency (EQE) under a driving current density of 10 mA/cm^2) OLEDs on polymer TCE. The results are shown in Figure 4c–d [47], comparing with those made from the ITO substrate. Moreover, from the optical study from Cai et al, the PEDOT:PSS anode could be more beneficial to the light outcoupling than ITO substrates, due to the match of the refractive index [48]. At the current stage, however, these polymers still exhibit a lower conductivity and transmission than CNTs and metallic nanowire networks. In order to achieve a better device performance compared with ITO and other new materials, the properties and stability of the conductive polymers need to be further improved before they can be mass produced as the next generation TCE.

5. GRAPHENE

Graphene is another promising candidate as a TCE in OLEDs. A flexible two-dimensional sheet of sp^2-hybridized carbon atoms has very high conductivity and is nearly transparent. A single layer of graphene showed a sheet resistance of 125 ohm/sq and 97.4% transmission at the 550 nm wavelength, which is superior to the ITO substrate and other reported TCEs (Figure 5a) [49].

Wu et al. demonstrated solution processed graphene oxide thin films fabricated using Hummers' method and the ability to further reduce them to graphene thin films [12,50]. The graphene films were used as the TCE for OLEDs with a film thickness of about 7 nm, a sheet resistance of about 800 ohm/sq and 82% transmission at 550 nm, which is lower than the theoretical value, due to the existence of multiple grain boundaries, lattice defects and oxidative traps formed during the fabrication process. The device performance with a solution processed graphene electrode showed a turn-on voltage of 4.5 V and a luminance of 300 cd/m^2 at 11.7 V, which is comparable to those of ITO substrates, which had a turn-on voltage of 3.8 V and a luminance of 300 cd/m^2 at 9.9 V bias (Figure 5b,c). Han et al. recently reported an extremely efficient flexible OLED by introducing a work function tunable layer [51]. The mismatch of the electrical band alignment between graphene (work function ~4.4 eV) and the hole transport layer, like N,N′-Di(1-naphthyl)-N,N′-diphenyl-(1,1′-biphenyl)-4,4′-diamine (NPB) (work function ~5.4 eV), needs to be reduced for efficient hole injection. The authors incorporated a self-organized gradient hole injection layer (GraHIL), which was composed of PEDOT:PSS and tetrafluoroethylene-perfluoro-3,6-dioxa-4-methyl-7-octenesulphonic acid copolymer, one of the perfluorinated ionomers (PFIs). The GraHIL provided a work function gradient throughout the hole injection layer and, in turn, improved the charge injection efficiency. Based on a graphene/GraHIL/NPB/Alq_3/LiF/Al structure, the luminance efficiencies with doped graphene TCE as the anode

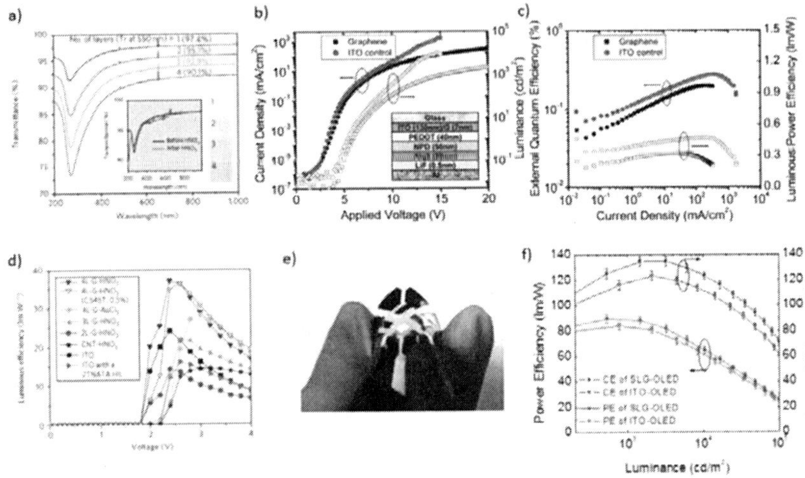

Figure 5. (**a**) Transmittance of the roll-to-roll layer-by-layer transferred graphene films on quartz substrates. The inset shows the transmittance spectra of graphene films with and without HNO_3 doping and the optical images for the corresponding number of transferred layers (1×1 cm^2) [49]; (**b**) Current density (filled symbols) and luminance (open symbols) vs. applied forward bias for an OLED on graphene (squares) and ITO (circles), with an OLED device structure of anode/PEDOT:PSS/NPD/Alq3/LiF/Al; (**c**) External quantum efficiency (EQE) (filled symbols) and luminous power efficiency (open symbols) for an OLED on graphene film (squares) and ITO glass (circles) [12]; (**d**) Power efficiencies of OLED devices using various graphene layers (doped with HNO_3 or $AuCl_3$) and ITO as the anode; (**e**) Photograph of a flexible fluorescent green OLED with a four-layered graphene anode (4L-G) doped with HNO3 (4L-G-HNO3) [51]; (**f**) Power efficiency and current efficiency of white OLED (WOLED) based on ITO and single layer graphene electrode [52].

(37.2 lm/W in fluorescent OLEDs, 102.7 lm/W in phosphorescent OLEDs) is higher than the devices using conventional ITO as the TCE (24.1 lm/W in fluorescent OLEDs, 85.6 lm/W in phosphorescent OLEDs, shown in the Figure 5d,e). Li et al. showed that white OLEDs (WOLEDs) on graphene electrode can indeed exhibit performance satisfying general lighting requirements [52]. WOLEDs on graphene with a power efficiency of 80 lm/W at a high brightness of 3000 cd/m^2 are demonstrated (Figure 5f). It is also found that graphene electrodes have the advantage of light extraction over ITO. White ITO has significant light reflection at both the top and bottom interface and also has significant light trapping. In the waveguide mode, Graphene, on the other hand, is so thin that it is optically negligible. There is almost no light reflection and trapping in the graphene layer. Ultimately, more light can be coupled out of the graphene OLED than the ITO OLED [52].

The above high-performance OLEDs on grapheme were all demonstrated using polymer interface layers. In contrast, OLEDs on graphene only electrodes without a polymer interface layers were also reported. In the early work, Sun et al. applied multilayered graphene from chemical vapor deposition as the anode for an OLED device and showed working devices without using the polymer interface layer, but with much less performance compared with the ITO reference device. The high leakage current and low efficiency were attributed to the surface roughness and work function mismatch [53]. Hwang et al. found that using weak plasma treatment to multilayered graphene can improve the hole injection efficiency and demonstrated high OLED device performance on graphene without polymer interface layers [54]. These works clearly indicate graphene's capability of charge conduction and injection. After optimization, graphene can serve as an excellent electrode by itself or together with a polymer layer.

Besides the high device performance, the stability of graphene is also excellent. Due to the inert material properties, it is not reactive with most chemicals. In addition, it does not have the problem of device degradation caused by electrode material diffusion into OLEDs, which can happen for Ag and ITO electrodes. The upcoming challenge for the grapheme electrode is to further improve the electrical and optical properties and to make the fabrication process faster, easier and more cost effective.

6. CONCLUSIONS

Four different types of TCE materials are discussed in this review in terms of their electrical and optical performance as a TCE and the resultant OLED device performance. A comparison of their current status is summarized in Table 1. The performance of CNT TCEs is mainly limited by the contact resistance between adjacent wires, while it provides excellent mechanical properties and stability. The purity and type of CNT will also significantly affect the device performance, due to the difference in conductivity and surface roughness. Metallic nanowire TCEs exhibit better sheet resistance and transmission compared to their CNT counterparts, which resulted in highly efficient and stretchable OLED devices. Optical enhancement by the metallic nanowires is another major advantage for their future application and needs further investigation. Better control of the dimension and alignment of the metallic nanowires also needs to be further investigated. Conducting polymers have a longer developing history comparing with the other mentioned techniques. Their properties are becoming more and more competitive with traditional ITO glass showing similar conductivity values. They can also be tuned to provide optimal refractive index matching for efficient light outcoupling in OLEDs. The search for methods to precisely control the morphologies of the conducting polymers and to further improve the electrical and optical properties is ongoing. Graphene is the latest material to be applied as the TCE for OLEDs. The high transparency and the absence of light trapping from the ultrathin graphene sheet, the high in-plane conductivity and its ability to be solution processed make it a promising candidate as the next generation TCE. Its potential as the TCE in OLEDs has been proven.Outstanding OLED performance comparable to those on ITO glass has been achieved. Surface modification, the reduction of grain boundaries and defects for roll-to-roll processed graphene are some of the key routes towards further improvement.

Although most of these new TCEs still have many challenges in order to replace ITO completely, their performance is improving very rapidly. Even in the current stage, very low sheet resistance can be obtained by combining the new TCEs with a metal grid for large area applications [55]. Light extraction is an aspect that new TCEs can improve over ITO. The light reflection and trapping in most new TCEs are reduced comparing with ITO, due to the porous nature and scattering effect of CNTs and metal nanowires, the tunable index of the polymer, and the ultra-thin thickness of graphene, respectively. Because of this, light-extraction methods can be much simplified, and more light can be coupled out ultimately. In addition, new TCEs do enable many new attractive features, such as high mechanical flexibility and stretchability. High efficiency or reasonably good efficiency devices are demonstrated for most of these TCEs, indicating that the process integration of new TCEs into high performance OLED devices will not be a fundamental hurdle for adopting these new materials. Low cost, high throughput and reliable process development is the key to commercially viability. Once a successful process is established, the impact of a new TCE is not only limited to the field of OLED devices, but can be readily applied to various other opto electronic devices. A myriad of applications can result from the development of one alternative material. These materials and related techniques are very likely to have a significant impact on optoelectronic research and the industry in the next few decades.

Table 1. Comparison of the TCE materials.

TCE materials for OLEDs	ITO	CNTs	Metal Nanowires	Conductive Polymers	Graphene
Conductivity/ Transparency (at 550 nm)	10 Ω/sq at 90%	180 Ω/sq at 85%	9.7 Ω/sq at 89% 30 Ω/sq at 93%	42 Ω/sq at 82% 240 Ω/sq at 97%	125 Ω/sq at 97% 30 Ω/sq at 90%
Light Reflection	High	Low	Low	Low	No reflection
Light Trapping	High	Low	Low	Low	No Trapping
Material Cost	High	Low	Medium	Low	Low
Process Cost	High	Potentially Low	Potentially Low	Low	Potentially Low
Stability	Good	Excellent	Medium	Medium	Excellent
Flexibility	Poor	Flexible	Flexible	Highly flexible	Highly flexible
Key OLED Performance Demonstrated	>100 lm/W (white)	10 cd/A at 1000 cd/m (green)	54 lm/W similar to ITO control device (white)	12% EQE 900 hours (green)	103 lm/W (green) 80 lm/W at 3000 cd/m^2 (white)
Advantages over ITO		Solution process Flexible Stretchable	Solution process Flexible Stretchable Angle uniformity	Solution process Flexible Light extraction	Highly flexible Ultra-thin Light extraction
Challenges to Replace ITO		Conductivity Roughness Cost	Stability Cost	Conductivity Stability	Conductivity Cost
Reference	[56]	[23,24,57,58]	[29,30,59,60]	[45–48]	[12,49,61]

ACKNOWLEDGMENTS

The authors would like to thank Xiaodan Zhu for proof reading.

CONFLICTS OF INTEREST

The authors declare no conflict of interest.

REFERENCES

1. Tsutsui, T.; Takada, N. Progress in emission efficiency of organic light-emitting diodes: Basic understanding and its technical application. Jap. J. Appl. Phys. 2013, 52, 110001.

2. Grimsdale, A.C.; Leok Chan, K.; Martin, R.E.; Jokisz, P.G.; Holmes, A.B. Synthesis of light-emitting conjugated polymers for applications in electroluminescent devices. Chem. Rev. 2009, 109, 897–1091.

3. Li, J.; Liu, D. Dendrimers for organic light-emitting diodes. J. Mater. Chem. 2009, 19, 7584–7591.

4. Rehmann, N.; Hertel, D.; Meerholz, K.; Becker, H.; Heun, S. Highly efficient solution-processed phosphorescent multilayer organic light-emitting diodes based on small-molecule hosts. Appl. Phys. Lett. 2007, 91, 103507.

5. Gustafsson, G.; Cao, Y.; Treacy, G.M.; Klavetter, F.; Colaneri, N.; Heeger, A.J. Flexible light-emitting diodes made from soluble conducting polymers. Nature 1992, 357, 477–479.

6. Islam, A.; Rabbani, M.; Bappy, M.H.; Miah, M.A.R.; Sakib, N. A Review on Fabrication Process of Organic Light Emitting Diodes. ICIEV 2013.

7. Tang, C.W.; VanSlyke, S.A. Organic electroluminescent diodes. Appl. Phys. Lett. 1987, 51, 913–915.

8. Gordon, R.G. Criteria for choosing transparent conductors. MRS Bull. 2000, 25, 52–57.

9. Hellstrom, S.L.; Ugolotti, J.; Britovsek, G.J.P.; Jones, T.S.; White, A.J.P. The effect of fluorination on the luminescent behaviour of 8-hydroxyquinoline boron compounds. New J. Chem. 2008, 32, 1379–1387.

10. Peng, C.; Jia, Z.; Neilson, H.; Li, T.; Lou, J. In situ electro-mechanical experiments and mechanics modeling of fracture in indium tin oxide-based multilayer electrodes. Adv. Eng. Mater. 2012, 15, 250–256.

11. Kim, N.-R.; Lee, J.-H.; Lee, Y.-Y.; Nam, D.-H.; Yeon, H.-W.; Lee, S.-Y.; Yang, T.-Y.; Lee, Y.-J.; Chu, A.; Nam, K.T.; et al. Enhanced conductivity of solution-processed indium tin oxide nanoparticle films by oxygen partial pressure controlled annealing. J. Mater. Chem. C 2013, 1, 5953–5959.

12. Wu, J.; Agrawal, M.; Becerril, H.C.A.; Bao, Z.; Liu, Z.; Chen, Y.; Peumans, P. Organic light-emitting diodes on solution-processed graphene transparent electrodes. ACS Nano 2009, 4, 43–48.

13. Zhang, M.; Fang, S.; Zakhidov, A.A.; Lee, S.B.; Aliev, A.E.; Williams, C.D.; Atkinson, K.R.; Baughman, R.H. Strong, transparent, multifunctional, carbon nanotube sheets. Science 2005, 309, 1215–1219.

14. Hu, L.; Hecht, D.S.; Gruner, G. Carbon nanotube thin films: Fabrication, properties, and applications. Chem. Rev. 2010, 110, 5790–5844.

15. De Volder, M.F.L.; Tawfick, S.H.; Baughman, R.H.; Hart, A.J. Carbon nanotubes: Present and future commercial applications. Science 2013, 339, 535–539.

16. Dresselhaus, M.; Dresselhaus, G.; Avouris, P.; Ajayan, P.; Zhou, O. Applications of Carbon Nanotubes. In Carbon Nanotubes; Springer: Berlin/Heidelberg, Germany, 2001; Volume 80, pp. 391–425.

17. Martel, R.; Schmidt, T.; Shea, H.R.; Hertel, T.; Avouris, P. Single- and multi-wall carbon nanotube field-effect transistors. Appl. Phys. Lett. 1998, 73, 2447–2449.

18. Zhao, J.; Han, J.; Lu, J.P. Work functions of pristine and alkali-metal intercalated carbon nanotubes and bundles. Phys. Rev. B 2002, 65, 193401.

19. Suzuki, S.; Bower, C.; Watanabe, Y.; Zhou, O. Work functions and valence band states of pristine and cs-intercalated single-walled carbon nanotube bundles. Appl. Phys. Lett. 2000, 76, 4007–4009.

20. Zhang, D.; Ryu, K.; Liu, X.; Polikarpov, E.; Ly, J.; Tompson, M.E.; Zhou, C. Transparent, conductive, and flexible carbon nanotube films and their application in organic light-emitting diodes. Nano Lett. 2006, 6, 1880–1886.

21. Li, J.; Hu, L.; Wang, L.; Zhou, Y.; Gruner, G.; Marks, T.J. Organic light-emitting diodes having carbon nanotube anodes. Nano Lett. 2006, 6, 2472–2477.

22. Ou, E.C.W.; Hu, L.; Raymond, G.C.R.; Soo, O.K.; Pan, J.; Zheng, Z.; Park, Y.; Hecht, D.; Irvin, G.; Drzaic, P.; et al. Surface-modified nanotube anodes for high performance organic light-emitting diode. ACS Nano 2009, 3, 2258–2264.

23. Yu, Z.; Hu, L.; Liu, Z.; Sun, M.; Wang, M.; Gruner, G.; Pei, Q. Fully bendable polymer light emitting devices with carbon nanotubes as cathode and anode. Appl. Phys. Lett. 2009, 95, 203304.

24. Yu, Z.; Niu, X.; Liu, Z.; Pei, Q. Intrinsically stretchable polymer light-emitting devices using carbon nanotube-polymer composite electrodes. Adv. Mater. 2011, 23, 3989–3994.

25. Hu, L.; Wu, H.; Cui, Y. Metal nanogrids, nanowires, and nanofibers for transparent electrodes. MRS Bull. 2011, 36, 760–765.

26. Lee, J.-Y.; Connor, S.T.; Cui, Y.; Peumans, P. Solution-processed metal nanowire mesh transparent electrodes. Nano Lett. 2008, 8, 689–692.

27. Yu, Z.; Zhang, Q.; Li, L.; Chen, Q.; Niu, X.; Liu, J.; Pei, Q. Highly flexible silver nanowire electrodes for shape-memory polymer light-emitting diodes. Adv. Mater. 2011, 23, 664–668.

28. Gaynor, W.; Hofmann, S.; Christoforo, M.G.; Sachse, C.; Mehra, S.; Salleo, A.; McGehee, M.D.; Gather, M.C.; Lüssem, B.; Müller-Meskamp, L.; et al. Color in the corners: ITO-free white oleds with angular color stability. Adv. Mater. 2013, 25, 4006–4013.

29. Liang, J.; Li, L.; Niu, X.; Yu, Z.; Pei, Q. Elastomeric polymer light-emitting devices and displays. Nat. Photon. 2013, 7, 817–824.

30. Song, T.-B.; Chen, Y.; Chung, C.-H.; Yang, Y.; Bob, B.; Duan, H.-S.; Li, G.; Tu, K.-N.; Huang, Y. Nanoscale joule heating and electromigration enhanced ripening of silver nanowire contacts. ACS Nano ASAP 2014.

31. Karim, S.; Toimil-Molares, M.E.; Balogh, A.G.; Ensinger, W.; Cornelius, T.W.; Khan, E.U.; Neumann, R. Morphological evolution of au nanowires controlled by rayleigh instability. Nanotechnology 2006, 17, 5954–5959.

32. Hsu, S.-Y.; Lee, M.-C.; Lee, K.-L.; Wei, P.-K. Extraction enhancement in organic light emitting devices by using metallic nanowire arrays. Appl. Phys. Lett. 2008, 92, 013303.

33. Cao, Y.; Treacy, G.M.; Smith, P.; Heeger, A.J. Solutioncast films of polyaniline: Optical quality transparent electrodes. Appl. Phys. Lett. 1992, 60, 2711–2713.

34. Yang, Y.; Heeger, A.J. Polyaniline as a transparent electrode for polymer lightemitting diodes: Lower operating voltage and higher efficiency. Appl. Phys. Lett. 1994, 64, 1245–1247.

35. Fehse, K.; Schwartz, G.; Walzer, K.; Leo, K. Combination of a polyaniline anode and doped charge transport layers for high-efficiency organic light emitting diodes. J. Appl. Phys. 2007, 101, 124509.

36. Bayer, A.G. Eur. Patent 440957, 1991.

37. Cao, Y.; Yu, G.; Zhang, C.; Menon, R.; Heeger, A.J. Polymer light-emitting diodes with polyethylene dioxythiophene-polystyrene sulfonate as the transparent anode. Synth. Met. 1997, 87, 171–174.

38. Kim, J.Y.; Jung, J.H.; Lee, D.E.; Joo, J. Enhancement of electrical conductivity of poly(3,4-ethylenedioxythiophene)/poly(4-styrenesulfonate) by a change of solvents. Synth. Met. 2002, 126, 311–316.

39. Ouyang, J.; Chu, C.W.; Chen, F.C.; Xu, Q.; Yang, Y. High-conductivity poly(3,4-ethylenedioxythiophene):Poly(styrene sulfonate) film and its application in polymer optoelectronic devices. Adv. Funct. Mater. 2005, 15, 203–208.

40. Fehse, K.; Walzer, K.; Leo, K.; Lövenich, W.; Elschner, A. Highly conductive polymer anodes as replacements for inorganic

materials in high-efficiency organic light-emitting diodes. Adv. Mater. 2007, 19, 441–444.

41. Xia, Y.; Zhang, H.; Ouyang, J. Highly conductive PEDOT:PSS films prepared through a treatment with zwitterions and their application in polymer photovoltaic cells. J. Mater. Chem. 2010, 20, 9740–9747.

42. Xia, Y.; Ouyang, J. PEDOT:PSS films with significantly enhanced conductivities induced by preferential solvation with cosolvents and their application in polymer photovoltaic cells. J. Mater. Chem. 2011, 21, 4927–4936.

43. Yoo, J.E.; Lee, K.S.; Garcia, A.; Tarver, J.; Gomez, E.D.; Baldwin, K.; Sun, Y.; Meng, H.; Nguyen, T.-Q.; Loo, Y.-L. Directly patternable, highly conducting polymers for broad applications in organic electronics. Proc. Natl. Acad. Sci. USA 2010, 107, 5712–5717.

44. Kim, Y.H.; Sachse, C.; Machala, M.L.; May, C.; Müller-Meskamp, L.; Leo, K. Highly conductive PEDOT:PSS electrode with optimized solvent and thermal post-treatment for ito-free organic solar cells. Adv. Funct. Mater. 2011, 21, 1076–1081.

45. Vosgueritchian, M.; Lipomi, D.J.; Bao, Z. Highly conductive and transparent PEDOT:PSS films with a fluorosurfactant for stretchable and flexible transparent electrodes. Adv. Funct. Mater. 2012, 22, 421–428.

46. Xia, Y.; Sun, K.; Ouyang, J. Solution-processed metallic conducting polymer films as transparent electrode of optoelectronic devices. Adv. Mater. 2012, 24, 2436–2440.

47. Kim, Y.H.; Lee, J.; Hofmann, S.; Gather, M.C.; Müller-Meskamp, L.; Leo, K. Achieving high efficiency and improved stability in ito-free transparent organic light-emitting diodes with conductive polymer electrodes. Adv. Funct. Mater. 2013, 23, 3763–3769.

48. Cai, M.; Ye, Z.; Xiao, T.; Liu, R.; Chen, Y.; Mayer, R.W.; Biswas, R.; Ho, K.-M.; Shinar, R.; Shinar, J. Extremely efficient indium-tin-oxide-free

green phosphorescent organic light-emitting diodes. Adv. Mater. 2012, 24, 4337–4342.

49. Bae, S.; Kim, H.; Lee, Y.; Xu, X.; Park, J.-S.; Zheng, Y.; Balakrishnan, J.; Lei, T.; Kim, H.R.; Song, Y.I.; et al. Roll-to-roll production of 30-inch graphene films for transparent electrodes. Nat. Nanotechnol. 2010, 5, 574–578.

50. Hummers, W.S.; Offeman, R.E. Preparation of graphitic oxide. J. Am. Chem. Soc. 1958, 80, 1339–1339.

51. Han, T.-H.; Lee, Y.; Choi, M.-R.; Woo, S.-H.; Bae, S.-H.; Hong, B.H.; Ahn, J.-H.; Lee, T.-W. Extremely efficient flexible organic light-emitting diodes with modified graphene anode. Nat. Photon. 2012, 6, 105–110.

52. Li, N.; Oida, S.; Tulevski, G.S.; Han, S.-J.; Hannon, J.B.; Sadana, D.K.; Chen, T.-C. Efficient and bright organic light-emitting diodes on single-layer graphene electrodes. Nat. Commun. 2013, 4, 2294.

53. Sun, T.; Wang, Z.L.; Shi, Z.J.; Ran, G.Z.; Xu, W.J.; Wang, Z.Y.; Li, Y.Z.; Dai, L.; Qin, G.G. Multilayered graphene used as anode of organic light emitting devices. Appl. Phys. Lett. 2010, 96, 133301.

54. Hwang, J.; Choi, H.K.; Moon, J.; Kim, T.Y.; Shin, J.-W.; Joo, C.W.; Han, J.-H.; Cho, D.-H.; Huh, J.W.; Choi, S.-Y.; et al. Multilayered graphene anode for blue phosphorescent organic light emitting diodes. Appl. Phys. Lett. 2012, 100, 133304.

55. Zhu, Y.; Sun, Z.; Yan, Z.; Jin, Z.; Tour, J.M. Rational design of hybrid graphene films for high-performance transparent electrodes. ACS Nano 2011, 5, 6472–6479.

56. Ellmer, K. Past achievements and future challenges in the development of optically transparent electrodes. Nat. Photon. 2012, 6, 809–817.

57. Nirmalraj, P.N.; Lyons, P.E.; De, S.; Coleman, J.N.; Boland, J.J. Electrical connectivity in single-walled carbon nanotube networks. Nano Lett. 2009, 9, 3890–3895.

58. Gruner, G. Carbon nanotube films for transparent and plastic electronics. J. Mater. Chem. 2006, 16, 3533–3539.

59. Chung, C.-H.; Song, T.-B.; Bob, B.; Zhu, R.; Duan, H.-S.; Yang, Y. Silver Nanowire Composite Window Layers for Fully Solution-Deposited Thin-Film Photovoltaic Devices. Adv. Mater. 2012, 24, 5499–5504.

60. Leem, D.-S.; Edwards, A.; Faist, M.; Nelson, J.; Bradley, D.D.C.; De Mello, J.C. Efficient organic solar cells with solution-processed silver nanowire electrodes. Adv. Mater. 2011, 23, 4371–4375.

61. Eda, G.; Fanchini, G.; Chhowalla, M. Large-area ultrathin films of reduced graphene oxide as a transparent and flexible electronic material. Nat. Nanotechnol. 2008, 3, 270–274.

7

CHAPTER

CONDUCTING POLYMER-BASED NANOHYBRID TRANSDUCERS: A POTENTIAL ROUTE TO HIGH SENSITIVITY AND SELECTIVITY SENSORS

Seon Joo Park [1,†], Oh Seok Kwon [1,2,†], Ji Eun Lee [3], Jyongsik Jang [1,*] and Hyeonseok Yoon [3,4,*]

[1] *School of Chemical and Biological Engineering, Seoul National University, Seoul 151-742, Korea*

[2] *Department of Chemical and Environmental Engineering, School of Engineering and Applied Science, Yale University, New Haven, CT 06511, USA*

[3] *Department of Polymer Engineering, Graduate School, Chonnam National University, Gwangju 500-757, Korea*

[4] *Alan G. MacDiarmid Energy Research Institute, School of Polymer Science and Engineering, College of Engineering, Chonnam National University, Gwangju 500-757, Korea*

ABSTRACT

The development of novel sensing materials provides good opportunities to realize previously unachievable sensor performance. In this review, conducting polymer-based nanohybrids are highlighted as innovative transducers for high-performance chemical and biological sensing devices. Synthetic strategies of the nanohybrids are categorized into four groups:

(1) impregnation, followed by reduction; (2) concurrent redox reactions; (3) electrochemical deposition; (4) seeding approach. Nanocale hybridization of conducting polymers with inorganic components can lead to improved sorption, catalytic reaction and/or transport behavior of the material systems. The nanohybrids have thus been used to detect nerve agents, toxic gases, volatile organic compounds, glucose, dopamine, and DNA. Given further advances in nanohybrids synthesis, it is expected that sensor technology will also evolve, especially in terms of sensitivity and selectivity.

Keywords: conducting polymer; nanohybrids; transducers; chemical sensors; biosensors

1. INTRODUCTION

Of organic materials, only conducting polymers have electrical and optical properties which are similar to those of inorganic semiconductors or sometimes metals. Thus, conducting polymers have been widely used to fabricate versatile chemical and biological sensors. Conducting polymers themselves are very sensitive to their surrounding environments, which makes them suitable for various sensor transducers. However, the sensitivity and selectivity of conducting polymer-based sensors still leave room for improvement. More specifically, there are several important factors which need to be improved for further success of conducting polymer-based sensors:

- sensitivity
- selectivity
- surface area
- environmental stability
- surface properties

The concept of using organic/inorganic hybrids as sensor transducers has been introduced to improve the important factors listed above [1–3]. There is a rich variety of possible combinations of organic-inorganic materials. Conducting polymers as the organic component can be also hybridized with one or more of the following materials:

- metal nanoparticles: gold, silver, platinum, silicon, palladium
- inorganic compounds: oxides, halides
- carbon nanomaterials: carbon nanotubes, fullerene, graphene

Polymers have traditionally held advantages over metal or inorganic materials, such as easy synthesis and processing, chemical and structural diversity, low weight, and flexibility [4]. Conducting polymers are a fascinating class of materials that combine the advantages of organic polymers with the electrical and optical properties of metal or inorganic semiconductors [5]. Accordingly, conducting polymers may be the best candidate for fabricating functional organic/inorganic hybrids. Moreover, conducting polymer-based hybrids are expected to have several synergistic properties between the polymer and the inorganic components, making them promising candidates for application in several fields such as catalysis, sensors, electronics, optoelectronics, and medicine [6,7]. On the other hand, nanostructured materials show unique properties depending on the size and shape [8,9]. In terms of sensor applications, moreover, they have beneficial advantages such as high surface area and small dimensions [10–12]. The enlarged surface area enhances the interactions between the materials and analytes, which leads to high sensitivity, and the small dimensions facilitate adsorption/desorption kinetics for analytes in the materials, which allows a rapid response time and high signal reproducibility [13]. In this review article, we focus on addressing synthesis strategies for obtaining conducting polymer-based nanohybrids and notable recent examples of sensor applications. In fact, many kinds of polymers including conducting polymers have been combined with carbon nanomaterials, such as fullerene, carbon nanotubes, graphene, and graphene oxide, to make diverse functional composites, because the synthesis routes and properties of the carbon nanomaterials have been well studied [14]. However, since there are already a couple of comprehensive review articles about carbon nanomaterials-based polymer composites, providing abundant information on the corresponding synthesis methods and applications, from a materials perspective, herein a considerable fraction will therefore be devoted to nanostructured hybrids that contain inorganic compounds such as metals or metal oxides/halides, and not carbon species.

2. SYNTHETIC ROUTES TO CONDUCTING POLYMER-BASED NANOHYBRIDS

A variety of methods have been used to prepare conducting polymer-based nanohybrids. A review article published in 2000 provides an excellent overview of conducting polymer nanocomposites [6]. Here, mainly recent research examples since 2000 will be discussed. The main synthesis strategies can be classified into the following four categories: (i) impregnation of metal precursors, followed by reduction; (ii) concurrent redox reactions; (iii) electrochemical deposition; (iv) seeding approach.

(i) Impregnation, followed by reduction

Various functional groups such as –COOH, –CN, $-NH_2$, –SH can interact with metal cations through ion-ion or ion-dipole interactions. Conducting polymers containing either nitrogen or sulfur atoms in their repeating unit are attractive candidates for the preparation of metal/polymer hybrids. For example, polypyrrole (PPy) and polyaniline (PANI) have a nitrogen atom in their heterocyclic ring, and polythiophene (PTh) and poly(3,4-ethylenedioxythiophene) (PEDOT) have a sulfur atom in their heterocyclic ring. Thus, Ag, Au, and Pd nanoparticles could be readily deposited on the surface of PANI, PPy and PEDOT nanostructures without any dispersing or reducing agents [15–18]. On the other hand, monomers with a functional side group can be used to introduce desired functionality to the resulting polymers. Of course, it should be noted that the chemical/physical properties of monomers with a functional side group are different those of unfunctionalized, pristine monomers. Thus, it is important to judiciously choose detailed synthesis conditions for the successful polymerization. There are several good examples of using pyrrole derivatives to fabricate conducting polymer nanostructures with surface functional groups. First, pyrrole has been copolymerized with pyrrole-3-carboxylic acid (P3CA) in emulsion systems to fabricate tubular nanostructures [19–21]. The introduction of carboxylic groups was confirmed by immobilizing a fluorescent dye onto the nanotubes. As shown in Figure 1, the nanotubes exhibited cyan emission with uniform distribution over their surfaces, indicating that the carboxylic groups were thoroughly introduced into the polymer nanotubes.

There is a notable example of fabricating metal nanoparticles-decorated PPy nanotubes using the above strategy [22]. Carboxylated pyrrole, P3CA monomers were chemically polymerized by ferric chloride within the cylindrical pores of a membrane template and magnetic phases were readily introduced into the nanotubes by precipitation of residual iron salt complexes during the template removal process. The main synthesis steps are represented in Figure 2. Magnetic carboxylated PPy (MCPPy) nanotubes were prepared by vapor-deposition polymerization (VDP) of P3CA monomers in porous aluminum template and subsequent chemical reduction of Fe(III) during the template etching process. The magnetic CPPy (MCPPy) nanotubes provide the anchoring sites for metal cations, thus making it possible to generate Pd nanoparticles on the nanotube surface at room temperature. Figure 3a shows transmission electron microscopy (TEM) image of MCPPy nanotubes with a diameter of ca. 100 nm and a wall thickness of 10–20 nm. The electron diffraction pattern indicated the presence of magnetites because only PPy is amorphous. Figure 3b–d exhibit TEM images of Pd nanoparticles-decorated MCPPy NTs prepared with different palladium salt ($PdCl_2$) concentrations. As the palladium salt concentration increases, the diameter of the formed Pd nanoparticles turned out to increase from 3.6 up to 9.8 nm. It is considered that rapid nucleation relative to growth lead to a small particle size at low concentrations of Pd salt. Figure 3b–d demonstrate that relatively uniform Pd nanoparticles are formed on the surface of the MCPPy nanotubes. The Pd nanoparticles have a narrow size distribution (the inset histograms).

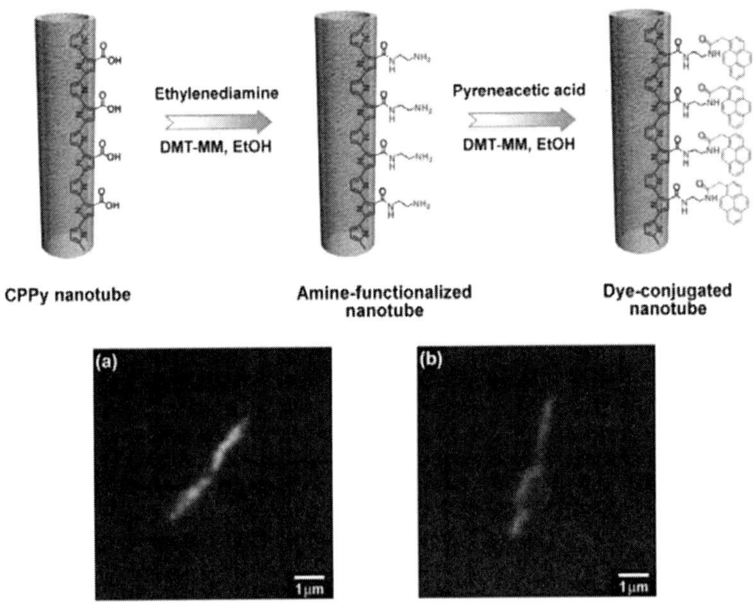

Figure 1. Schematic illustration of reaction steps for the covalent immobilization of fluorescent dye molecules on the surface of carboxylated PPy (CPPy) nanotubes: the first step involves the coupling reaction of CPPy nanotubes with ethylenediamine as a diamino linker, and then pyreneacetic acid molecules are covalently bonded to the nanotube surface by a condensation reaction between the carboxylic groups and the surface amino groups. The fluorescent dye molecules were tracked in the samples using confocal laser scanning electron microscopy (CLSM) (λ_{exc} = 458 nm): CSLM images of the nanotubes synthesized with the P3CA-to-pyrrole molar ratios of (**a**) 1:15 and (**b**) 1:30. With permission from [19]; Copyright 2008, Wiley-VCH Verlag GmbH & Co. KGaA.

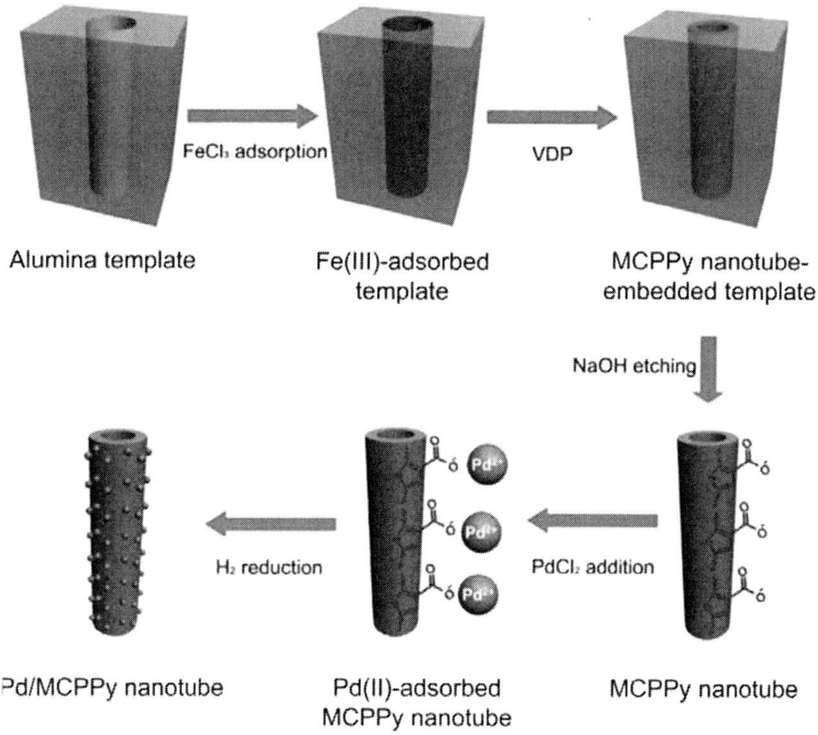

Figure 2. Schematic illustration of the fabrication of Pd nanoparticles-decorated MCPPy nanotubes. With permission from [22]; Copyright 2006, Wiley-VCH Verlag GmbH & Co. KGaA.

Silica nanoparticles were also attached to PPy nanotubes through a chemical binding between the amino group of silica nanoparticles and the carboxylic group of the nanotubes [23]. Furthermore, the carboxylic group can be modified to thiol group, which can be used as nucleation sites for the formation of metal or inorganic semiconductor nanoparticles. Typically, PbSe quantum dots could be decorated on the thiolated PPy nanotubes for photovoltaic application [24].

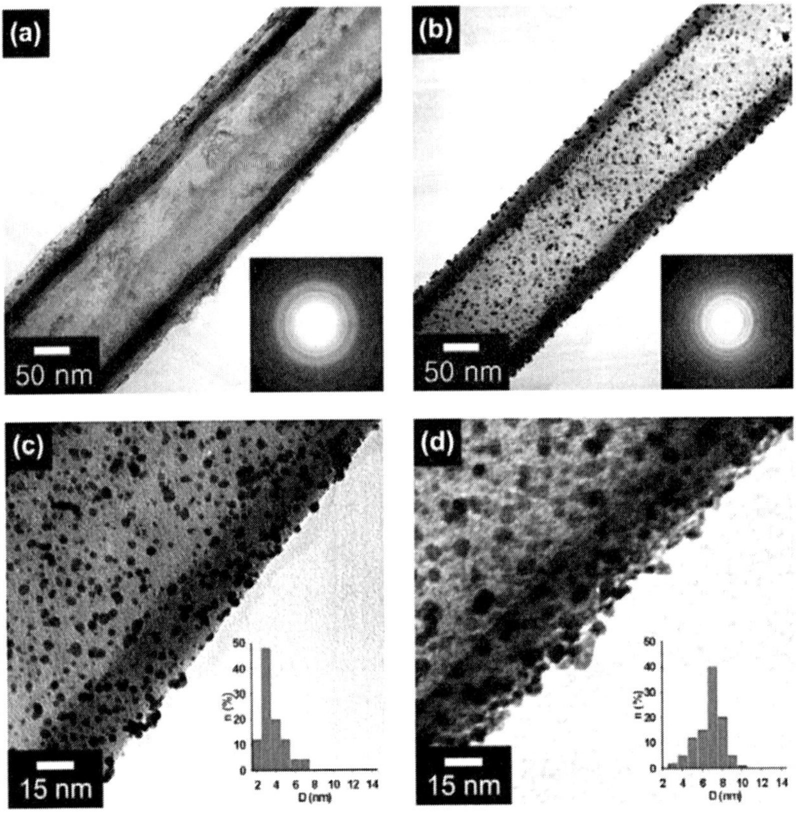

Figure 3. TEM images of (**a**) a MCPPy nanotube and (**b**) a Pd nanoparticles-decorated MCPPy nanotube prepared with a dilute $PdCl_2$ solution. The inset images are the electron diffraction patterns; (**c**) and (**d**) magnified TEM images of Pd nanoparticles-decorated MCPPy nanotubes obtained with the palladium salt concentrations of 10^{-4} and 10^{-3} M, respectively. The Pd nanoparticle size is illustrated in the inset histogram. With permission from [22]; Copyright 2006, Wiley-VCH Verlag GmbH & Co. KGaA.

(ii) Concurrent redox reactions

Most conducting polymers such as PPy, PANI, and PTh can be synthesized by oxidative polymerization using appropriate oxidizing agents. Normally, metal salts such as copper chloride, ferric chloride, ferric perchlorate, and ferric nitrate are widely using to obtain conducting polymers via chemical oxidation polymerization. Similarly, metal precursors containing iron, gold, silver, and platinum can be also used as oxidizing agents to achieve the oxidation polymerization of conducting polymer monomers, which allows the simultaneous formation of the metal and conducting polymer. Various nanoparticles have been synthesized using this approach, such as AgCl/PPy/chitosan nanospheres [25], gold/PANI hollow nanoparticles [26], gold/PEDOT nanoparticles [27], platinum/PEDOT nanoparticles [28], and silver/PPy nanostructures [29].

As a recent example, silver-PPy core-shell nanoparticles were readily obtained through the concurrent reduction/oxidation of silver nitrate/pyrrole in the presence of starch stabilizer [30]. The overall formation mechanism of the core-shell nanoparticles is described in Figure 4. First, pyrrole monomers are oxidized by silver cations, yielding PPy and silver atoms simultaneously (see Figure 4a). The polar groups of starch used as a steric stabilizer can act as nucleation sites to generate the core-shell nanoparticles at the initial stage (stage I). At the nucleation sites, silver cations are reduced to atomic silver by pyrrole monomers, whereas the pyrrole monomers are oxidized to pyrrole radical cations (stage II). Thus, the silver nanoseeds are being formed at the nucleation sites and coupling of pyrrole radical cations results in PPy short chains. Owing to the so-called common ion adsorption effect, additional silver cations are adsorbed onto the nanoseed surface that lead to the formation of a PPy shell through subsequent growth of surrounding PPy short chains (stage III). The resulting core-shell nanoparticles were ca. 36 nm in core diameter and ca. 13 nm in shell thickness, respectively. Importantly, those sequential reaction steps for yielding the core-shell nanoparticles are not complicated at all. The two reactants, pyrrole and silver nitrate are subjected to the successive reactions in a reactor to yield the final product, often referred to as one-pot synthesis.

The availability of this concurrent oxidation/reduction approach was also demonstrated in an interfacial polymerization system. Gold/PEDOT co-axial nanocables were fabricated without any special template by interfacial oxidative polymerization of 3,4-ethylenedioxythiophene (EDOT) in organic solution with a gold precursor, $HAuCl_4$ in aqueous solution [31]. The nanocables had lengths of several micrometers, outer diameters ca. 50 nm, and central core ca. 30 nm. It is known that PEDOT has low band gap, high environmental stability, and high optical transparency in the oxidized, conductive state.

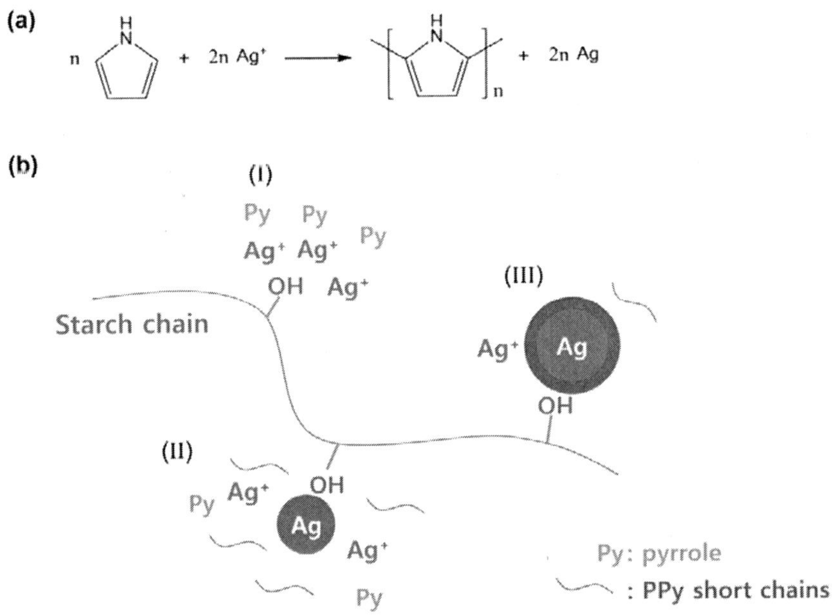

Figure 4. Schematic illustration of the formation mechanism of silver-PPy nanoparticles: (**a**) the scheme describing the chemical reaction between pyrrole and silver cation; (**b**) the reaction process could be divided into three stages (I, II, and III). Adapted with permission from [30]; Copyright 2012, American Chemical Society.

In the case of PANI, hydrogen peroxide was used as an oxidizing agent both to reduce a gold precursor ($HAuCl_4$) to gold atoms and to polymerize aniline monomer, which is probably due to the relatively high oxidation potential of aniline monomer [32]. A new synthetic approach was used: namely, aniline is introduced in the vapor phase to an aqueous micelle solution where the reduction of the gold precursor is being performing by hydrogen peroxide. As a result, gold/PANI composite nanoparticles could be prepared, with the diameter of greater than 100 nm and the electrical conductivity of ca. 1.1 $S \cdot cm^{-1}$.

(iii) Electrochemical deposition

In contrast with most polymers, the conductivity of conducting polymers makes them eligible for various electrochemical tool applications. As a typical example, conducting polymers can be readily obtained through electrochemical oxidation polymerization on electrodes. Electrochemical deposition involves a subsequent reduction of metal ions deposited from a

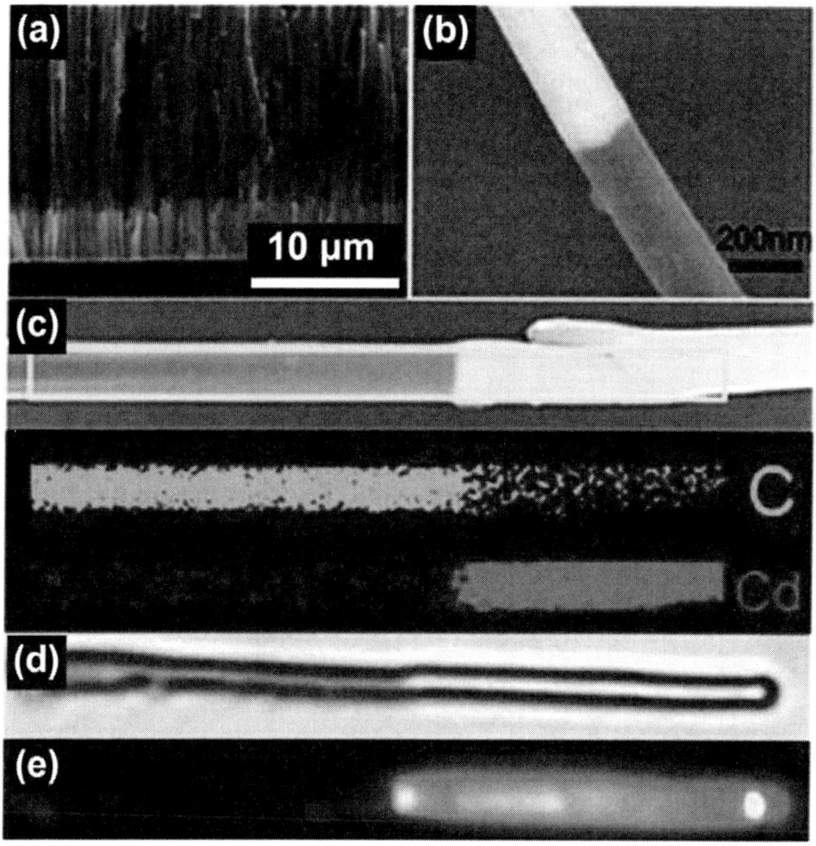

Figure 5. SEM images of CdS-PPy heterojunction nanowires: (**a**) side view image of CdS-PPy nanowires inside the alumina membrane template; (**b**) typical image of a single CdS-PPy nanowire; (**c**) element mapping of a single CdS-PPy nanowire; (**d**) optical image of a single CdS-PPy nanowire; (**e**) fluorescent image of a single CdS-PPy nanowire excited by a 405 nm laser. Adapted with permission from [33]; Copyright 2008, American Chemical Society.

solution onto an electrically conductive substrate. When conducting polymers are used as the conductive substrate, therefore, a variety of inorganic compounds can be introduced into the polymers using electrochemical deposition. Most related studies have used electrochemical co-deposition, for example, such as the electrochemical polymerization of a conducting polymer on an electrode, followed by the electrochemical deposition of an inorganic component. Li and co-workers reported the fabrication of CdS-PPy heterojunction nanowires by electrochemical co-deposition in porous alumina template (Figure 5) [33]. First, CdS was deposited into the cylindrical nanopores of the alumina membrane using $CdCl_2$ and element sulfur as precursors in a dimethyl sulfoxide solution at a current density of 2.5 mA·cm^{-1}. Subsequently, PPy was deposited in 0.1 M $LiClO_4$ acetonitrile solution by applying a voltage of 0.85 V (vs. SCE). Scanning electron microscopy (SEM) images and element mapping demonstrated the formation of the heterojunction nanowires with diameters of 200–400 nm. Similarly, CdS-PANI nanowires and gold-PPy heterojunction nanorods have been also synthesized [34,35].

PPy/titania nanotubes were also obtained through sequential electrochemical processes [36]. Highly ordered titania nanotube array was first fabricated by an electro-oxidation of titanium sheet through an electrochemical anodization process and PPy was then deposited into the well-aligned titania nanotubes through a voltammetry deposition process, in which TiO_2/Ti acted as a working electrode. The inner diameter, wall-thickness, and length of the titania nanotubes were 120–150 nm, 10–20 nm, and ca. 1 μm, respectively. The wall-thickness of the titania nanotubes was found to increase up to 50–80 nm after PPy deposition.

(iv) Seeding approach

The so-called seeding approach has been widely used to produce micro- and nano-particles [37]. For example, magnetite (Fe_3O_4)/PEDOT core-shell nanoparticles could be synthesized through a seeding approach [38]. Magnetite nanoparticles were first prepared by a chemical precipitation method and were then used as nanoseeds for the polymerization of EDOT monomers. The magnetite nanoseeds were acid-etched, resulting in iron cations that can initiate the surface-confined oxidation polymerization. During this process, the polymerization kinetics affects highly the morphology of the final product at the nanometer regime. The polymerization rate of EDOT is relatively slow, which makes it possible to form ultrafine core-shell nanoparticles (ca. 13 nm in diameter) without interparticle aggregation.

One-dimensional nanostructures could be also obtained through seeding approaches. The Manohar and Zhang groups have also reported interesting results of fabricating PPy and PANI nanofibers using several different kinds of nanofiber seeds [39–41]. Nanofiber seeds such as V_2O_5 nanofibers and single-walled carbon nanotubes (SWNTs) acted as a reactive template that chemically interact with the monomer prior to the addition of initiator (Figure 6). Employing such nanofibers seeds has resulted in the formation of PPy or PANI composite nanofibers with ca. 20–50 nm, in which the morphology of the final product was dependent on the species, dimensions, and surface properties of seeds.

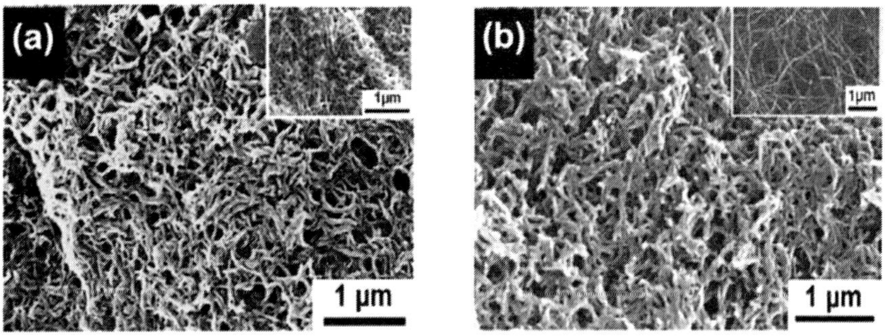

Figure 6. SEM images of PANI nanofibers synthesized with nanofiber seeds (insets): (**a**) single-walled carbon nanotubes; (**b**) V_2O_5 nanofibers. Adapted with permission from [39]; Copyright 2012, American Chemical Society.

3. CONDUCTING POLYMER NANOHYBRID-BASED SENSORS

3.1. Chemiresistive Gas Sensors

Various chemical species that can be harmful to human society or be used as diagnostic markers for disease, such as chemical weapons, toxic gases, and metabolites/volatile organic compounds (VOCs), have been found from the nature. For examples, nerve gas agents with odorless and colorless characteristics such as sarin, VX, soman, and tabun have are used as chemical weapons in the past wars. Organic compounds including ammonia, hydrogen chloride, and various hydroxyl acids are also characterized with colorless gas with a pungent and suffocating odor that is recognized as one of the primary irritants to human body. Moreover, metabolites compounds with extremely low concentrations (part-per-billions or part-per-trillions) in exhaled breath provide various disease informations or utilized for diagnostic markers. The level of ethane (ca. 800 pmol·L^{-1}) and pentane in breath indicate high possibility in liver diseases and hepatic disease is marked by the presence of ammonia (ca. 23 μg %).

To detect or discriminate these chemical species, excellent technologies have been developed using various conducting polymer nanomaterials [10,11]. Multidimensional conducting polymer nanotubes were fabricated by electrospinning and vapor deposition polymerization processes, leading to the successful high-performance chemical nerve agent sensors [42]. Also, one-dimensional conducting polymer-based chemiresistive gas sensors were designed using microemulsion process that is one of the template-free methods, leading to the excellent chemiresistive transistor in sensor applications [43]. Although they displayed significant results in detectable sensing limitation, the advanced nanomaterials based on conducting polymers should be continuously developed for the current limitations such as low sensitivity and selectivity, response/recovery time, and complex pre-/post-process.

For the high-performance sensing devices, several important factors in sensing geometry design should be considered with as follows: (i) enlarged-surface areas, (ii) specific selectivity, and (iii) facile and portable geometry. From these fundamental conditions, inorganic/organic hybrid nanomaterials based on conducting polymers are one of the suitable transistors in chemiresistive gas sensor applications. It is caused that the conducting polymer nanohybrids can provide enhanced interaction and target selectivity by attaching or adsorbing pristine metal or metal-oxide nanomaterials.

3.1.1. Nerve Agent Gas Detection

Nerve agents as organophosphate compounds induce the inhibition of serine protease in the nervous system and serious effects in human by reducing the breakdown of acetylcholine. There are various nerve agents. Especially, dimethyl methylphosphonate (DMMP), which is one of the diverse nerve gas simulants, has similar chemical structures with sarin gas that is used as a chemical weapon owing to its extreme potency as a nerve agent. Therefore, diverse technologies, including colorimetric assay, surface acoustic wave, and enzymatic assay, have been developed to detect DMMP gas. Representative researches toward nerve agent gas detection have been made with conducting polymers or carbon nanomaterials because of their excellent conductivity and facile functionalization, showing the high-performance sensing properties [42,44]. Especially, DMMP, which contains phosphoryl groups, can act as hydrogen-bond basicity, so that it easily can accept a proton from functionalized conducting polymer nanomaterials or nanohybirds. Recently, there was an interesting research using conducting polymer/metal hybrids without functional groups. Tiwari et al. fabricated PPy nanohydrids using electrochemical incorporation in presence of a cationic surfactant, cetyltrimethylammonium bromide (CTAB) [45]. Multidimensional PPy/copper phthalocyanine (CuPc)/CTAB/$NaClO_4$ with unique flake-like CuPc substructure on the indium tin oxide substrate was demonstrated with infrared (IR) spectroscopy, cyclic voltammetry, and energy dispersive X-ray (EDX) analysis. Moreover, the flake-like CuPc can provide the enlarged surface area, leading to the higher sensitivity than that of flat structures. The sensing geometry using PPy nanohybrids with flake-like CuPc was developed in order to detect DMMP and toxic gases, which was connected to an acquisition system. This sensor showed significant frequency shift under various DMMP concentrations (Figure 7). Although the sensitivity from the PPy/copper nanohybrids was low compared to multidimensional conducting polymer nanostructures, they showed a possibility of detecting DMMP gases using metal nanohybrids.

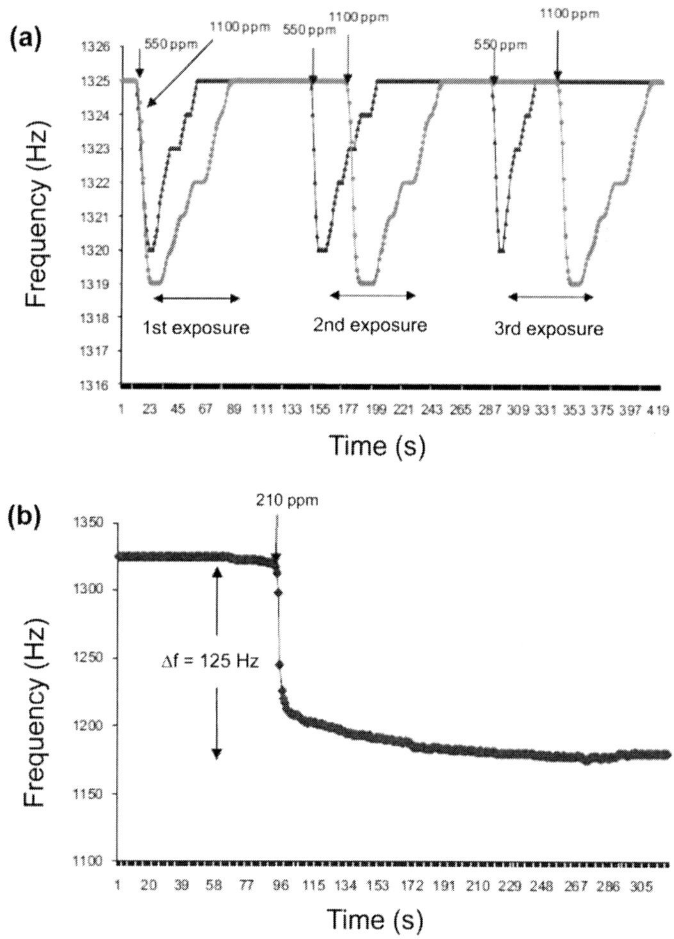

Figure 7. Responses of DMMP gas sensor toward (**a**) methanol and (**b**) DMMP. Adapted with permission from [45]; Copyright 2010, Elsevier B.V.

3.1.2. Toxic Gas Detection

To protect environment and human health from toxic gases, the sensing performance of toxic gas detectors has become increasingly important because low concentration of toxic gas can be even fatal. For example, hydrogen sulfide (H_2S) from industrial processes is one of the highly toxic and corrosive gases and low concentrations (above 250 ppm) can lead to death [46]. Also, it can have potential use in terrorist attacks since it is colorless, flammable, and heavier than air [47]. Nitrogen oxides (NO, NO_2, and N_2O_4) that produce the combustion of fuels can induce toxic acid rain, photochemical smog, and depletion of the ozone layer [48].

In the ammonia gas sensing, the most important issue is its fatal dose defined from Occupational Safety and Health Administration (U.S. Department of Labor, Washington, DC, USA) (15 min exposure limit for NH_3 gas of 35 ppm by volume in the ambient air), leading to promoting the fabrication of advanced sensing geometries. From these concerns, various sensing materials which are inorganic or organic materials have been developed. Although they have attractive advantages including facile fabrication process, easy-controllable morphology, and low-cost, the limitations including high-temperature operation, slow response/recovery time, and low sensitivity remain as challenges.

Hydrogen sulfide can interact with PANI via a doping mechanism, which may result in a measurable change in PANI conductivity. However, H_2S as a weak acid is partially doped into the polymer backbone, allowing only low sensitivity. To overcome this drawback, Virji et al. demonstrated the concept by incorporating transition-metal chlorides into PANI nanofibers and utilized as transistor for H_2S detection (10 ppm) [47]. This is based on the formation of metal sulfide precipitates in metal salt solutions, giving rise to strong acids as the by-product. They provided experimental data to compare the sensing properties between pristine PANI and PANI/metal salts. Moreover, they prepared PANI nanofibers with various metal salts and characterized doping process using coordination of metal cations to the nitrogen atoms. The chemical sensors based on the PANI hybrids were measured the resistance changes upon the exposure to H_2S (10 ppm). Also, they characterized the formation of sulfur in the films with EDX analysis. From the experimental results, the $CuCl_2$ as metal source showed the strongest sensitivity compared to Zn and Cd metal salts. Compared to the effect of incorporated transition-metals, in sensing H_2S using metal nanohybrids, another one of the important factors is a temperature because the activity of the metal substrates can be sensitive on temperature changes. Geng et al. also prepared PPy nanohybrids mixing mechanically with WO_3 and showed excellent H_2S detection at low temperature (90 °C) compared with conventional H_2S sensors based on hybrids [49]. The PPy

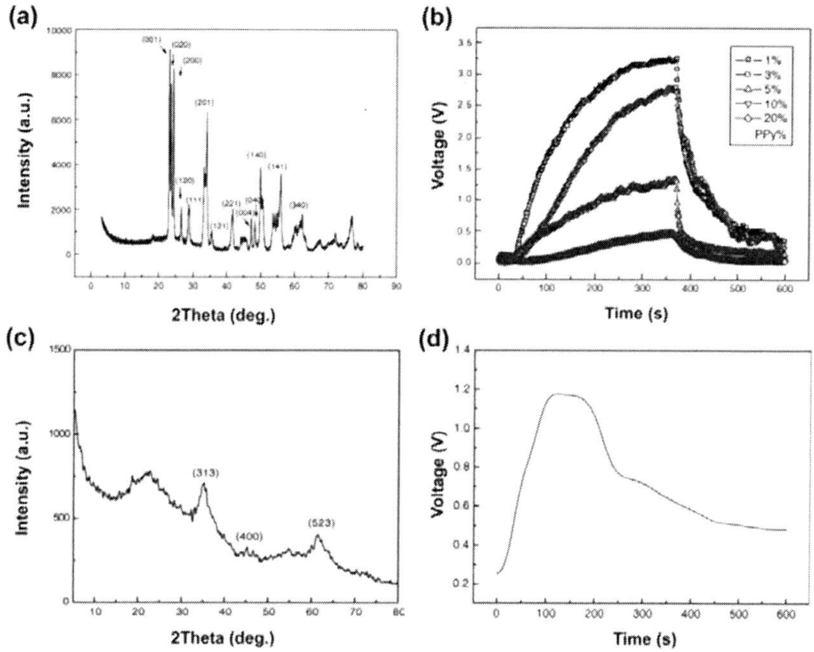

Figure 8. (**a**) XRD pattern of PPy/WO_3 nanohybrids and (**b**) their response-recovery curves to H_2S at 90 °C. Adapted with permission from Ref. [49]; Copyright 2006, Elsevier B.V. (**c**) XRD pattern (the three peaks correspond to γ-Fe_2O_3 phase) of PPy/γ-Fe_2O_3 nanohybrids annealed at 150 °C and (**d**) their response curve (voltage vs. time) to 0.1% H_2S at 90 °C. Adapted with permission from [50]; Copyright 2006, Elsevier B.V.

nanohybrids were characterized using HRTEM and X-ray diffraction (XRD) pattern, resulting that the average size of WO_3 particles was ca. 29 nm calculated with Scherrer equation (Figure 8a). The voltage of the nanohybrid-based sensors was monitored by using a conventional circuit in which the sensor was connected with an external resistor in the testing circuit. In the sensing properties, the PPy nanohybrids showed stable reversibility and high sensitivity depending on the amount of PPy mass percent when operated at 90 °C. Moreover, they demonstrated that the mixing ratios of organic-inorganic hybrids can affect the sensing properties (Figure 8b). In addition, Geng et al. prepared PPy/γ-Fe_2O_3 nanohybrids using chemical oxidation with $Fe(NO_3)_3 \cdot 9H_2O$ and used it for the detection of toxic gases [50]. The PPy-based nanohybrids were prepared by simultaneous gelling and polymerization under methoxy ethanol and $Fe(NO_3)_3 \cdot 9H_2O$. The nanohybrids were characterized by IR spectroscopy, thermogravimetric and differential thermal analysis, XRD analysis (Figure 8c), and electron microscopy. From these characterizations, the optimized annealing temperature and mole ratios of oxidation were selected and the operating temperature of the PPy/γ-Fe_2O_3 nanohybrids was studied at 90 °C. Using the PPy nanohybrids as transducer, they showed the sensitivity difference toward 0.1% H_2S depending on the different mole ratios of hybrid composition and also demonstrated the maximized annealing temperature vs. H_2S sensing properties in the PPy/γ-Fe_2O_3 nanohybrid system (Figure 8d). From these results, H_2S sensors using conducting polymer/metals can be optimized with their kinds, detecting temperature, and annealing temperature.

Metal oxides such as SnO_2, WO_3, and ZnO have been utilized in monitoring NO_2 gas [51–53]. However, they have low selectivity and sensitivity at ambient temperature. Therefore, organic-metal nanohybrids were developed to overcome their limitations. Kong et al. synthesized PTh/SnO_2 nanocomposites by the in situ chemical oxidative polymerization and applied it for NO_x detection [54]. The SnO_2 particles were coated with PTh shell which was several thickness in nanoscale (Figure 9a). They found that the sensitivity and selectivity can increase by controlling PTh mass percent (1%, 5%, 10%, 20%, and 30%) in PTh/SnO_2 nanocomposites. Based on the effects of p-n heterojunction between PTh and SnO_2, the minimum detectable level (MDL) of nanocomposites toward NO_x gas was ca. 10 ppm at room temperature, which was higher than pure SnO_2. Moreover, the stability of PTh/SnO_2 (PTh mass: 5%, 10%, and 20%) sensor was calculated during two months at room temperature, by which the enhanced stability was confirmed (Figure 9b). This NO_x sensor using PTP/SnO_2 nanocomposite operated high temperature (~90 °C), while WO_3-based transistors can detect NO_x at low temperature. Using similar fabrication process, PTh/WO_3 nanohybrids were also designed by Huang et al. and characterized with TEM, XRD analysis, and thermo-gravimetric analysis [55]. This nanohybrid showed higher thermal stability than pure PTh, leading to the utilization as sensing materials in chemical sensors. The MDL of the PTh/WO_3 nanohybrids toward NO_2 gas was a ppm level at low temperature. They demonstrated several critical factors, including the operating temperature and PTh contents, in conducting polymer/metal oxide-based chemical sensors. In this study, the 10 wt% PTh/WO_3 nanohybrid showed the highest response at low operating temperature of 70 °C. Moreover, the NO_x sensor at room temperature was demonstrated using PANI/In_2O_3 nanocomposites and they displayed high sensitivity. Sadek et al. created excellent NO_2 gas sensor using PANI/In_2O_3 nanofiber composites [56]. The nanohybrids were fabricated by chemical oxidative polymerization of aniline in the presence of finely divided In_2O_3. The high-performance sensing geometry was developed by deposition of nanohybrids on surface acoustic wave (SAW), leading to the fast response and recovery times with good repeatability at room temperature. The conductivity of PANI/In_2O_3 increased with NO_2 exposure, which indicates that NO_2 gas can act as a dopant for the PANI/In_2O_3. Interestingly, based on this sensing mechanism, the MDL of PANI/In_2O_3 nanohybrid was ca. 510 ppb and rapid response/recovery time was calculated.

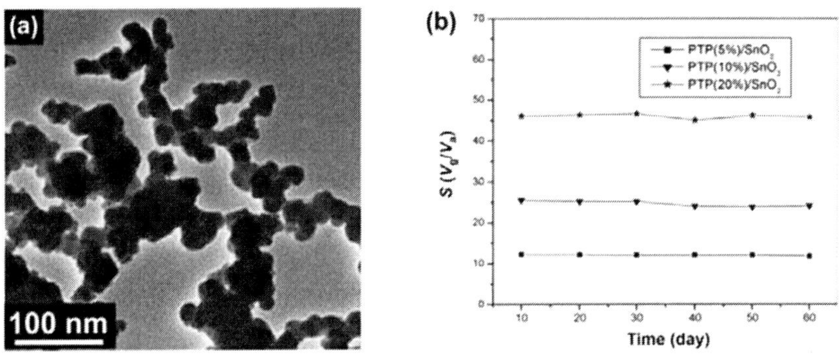

Figure 9. (**a**) TEM image of PTP (20%)/SnO_2 composites; (**b**) The stability of PTP/SnO_2 sensor to 100 ppm NO_x. Adapted with permission from [54]; Copyright 2012, American Chemical Society.

For the detection of NH_3 gas, various materials and technologies have been explored. Among many different strategies, the combination of organic-inorganic nanomaterials displayed the most excellent sensing performances toward NH_3 gas. Yang et al. provided a facile method in fabricating Ag/AgCl-decorated PPy (PPy/Ag-AgCl) nanotubes [57]. PPy nanotubes were fabricated through assembling process on the reactive self-degraded template with complex of ferric chloride and methyl orange. For uniform decoration of Ag and AgCl nanoclusters, PPy nanotubes were immersed in $AgNO_3$ solution, which resulted in PPy nanotube hybrids, as shown in Figure 10a. The PPy nanotube hybrids were utilized as chemiresistors in chemical sensor applications and showed significant conductivity changes when NH_3 gas (100 ppm) was exposed into the nanotube hybrids (Figure 10b). Moreover, the response time was ca. 150 s to reach the maximum and stable reproducible responses were also displayed. Several researchers also demonstrated that nanostructured metals deposited or chemically connected on the surface of the conducting polymers exhibits size-dependent sensing properties. Hong et al. reported the fabrication of a nanohybrids consisting of PPy and Pd through thermal dynamic refluxing of

Pd salt and subsequent encapsulating of formed Pd nanoparticles with a PPy shell via vapor deposition polymerization [58]. The nanohybrids designed with synthetic variables, including reaction time and polymer additives, were characterized by TEM and atomic force microscopy, and then they were utilized as the chemiresistor in NH_3 sensing system. The electrical responses from Pd-encapsulated PPy nanoparticles toward NH_3 were highly fast and stable at the range of the 50–200 ppm NH_3 gases. Moreover, the response and recovery time were 14 s and 148 s for 1000 ppm NH_3, respectively. Judging from these results, the size of Pd nanoparticles and the morphology of the hybrid films are important to achieve high-performance ammonia sensors based on the nanohybrids. As advanced nanohybrids for NH_3 sensing materials, Park et al. constructed silver nanoparticles-decorated PEDOT nanotubes using a modified concurrent redox reaction in porous alumina membrane [59]. To fabricate the PEDOT nanotube hybrids, the alumina template was immersed into $Fe(NO_3)_3$ and $AgNO_3$ solutions and vaporized EDOT monomer was introduced into the template membrane under 10^{-2} Torr at 60 °C for 3 h. Finally, PEDOT nanohybrids were successfully obtained after etching the template membrane. To characterize these nanohybrids, various analysts including TEM, SEM, and XRD analysis were employed (Figure 10c). The diameter and wall-thickness of PEDOT nanohybrids were ca. 100 and ca. 25 nm, and the size of the decorated silver nanoparticles was ca. 10 nm. The PEDOT nanotube hybrids were utilized as the chemiresistor in ammonia sensor by virtue of the enhanced conductivity and high affinity toward NH_3 by silver nanoparticles. The highest sensitivity and rapid response time toward NH_3 (1 ppm and less than 1s) can be obtained from PEDOT nanohybrids owing to the silver nanoparticles dispersed on the surface of PEDOT nanotubes. Moreover, the optimal deposited amount of silver nanoparticles were also determined by adjusting $AgNO_3$ concentration (Figure 10d). The sensing geometry was constructed by the deposition of nanohybrids on the electrodes and induced the resistance changes when NH_3 was adsorbed on the surface of nanohybrids. They showed comprehensive sensing properties using experimental date: (i) the sensitivity and selectivity changes dependence upon amount of Ag NPs on the surface of PEDOT nanotubes, and (ii) response and recovery time upon increasing NH_3 concentrations.

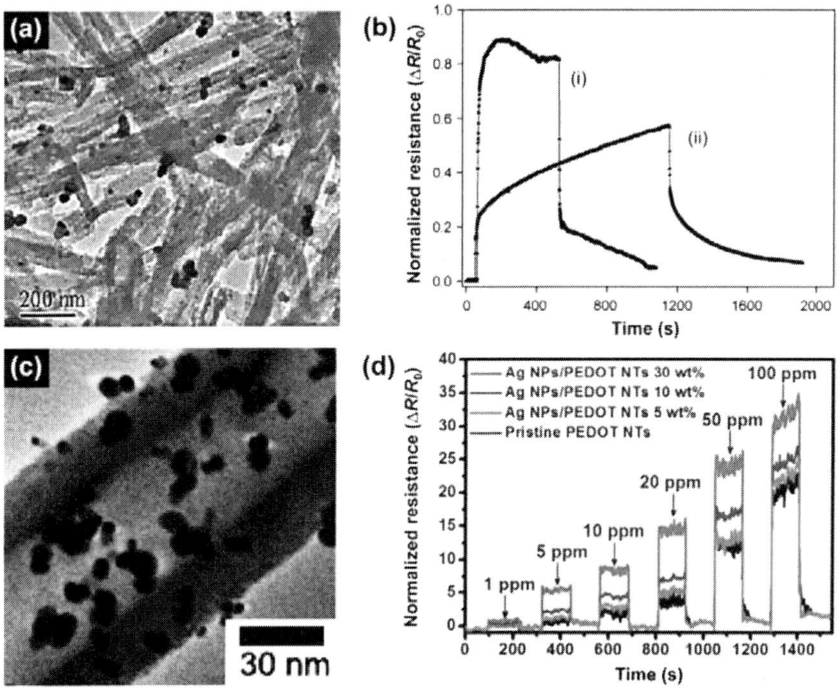

Figure 10. (**a**) TEM image of PPy/Ag-AgCl nanotubes; (**b**) Response of (i) PPy nanotubes and (ii) PPy/Ag-AgCl nanotubes upon exposure to 100 ppm of NH_3. Adapted with permission from Ref. [58]; Copyright 2010, Elsevier B.V; (**c**) TEM images of silver/PEDOT nanotubes; (**d**) The sensing performances toward 1–100 ppm NH_3 gases. Adapted with permission from [59]; Copyright 2012, Royal Society of Chemistry.

In the detection of ammonia gas, the p-n junctional type as transistor was introduced for the first time. Gong et al. fabricated unique conducting polymer nanograins using PANI and electrospun titania fibers, leading to the ultrasensitive NH_3 gas sensor (Figures 11a,b) [60]. They introduced p-n heterojunctions combining p-type PANI and an electrospun n-type semiconductive titania nanofiber using immersing process. Moreover, the chemiresistive ammonia sensor based on titania/PANI nanohybrids was developed with electric current switches that can be operated by the resistance change from p-n heterojunction when NH_3 gas is absorbed by PANI nanograins (Figure 11c). Surprisingly, the MDL of chemiresistive sensor was ca. 50 ppt NH_3 gas in air, which is 1000 times more sensitive than the best PANI sensor reported in the literature.

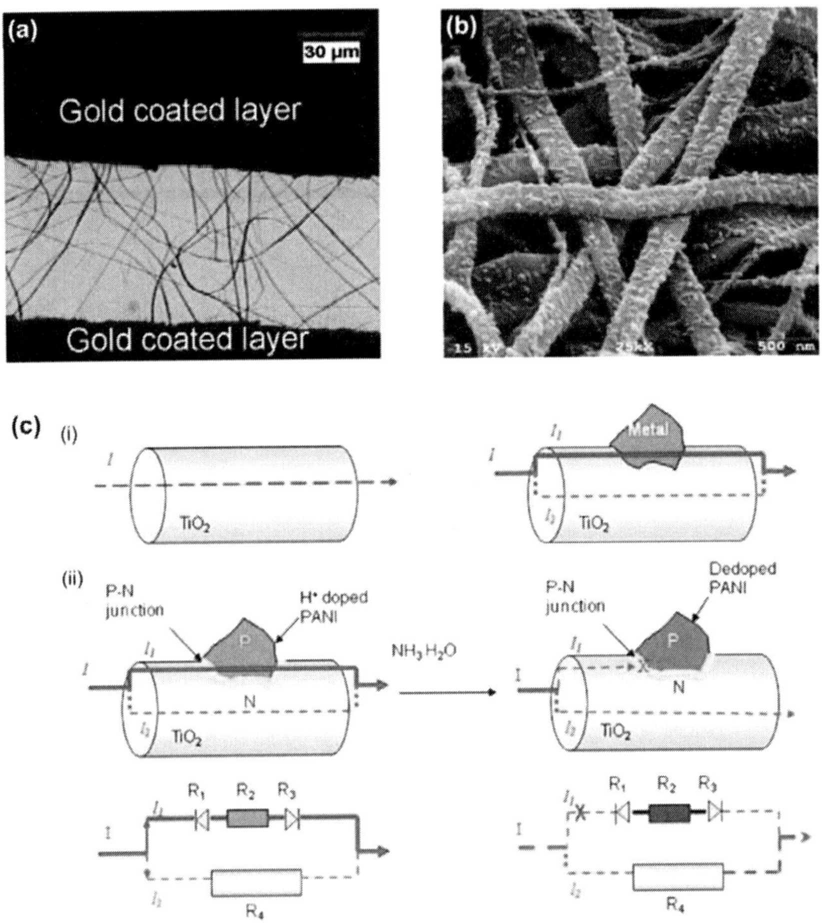

Figure 11. (**a**) An optical microscope image of the sensor and (**b**) SEM image of titania/PANI nanofiber hybrids. (**c**) Scheme of nanosized p-n heterojunction as a switch to control the electric current flow in titania nanofibers. Adapted with permission from [60]; Copyright 2010, American Chemical Society.

3.1.3. Volatile Organic Compound Detection

There are a variety of VOCs, including alkanes and benzene derivatives, to which we are exposed our daily life. Moreover, extremely low concentrations (ppb to ppt) of metabolites or VOCs can be used as diagnostic markers for diseases [61]. Contents over certain amount of ethanol (ca. 800 pmol·L^{-1}) can be diagnosed as liver diseases and, in case of lung cancer, 22 VOCs can be found in the breath from patients [62]. Therefore, various detection tools, such as gas/liquid chromatography, mass spectroscopy, and chemical treatments, have been explored to discriminate exhaled breath. Although the detection tools have individual advantages, their limitations in terms of sensitivity and selectivity still remained as challenges because the chemical structures of VOCs are almost similar or slightly have atomic-differences. Recently, many researchers have tried to overcome the limitations using conducting polymer nanohybrids owing to their specificity toward alkanes. Athawale et al. successfully obtained Pd-PANI nanohybrids using oxidation polymerization with aniline solution containing Pd nanoparticles [63]. The nanohybrids can induce specific interaction between the methanol molecules with the imine nitrogen. Specifically, the positive charge on the imine nitrogen is reduced by interacting with methanol in the presence of Pd nanoparticles, resulting in that the imine nitrogen was converted to amine. Based on this sensing mechanism, the Pd-PANI nanohybrids had highly selective and sensitive to methanol vapors (Figure 12). Interestingly, their specific selectivity was also demonstrated by exposing to mixtures including methanol-ethanol and methanol-isopropanol. Moreover, the methanol sensors based on the nanohybrids displayed a stable response for a long time.

Choudhury also demonstrated highly specific ethanol sensors using silver/PANI nanohybrids that were prepared by in situ oxidative polymerization of aniline monomer in the presence of different contents of silver nanoparticles [64]. The formation of the nanohybrids can be easily controlled by adjusting silver concentrations owing to the aggregation effect. They showed comparison analysis between pure PANI and silver/PANI nanohybrids. The most outstanding feature was their conductivity difference, which affects the sensitivity toward ethanol sensitivity. The PANI nanohybrids had superior ethanol sensing capacity compare to pure PANI. Moreover, the sensitivity showed linear relationship between ethanol and silver concentrations. The PANI nanohybrid sensors also provided reproducible sensing properties.

3.2. Biosensors

One of the great goals in developing the biological sensing materials is to fabricate rapid, accurate, and portable sensing geometry or diagnostic technologies for the prevention of fatal disease. Various sensing nanomaterials such as graphene, carbon nanotubes, metal transistors, and nanohybrids have been developed. Especially, conducting polymer-based nanohybrids have been studied to detect directly or indirectly related-biomolecules, such as hydrogen peroxide, glucose, dopamine, DNA, and proteins, by virtue of their electrical conductivity, affinity toward biomolecules, and biocompatibility.

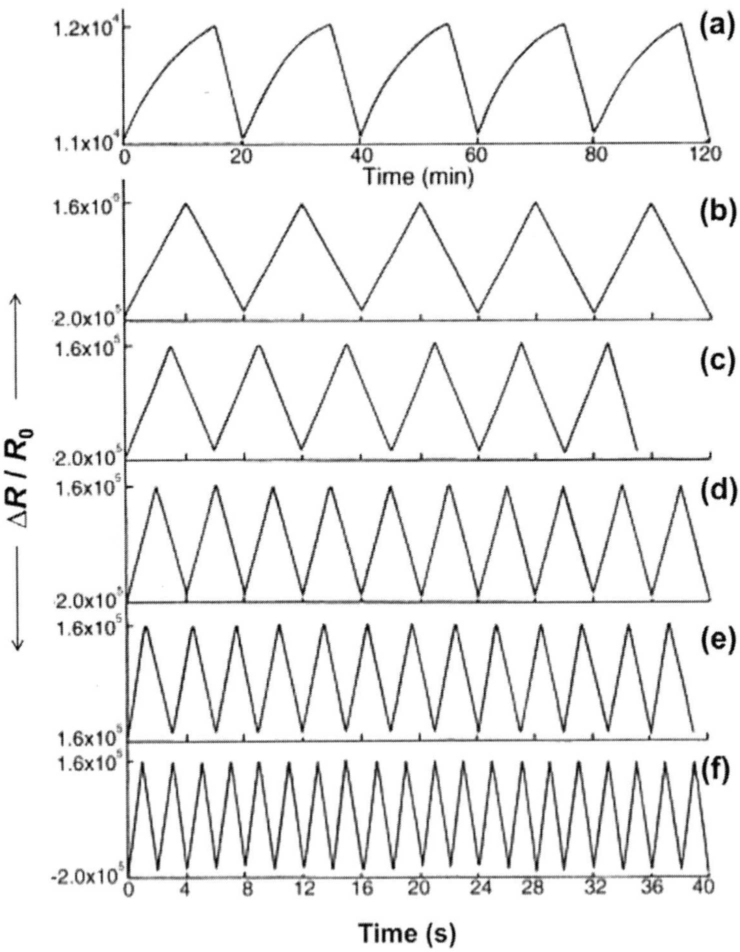

Figure 12. (**a**) Response curves of (**a**) PANI_only and (**b–f**) Pd-PANI nanohybrids toward different methanol concentrations: (**a**) 2000 ppm; (**b**) 1 ppm; (**c**) 5 ppm; (**d**) 10 ppm; (**e**) 100 ppm; and (**f**) 2000 ppm. Adapted with permission from [63]; Copyright 2006, Elsevier B.V.

3.2.1. Glucose Detection

Incorporation of biomolecules as the recognition element into signal transducers can enable to detect the target molecules. A variety of biosensors based on transducers conjugated with bioreceptors such as antibodies, aptamers, cells, and enzymes have been developed. In particular, for diagnosing diabetes, amperometric glucose biosensors have been constructed with glucose oxidase (GOx) electrodes that can induce enzymatic regeneration reaction, leading to quantitative information on the glucose level. First, the controlled loading amount of GOx can affect the sensing properties of glucose sensors. Xian et al. reported a novel glucose biosensor based on gold nanoparticles-conjugated PANI nanofiber hybrids immobilized with GOx [65]. The PANI nanofiber hybrids provided fascinating advantages, including enlarged surface areas, excellent conductivity, and many micron-sized gaps to easily immobilize GOx. Moreover, this enzyme-conjugated nanohybrids allow the rapid transmit electron and enhanced interactions toward various glucose concentrations. To characterize their sensing operation, first of all, several experiments were performed to determine the electrochemical response of H_2O_2 at the PANI nanofiber hybrids. PANI nanofiber hybrids displayed the most efficient sensing properties toward H_2O_2 because incorporation of gold nanoparticles in PANI nanofibers enabled the electron transfer between electrodes and H_2O_2. To optimize sensing operation, several experimental factors (pH and loading amount of GOx) were also characterized. Based on these results, the sensing properties of PANI nanohybrids were further examined to evaluate glucose sensor performance. In real samples, this glucose sensor showed rapid response time (less than 5 s) and significant current changes toward various glucose concentrations. Moreover, excellent selectivity from the real-sample was also displayed. Second, to obtain enhanced sensitivity, the measurement conditions are considered. Xu et al. also fabricated enzyme-immobilized conducting polymer nanohybrids to construct glucose biosensor [66]. They fabricated multiwalled-carbon nanotubes (MWNTs) coated with PANI and dendrimer-encapsulated Pt nanoparticles (Pt-DENs) by in site polymerization and immobilized GOx on the nanotube surface. The glucose biosensors using PANI nanohybrids was examined with important factors (pH and potentials), by which the optimized operating condition was determined from stable responses. Moreover, the attached Pt-DENs can accelerate electron transfer and aid the immobilization of enzyme, helping their sensing performances. To confirm the biosensing performance using PANI nanohybrids, the real-time responses of the nanohybrids-based electrode were demonstrated with the addition of various glucose concentrations. The significant current changes were displayed within 5 s. Its MDL was ca. 0.5 μM and saturated over 16 mM. Moreover, the influence of common interfering electroactive substances such as uric acid, acetaminophen, and ascorbic acid was also characterized for excellent selectivity. The PANI nanohybrids have shown excellent electrochemical properties owing to their electrochemical doping and re-doping behavior under acid and base condition. Especially, the stable electrical properties of the nanohybrids make them suitable for biosensing. Liu et al. reported the formation of novel PANI-coated Fe_3O_4 nanoparticles-MWNT nanohybrids by the co-precipitation of Fe(III) and Fe(II), followed by in site polymerization [67]. The PANI nanohybrids were characterized by IR spectroscopy, electron microscopy, XRD analysis and thermogravimetric analysis, which proved their successful fabrication. Also, the magnetic properties of PANI nanohybrids were confirmed by hysteresis loops of Fe_3O_4 nanoparticles and PANI-Fe_3O_4-MWNT nanohybrids. These nanohybrids were utilized for glucose biosensor by electrochemical doping of GOx. The glucose biosensor showed rapid dynamic equilibrium within 5–6 s and stable reproducibility toward 1 mM glucose.

Lastly, the metal density also becomes a one of important factors in glucose sensors. Guo et al. also reported a simple and facile procedure for fabricating PANI nanohybrids for high-efficiency electrocatalysts [68]. They prepared PANI nanofiber-supported high-density Pt nanoparticles at room temperature, which had a couple of benefits such as high-density of Pt nanoparticles without linkers, free-template at room temperature, and high-performance electrochemical devices. The PANI nanohybrids, which were fabricated by a simple wet-chemical approach, were combined with GOx to utilize as nanoprobes of glucose biosensor. The PANI nanohybrids-based glucose biosensor showed highly sensitive response toward various glucose concentrations owing to enhanced electron transfer. Recently, Bian et al. provided enzyme-free electrochemical biosensing toward hydrogen peroxide. This methodology provide cost-effective and easy-fabrication process. They fabricated Pt/PPy hollow hybrids using wet-chemical method via magnetite template and applied them into the H_2O_2 biosensors [69]. The cyclic measurements from Pt/PPy hollow nanosphere hybrids exhibited good electrocatalytic activity towards the reduction of H_2O_2 in a buffer solution (pH = 7.0). The response from hybrid-based biosensor was less than 2 s with linear range of 1.0–9.0 mM and the MDL was 1.2 μM (signal to noise: 3).

3.2.2. Dopamine Detection

Dopamine that belongs to excitatory chemical neurotransmitters is one of the most significant catecholoamines and affects to the central nervous and hormone systems. Abnormal dopamine concentration in human body can be used as diagnosis measurement of fatal diseases such as Parkinson's and Alzheimer's diseases. Therefore, efficient sensing methodologies have been designed to detect dopamine molecules. Feng et al. fabricated gold/PANI hollow nanosphere hybrids using polystyrene-sulfonated polystyrene core-shell gel particle templates [26]. Moreover, they controlled the shell-thickness and amount of gold nanoparticles decorated on the surface of PANI shell, leading to the advanced transistor in the range of the sensor applications (Figure 13a). Especially, the conductivity of the PANI nanohybrids was more than 3 times higher than that of the conventional PANI hollow sphere. Based on excellent electrical properties, the nanohybrids were applied for dopamine detection. Also, gold/PANI hybrid system can provide redox behavior in different pH buffer solutions because gold nanoparticles facilitate the electrocatalytic oxidation of dopamine, resulting to efficient dopamine biosensor. The MDL of gold/PANI nanohybrids was ca. 2 mM dopamine concentration, which is ca. three times higher than that in the PANI system (Figure 13b). Compared to conventional dopamine biosensors, Li et al. proposed a novel sensing platform based on gold nanoparticles/PANI/polydopamine hybrids [70]. The nanohybrids were synthesized by facile processes: (i) spontaneous self-polymerization of dopamine, (ii) the formation of gold nanoparticles with both polydopamine and aniline by reducing agent, leading to the excellent electroactivity in neural solution. Importantly, the role of the gold nanoparticles in the nanohybrids is to enhance catalytic ability. Under optimized conditions, the nanohybrid sensor showed rapid response (4 s) and ultrasensitive DML (4.0×10^{-7} M) to ascorbic acid. Recently, graphene-based dopamine sensor was introduced and displayed high sensitivity owing to its high conductivity and enlarged surface areas. Ultrasensitive dopamine detector was created by Wu and Shen [71]. They fabricated gold nanoparticles decorated PPy/graphene oxide nanohybrids using in-situ chemical oxidative polymerization. The gold@PPy/graphene oxide nanohybrids were characterized by electron microscopy and X-ray photoelectron spectroscopy (XPS). The nanohybrids displayed a crumpled and wrinkled surface, which can provide enhanced surface areas, and have uniform flower-like gold nanoparticles, leading to the ultrasensitive MDL (18.29 pM) with a linear range of 1–5000 nM for dopamine.

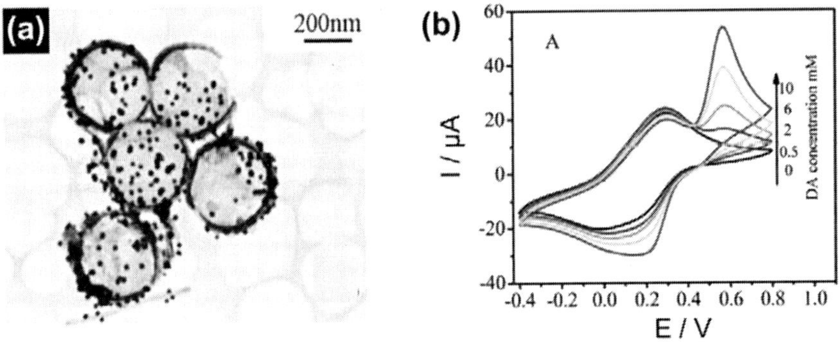

Figure 13. (**a**) TEM image and (**b**) cyclic voltammograms of gold/PANI hollow nanosphere nanohybrids toward various dopamine concentrations. Adapted with permission from [26]; Copyright 2006, American Chemical Society.

3.2.3. DNA Detection

Nucleic acids such as DNA or RNA can affect important protein information in human body. As a results, various DNA analysis have been reported with advanced sensing devices in related applications including infectious diseases, drug discoveries, and clinical diagnosis. One of the attractive DNA sensing systems is electrochemical detectors based on biomaterial-conjugated conductive geometries. Electrochemical DNA sensors showed direct and fast electrical-signals toward target biomolecules, leading to the simple sensing geometries without complex signal transduction equipment and expensive analyzers. Moreover, the hybridized transducers in DNA sensors showed more sensitive and selective sensing performances compared to pristine them. Especially, DNA detectors based on conducting polymer nanohybrids provided excellent electrical properties, resulting that high-performance DNA sensors were constructed. Tian et al. reported stable multilayer films using PANI/mercaptosuccinic-acid-capped gold (MSAG) nanoparticles hybrids by layer-by-layer method [72]. PANI can provide the excellent redox activity by molecular doping and shift their neutral conditions to significant redox position, leading to the excellent electrochemical sensing devices. In this study, the PANI layer was easily stimulated by modified MSAG under no polyelectrolytes conditions and its electroactivity was shifted to neutral pH without polyelectrolytes. Moreover, PANI hybrid layer was functionalized to introduce the amino-terminated DNA catcher probes (NH_2-DNA), which can be used for monitoring their hybridization with different DNA target strands under electrochemical methods. Based on the electrochemical impedance spectroscopy (EIS), the surface-charge density of the PANI hybrids increased from the hybridization process with target DNA, resulting that the electrostatic repulsion also increased. Therefore, the comparison between before and after the introduction of target DNA (500 nM) can be monitored with EIS spectra. Jang et al. reported excellent dual-functionalized conducting polymer nanohybrids substrates for molecular-probe and DNA-carrier [23]. They, for the first time, constructed bifunctionalized conducting polymer nanotube hybrids using VDP in ferric chloride-adsorbed alumina membranes with 100 nm pore diameter. Moreover, the conducting polymer nanotubes were also functionalized with carboxylic groups by the polymerization of pyrrole-2-carboxylic acid. Functionalized PPy nanotubes can be easily decorated a photoluminescent molecule, pyreneacetic acid, which silica nanoparticles were employed as the linker. The silica nanoparticles with surface amino groups were first attached to the PPy nanotubes and then pyreneacetic acid was linked with the silica nanoparticles. Furthermore, single-stranded DNA (ssDNA) with a terminal amino group was attached to the residual carboxyl groups of the nanotubes. The resulting PPy nanotube hybrids were characterized with IR spectroscopy, XPS, and electron microscopy. Interestingly, the size of the attached silica-NH_2 nanoparticles decreased with increasing number of the nanoparticles, resulting in that the total surface area of silica nanoparticles increased (Figure 14a–c). The PPy nanotubes hybrids with silica nanoparticles were found to have more sensitive responses than only PPy nanotubes owing to the enhanced surface area. The nanotube hybrids showed excellent PL intensity (Figure 14d) and different intensities depending on the size of the attached silica nanoparticles (7 nm > 12 nm > 11 nm). Feng et al. also fabricated DNA sensors using gold/PANI nanotubes hybridized membranes. The PANI nanohybrids were constructed on the glass carbon electrode and utilized as electrochemical sensing geometry to detect the immobilization and hybridization of DNA. The cyclic voltammetry, pulse voltammetry, and electrochemical impedance spectroscopy were introduced to characterize PANI nanohybrids. Based on the DNA probes using PANI nanohybrids, they showed excellent recognition of DNA hybridization under the phosphinothricin acetyltransferase gene (PAT). The DML toward the sequence-specific DNA was 3.1×10^{-13} mol·L^{-1}, which is high-performance DNA sensor compared to conventional DNA biosensors. Moreover, specific recognition and reproducibility of the DNA sensor using PANI nanohybrids were also evaluated.

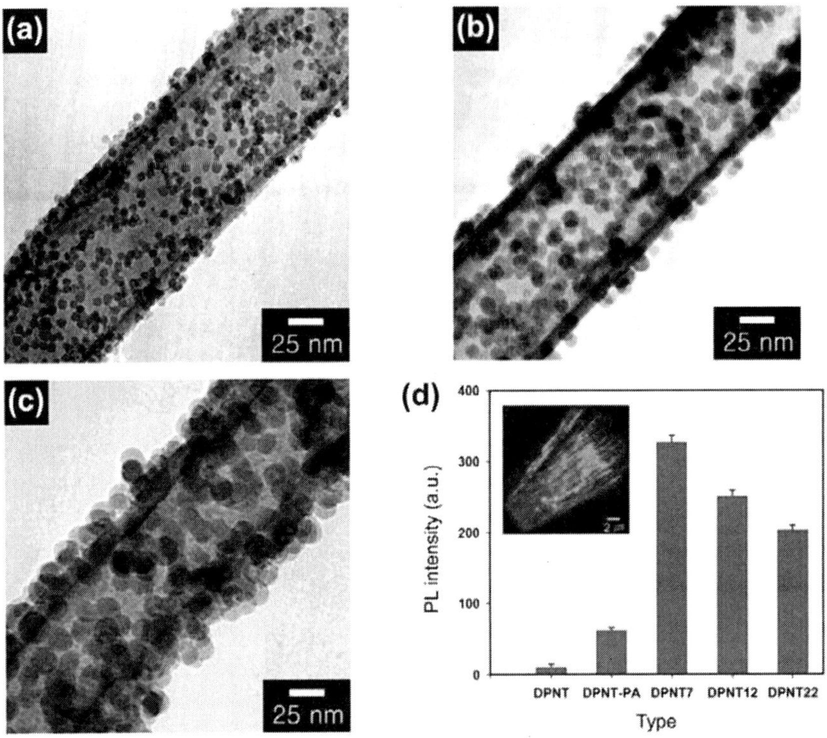

Figure 14. TEM images of PPy nanotubes with different sized amine-modified silica nanoparticles attached: (**a**) 7 nm; (**b**) 12 nm; and (**c**) 22 nm; (**d**) photoluminescence intensity of pyreneacetic acid conjugated with amine groups on the nanotube hybrids (DPNT: unconjugated PPy nanotube hybrids; DPNT-PA: the nanotube hybrid incubated with pyreneacetic acid in the absence of a condensing agent; DPNT7: the nanotube hybrids formed using 7 nm silica; DPNT22: the nanotube hybrids formed using 22 nm silica (inset: CLSM image of pyreneacetic acid conjugated with amine groups of the nanotube hybrids. Adapted with permission from [23]; Copyright 2006, Wiley-VCH Verlag GmbH & Co. KGaA.

4. CONCLUSIONS AND OUTLOOK

The purpose of this review article is to provide a better understanding for the recent research trend of conducting polymer-based nanohybrid sensors. Conducting polymer-based nanohybrids have the strong potential to provide distinct properties that were not found in the individual component and their bulk counterparts. However, on the other hand, it seems like that there has been relatively limited research on combining of conducting polymers with inorganic materials on nanoscale, except carbon nanomaterials, which is presumably due to the difficulties in obtaining well-defined polymer nanostructures. Although there have been successful trials that are highlighted in here, the absence of various material candidates is considered to retard the wide application of the conducting polymer nanohybrids to sensors. There is no doubt that new or improved chemical and physical properties from the conducting polymer nanohybrids can be exploited for fabricating high-performance sensors. Accordingly, we believe that further advance in synthesis techniques of conducting polymer nanohybrids will lead to a huge leap of sensor technology.

ACKNOWLEDGMENTS

This work was supported by the Basic Science Research Program (2012R1A1A1042024), the Global Research Laboratory (2013056090), and the National Research Foundation of Korea (NRF) grant (2011-0017125) of Korea (NRF), funded by the Ministry of Science, ICT (Information and Communications Technology) and Future Planning.

CONFLICTS OF INTEREST

The authors declare no conflict of interest.

REFERENCES

1. Janaky, C.; Visy, C. Conducting polymer-based hybrid assemblies for electrochemical sensing: A materials science perspective. Anal. Bioanal. Chem. 2013, 405, 3489–3511.

2. Hangarter, C.M.; Chartuprayoon, N.; Hernandez, S.C.; Choa, Y.; Myung, N.V. Hybridized conducting polymer chemiresistive nano-sensors. Nano Today 2013, 8, 39–55.

3. Liu, Z.; Zhang, L.; Poyraz, S.; Zhang, X.Y. Conducting polymer-metal nanocomposites synthesis and their sensory applications. Curr. Org. Chem. 2013, 17, 2256–2267.

4. Yoon, H.S.; Choi, M.J.; Lee, K.A.; Jang, J.S. Versatile strategies for fabricating polymer nanomaterials with controlled size and morphology. Macromol. Res. 2008, 16, 85–102.

5. Jang, J. Conducting polymer nanomaterials and their applications. Adv. Polym. Sci. 2006, 199, 189–259.

6. Gangopadhyay, R.; De, A. Conducting polymer nanocomposites: A brief overview. Chem. Mater. 2000, 12, 608–622.

7. Oh, W.K.; Kwon, O.S.; Jang, J. Conducting polymer nanomaterials for biomedical applications: Cellular interfacing and biosensing. Polym. Rev. 2013, 53, 407–442.

8. Park, H.W.; Kim, T.; Huh, J.; Kang, M.; Lee, J.E.; Yoon, H. Anisotropic growth control of polyaniline nanostructures and their morphology-dependent electrochemical characteristics. ACS Nano 2012, 6, 7624–7633.

9. Yoon, H.; Hong, J.Y.; Jang, J. Charge-transport behavior in shape-controlled poly (3,4-ethylenedioxythiophene) nanomaterials: Intrinsic and extrinsic factors. Small 2007, 3, 1774–1783.

10. Yoon, H.; Jang, J. Conducting-polymer nanomaterials for high-performance sensor applications: Issues and challenges. Adv. Funct. Mater. 2009, 19, 1567–1576.

11. Yoon, H. Current trends in sensors based on conducting polymer nanomaterials. Nanomaterials 2013, 3, 524–549.

12. Lu, X.F.; Zhang, W.J.; Wang, C.; Wen, T.C.; Wei, Y. One-dimensional conducting polymer nanocomposites: Synthesis, properties and applications. Prog. Polym. Sci. 2011, 36, 671–712.

13. Rajesh; Ahuja, T.; Kumar, D. Recent progress in the development of nano-structured conducting polymers/nanocomposites for sensor applications. Sens. Actuator B Chem. 2009, 136, 275–286.

14. Park, S.J.; Kwon, O.S.; Lee, S.H.; Song, H.S.; Park, T.H.; Jang, J. Ultrasensitive flexible graphene based Field-Effect Transistor (FET)-type bioelectronic nose. Nano Lett. 2012, 12, 5082–5090.

15. Tseng, R.J.; Huang, J.X.; Ouyang, J.; Kaner, R.B.; Yang, Y. Polyaniline nanofiber/gold nanoparticle nonvolatile memory. Nano Lett. 2005, 5, 1077–1080.

16. Zhang, X.Y.; Manohar, S.K. Narrow pore-diameter polypyrrole nanotubes. J. Am. Chem. Soc. 2005, 127, 14156–14157.

17. Zhang, X.Y.; Lee, J.S.; Lee, G.S.; Cha, D.K.; Kim, M.J.; Yang, D.J.; Manohar, S.K. Chemical synthesis of PEDOT nanotubes. Macromolecules 2006, 39, 470–472.

18. Qin, X.Y.; Lu, W.B.; Luo, Y.L.; Chang, G.H.; Sun, X.P. Preparation of Ag nanoparticle-decorated polypyrrole colloids and their application for H2O2 detection. Electrochem. Commun. 2011, 13, 785–787.

19. Yoon, H.; Kim, J.H.; Lee, N.; Kim, B.G.; Jang, J. A novel sensor platform based on aptamer-conjugated polypyrrole nanotubes for label-free electrochemical protein detection. ChemBioChem 2008, 9, 634–641.

20. Yoon, H.; Lee, S.H.; Kwon, O.S.; Song, H.S.; Oh, E.H.; Park, T.H.; Jang, J. Polypyrrole nanotubes conjugated with human olfactory receptors: High-performance transducers for FET-type bioelectronic noses. Angew. Chem. Int. Ed. 2009, 48, 2755–2758.

21. Yoon, H.; Jang, J. A field-effect-transistor sensor based on polypyrrole nanotubes coupled with heparin for thrombin detection. Mol. Cryst. Liq. Cryst. 2008, 491, 21–31.

22. Ko, S.; Jang, J. A highly efficient palladium nanocatalyst anchored on a magnetically functionalized polymer-nanotube support. Angew. Chem. Int. Ed. 2006, 45, 7564–7567.

23. Jang, J.; Ko, S.; Kim, Y. Dual-functionalized polymer nanotubes as substrates for molecular-probe and DNA-carrier applications. Adv. Funct. Mater. 2006, 16, 754–759.

24. Kim, J.S.; Kim, W.J.; Cho, N.; Shukla, S.; Yoon, H.; Jang, J.; Prasad, P.N.; Kim, T.D.; Lee, K.S. Synthesis and properties of quantum dot-polypyrrole nanotube composites for photovoltaic application. J. Nanosci. Nanotechnol. 2009, 9, 6957–6961.

25. Cheng, D.M.; Xia, H.B.; Chan, H.S.O. Facile fabrication of AgCl@polypyrrole-chitosan core-shell nanoparticles and polymeric hollow nanospheres. Langmuir 2004, 20, 9909–9912.

26. Feng, X.M.; Mao, C.J.; Yang, G.; Hou, W.H.; Zhu, J.J. Polyaniline/Au composite hollow spheres: Synthesis, characterization, and application to the detection of dopamine. Langmuir 2006, 22, 4384–4389.

27. Li, X.H.; Li, Y.C.; Tan, Y.W.; Yang, C.H.; Li, Y.F. Self-assembly of gold nanoparticles prepared with 3,4-ethylenedioxythiophene as reductant. J. Phys. Chem. B 2004, 108, 5192–5199.

28. Liu, Y.; Lu, N.; Poyraz, S.; Wang, X.L.; Yu, Y.J.; Scott, J.; Smith, J.; Kim, M.J.; Zhang, X.Y. One-pot formation of multifunctional Pt-conducting polymer intercalated nanostructures. Nanoscale 2013, 5, 3872–3879.

29. Wang, S.B.; Shi, G.Q. Uniform silver/polypyrrole core-shell nanoparticles synthesized by hydrothermal reaction. Mater. Chem. Phys. 2007, 102, 255–259.

30. Chang, M.; Kim, T.; Park, H.W.; Kang, M.; Reichmanis, E.; Yoon, H. Imparting chemical stability in nanoparticulate silver via a conjugated polymer casing approach. ACS Appl. Mater. Inter. 2012, 4, 4357–4365.

31. Lu, G.W.; Li, C.; Shen, J.Y.; Chen, Z.J.; Shi, G.Q. Preparation of highly conductive gold-poly(3,4-ethylenedioxythiophene) nanocables and their conversion to poly (3,4-ethylenedioxythiophene) nanotubes. J. Phys. Chem. C 2007, 111, 5926–5931.

32. Sarma, T.K.; Chattopadhyay, A. One pot synthesis of nanoparticles of aqueous colloidal polyaniline and its Au-nanoparticle composite from monomer vapor. J. Phys. Chem. A 2004, 108, 7837–7842.

33. Guo, Y.B.; Tang, Q.X.; Liu, H.B.; Zhang, Y.J.; Li, Y.L.; Hu, W.P.; Wang, S.; Zhu, D.B. Light-controlled organic/inorganic P-N junction nanowires. J. Am. Chem. Soc. 2008, 130, 9198–9199.

34. Park, S.; Lim, J.H.; Chung, S.W.; Mirkin, C.A. Self-assembly of mesoscopic metal-polymer amphiphiles. Science 2004, 303, 348–351.

35. Lin, H.W.; Liu, H.B.; Qian, X.M.; Lai, S.W.; Li, Y.J.; Chen, N.; Ouyang, C.B.; Che, C.M.; Li, Y.L. Constructing a blue light photodetector on inorganic/organic p-n heterojunction nanowire arrays. Inorg. Chem. 2011, 50, 7749–7753.

36. Xie, Y.B.; Du, H.X. Electrochemical capacitance performance of polypyrrole-titania nanotube hybrid. J. Solid State Electr. 2012, 16, 2683–2689.

37. Butterworth, M.D.; Bell, S.A.; Armes, S.P.; Simpson, A.W. Synthesis and characterization of polypyrrole-magnetite-silica particles. J. Colloid Interface Sci. 1996, 183, 91–99.

38. Shin, S.; Yoon, H.; Jang, J. Polymer-encapsulated iron oxide nanoparticles as highly efficient Fenton catalysts. Catal. Commun. 2008, 10, 178–182.

39. Zhang, X.Y.; Goux, W.J.; Manohar, S.K. Synthesis of polyaniline nanofibers by "nanofiber seeding". J. Am. Chem. Soc. 2004, 126, 4502–4503.

40. Zhang, X.Y.; Manohar, S.K. Bulk synthesis of polypyrrole nanofibers by a seeding approach. J. Am. Chem. Soc. 2004, 126, 12714–12715.

41. Liu, Z.; Poyraz, S.; Liu, Y.; Zhang, X.Y. Seeding approach to noble metal decorated conducting polymer nanofiber network. Nanoscale 2012, 4, 106–109.

42. Kwon, O.S.; Park, S.J.; Lee, J.S.; Park, E.; Kim, T.; Park, H.W.; You, S.A.; Yoon, H.; Jang, J. Multidimensional conducting polymer nanotubes for ultrasensitive chemical nerve agent sensing. Nano Lett. 2012, 12, 2797–2802.

43. Yoon, H.; Chang, M.; Jang, J. Formation of 1D poly(3,4-ethylenedioxythiophene) nanomaterials in reverse microemulsions and their application to chemical sensors. Adv. Funct. Mater. 2007, 17, 431–436.

44. Wang, F.; Gu, H.W.; Swager, T.M. Carbon nanotube/polythiophene chemiresistive sensors for chemical warfare agents. J. Am. Chem. Soc. 2008, 130, 5392–5393.

45. Tiwari, D.C.; Sharma, R.; Vyas, K.D.; Boopathi, M.; Singh, V.V.; Pandey, P. Electrochemical incorporation of copper phthalocyanine in conducting polypyrrole for the sensing of DMMP. Sens. Actuator B Chem. 2010, 151, 256–264.

46. Shirsat, M.D.; Bangar, M.A.; Deshusses, M.A.; Myung, N.V.; Mulchandani, A. Polyaniline nanowires-gold nanoparticles hybrid network based chemiresistive hydrogen sulfide sensor. Appl. Phys. Lett. 2009, 94, 083502:1–083502:3.

47. Virji, S.; Fowler, J.D.; Baker, C.O.; Huang, J.X.; Kaner, R.B.; Weiller, B.H. Polyaniline manofiber composites with metal salts: Chemical sensors for hydrogen sulfide. Small 2005, 1, 624–627.

48. Andringa, A.-M.P.C.; Katsouras, I.; Blom, P.W.M.; de Leeuw, D.M. NO2 detection and real-time sensing with field-effect transistors. Chem. Mater. 2014, 26, 773–785.

49. Geng, L.N.; Huang, X.L.; Zhao, Y.Q.; Li, P.; Wang, S.R.; Zhang, S.M.; Wu, S.H. H2S sensitivity study of polypyrrole/WO3 materials. Solid State Electron. 2006, 50, 723–726.

50. Geng, L.; Wang, S.R.; Zhao, Y.Q.; Li, P.; Zhang, S.M.; Huang, W.P.; Wu, S.H. Study of the primary sensitivity of polypyrrole/r-Fe2O3 to toxic gases. Mater. Chem. Phys. 2006, 99, 15–19.

51. Zhang, D.F.; Sun, L.D.; Xu, G.; Yan, C.H. Size-controllable one-dimensinal SnO2 nanocrystals: Synthesis, growth mechanism, and gas sensing property. Phys. Chem. Chem. Phys. 2006, 8, 4874–4880.

52. Jimenez, I.; Arbiol, J.; Dezanneau, G.; Cornet, A.; Morante, J.R. Crystalline structure, defects and gas sensor response to NO2 and H2S of tungsten trioxide nanopowders. Sens. Actuator B Chem. 2003, 93, 475–485.

53. Fan, F.Y.; Feng, Y.J.; Bai, S.L.; Feng, J.T.; Chen, A.F.; Li, D.Q. Synthesis and gas sensing properties to NO2 of ZnO nanoparticles. Sens. Actuator B Chem. 2013, 185, 377–382.

54. Kong, F.H.; Wang, Y.; Zhang, J.; Xia, H.J.; Zhu, B.L.; Wang, Y.M.; Wang, S.R.; Wu, S.H. The preparation and gas sensitivity study of polythiophene/SnO2 composites. Mater. Sci. Eng. B Adv. 2008, 150, 6–11.

55. Huang, J.; Kang, Y.F.; Yang, T.L.; Wang, Y.; Wang, S.R. Preparation of polythiophene/WO3 organic-inorganic hybrids and their gas sensing properties for NO2 detection at low temperature. J. Natl. Gas. Chem. 2011, 20, 403–407.

56. Sadek, A.Z.; Wlodarski, W.; Shin, K.; Kaner, R.B.; Kalantar-zadeh, K. A layered surface acoustic wave gas sensor based on a polyaniline/In2O3 nanofibre composite. Nanotechnology 2006, 17, 4488–4492.

57. Yang, X.M.; Li, L.A.; Zhao, Y. Ag/AgCl-decorated polypyrrole nanotubes and their sensory properties. Synth. Met. 2010, 160, 1822–1825.

58. Hong, L.J.; Li, Y.; Yang, M.J. Fabrication and ammonia gas sensing of palladium/polypyrrole nanocomposite. Sens. Actuator B Chem. 2010, 145, 25–31.

59. Park, E.; Kwon, O.S.; Park, S.J.; Lee, J.S.; You, S.; Jang, J. One-pot synthesis of silver nanoparticles decorated poly(3,4-ethylenedioxythiophene) nanotubes for chemical sensor application. J. Mater. Chem. 2012, 22, 1521–1526.

60. Gong, J.; Li, Y.H.; Hu, Z.S.; Zhou, Z.Z.; Deng, Y.L. Ultrasensitive NH3 gas sensor from polyaniline nanograin enchased TiO2 fibers. J. Phys. Chem. C 2010, 114, 9970–9974.

61. Kwon, O.S.; Park, S.J.; Yoon, H.; Jang, J. Highly sensitive and selective chemiresistive sensors based on multidimensional polypyrrole nanotubes. Chem. Commun. 2012, 48, 10526–10528.

62. Phillips, M.; Gleeson, K.; Hughes, J.M.B.; Greenberg, J.; Cataneo, R.N.; Baker, L.; McVay, W.P. Volatile organic compounds in breath as markers of lung cancer: A cross-sectional study. Lancet 1999, 353, 1930–1933.

63. Athawale, A.A.; Bhagwat, S.V.; Katre, P.P. Nanocomposite of Pd-polyaniline as a selective methanol sensor. Sens. Actuator B Chem. 2006, 114, 263–267.

64. Choudhury, A. Polyaniline/silver nanocomposites: Dielectric properties and ethanol vapour sensitivity. Sens. Actuator B Chem. 2009, 138, 318–325.

65. Xian, Y.Z.; Hu, Y.; Liu, F.; Xian, Y.; Wang, H.T.; Jin, L.T. Glucose biosensor based on Au nanoparticles-conductive polyaniline nanocomposite. Biosens. Bioelectron. 2006, 21, 1996–2000.

66. Xu, L.H.; Zhu, Y.H.; Yang, X.L.; Li, C.Z. Amperometric biosensor based on carbon nanotubes coated with polyaniline/dendrimer-encapsulated Pt nanoparticles for glucose detection. Mater. Sci. Eng. C Bio S 2009, 29, 1306–1310.

67. Liu, Z.; Wang, J.; Xie, D.H.; Chen, G. Polyaniline-coated Fe3O4 nanoparticle-carbon-nanotube composite and its application in electrochemical biosensing. Small 2008, 4, 462–466.

68. Guo, S.J.; Dong, S.J.; Wang, E.K. Polyaniline/Pt hybrid nanofibers: High-efficiency nanoelectrocatalysts for electrochemical devices. Small 2009, 5, 1869–1876.

69. Bian, X.J.; Lu, X.F.; Jin, E.; Kong, L.R.; Zhang, W.J.; Wang, C. Fabrication of Pt/polypyrrole hybrid hollow microspheres and

their application in electrochemical biosensing towards hydrogen peroxide. Talanta 2010, 81, 813–818.

70. Li, F.; Yang, L.M.; Zhao, C.; Du, Z.F. Electroactive gold nanoparticles/polyaniline/polydopamine hybrid composite in neutral solution as high-performance sensing platform. Anal. Methods 2011, 3, 1601–1606.

71. Qian, T.Y.C.; Zhou, X.; Wu, S.; Shen, J. Au nanoparticles decorated polypyrrole/reduced graphene oxide hybrid sheets for ultrasensitive dopamine detection. Sens. Actuator B Chem. 2014, 193, 759–763.

72. Tian, S.J.; Liu, J.Y.; Zhu, T.; Knoll, W. Polyaniline/gold nanoparticle multilayer films: Assembly, properties, and biological applications. Chem. Mater. 2004, 16, 4103–4108.

8
CHAPTER
CURRENT TRENDS IN SENSORS BASED ON CONDUCTING POLYMER NANOMATERIALS

Hyeonseok Yoon [1,2]

[1] *Alan G. MacDiarmid Energy Research Institute, Department of Polymer and Fiber System Engineering, Chonnam National University, 77 Yongbong-ro, Buk-gu, Gwangju 500-757, Korea; Tel.: +82-62-530-1778; Fax: +82-62-530-1779*

[2] *Department of Polymer Engineering, Graduate School, Chonnam National University, 77 Yongbong-ro, Buk-gu, Gwangju 500-757, Korea*

ABSTRACT

Conducting polymers represent an important class of functional organic materials for next-generation electronic and optical devices. Advances in nanotechnology allow for the fabrication of various conducting polymer nanomaterials through synthesis methods such as solid-phase template synthesis, molecular template synthesis, and template-free synthesis. Nanostructured conducting polymers featuring high surface area, small dimensions, and unique physical properties have been widely used to build various sensor devices. Many remarkable examples have been reported over the past decade. The enhanced sensitivity of conducting polymer nanomaterials toward various chemical/biological species and external stimuli has made them ideal candidates for incorporation into the design of sensors. However, the selectivity and stability still leave room for improvement.

Keywords: conducting polymers; nanomaterials; chemical sensors; biosensors; polypyrrole; polyaniline; poly(3,4-ethylenedioxythiophene)

1. INTRODUCTION

Conducting polymers have found a wide range of applications in the various fields of electronics, optics, energy devices, medicine, actuators, and composites as a viable alternative to metallic or inorganic semiconductor counterparts. In particular, there has recently been huge demand for developing flexible or wearable electronics, displays, and other devices, in which conducting polymers can ultimately be used as true flexible organic conductors or semiconductors. The most notable property of conducting polymers is their inherent electrical conductivity, which is closely connected to the charge transfer rate and electrochemical redox efficiency. Most conducting polymers act as semiconductors in terms of conductivity although several studies on metallic conducting polymers have been reported. Unlike their inorganic counterparts, a weak intermolecular overlap of electronic orbitals combined with a greater degree of disorder in conducting polymers result in narrow electronic bands and a low mobility of charge carriers. Consequently, conducting polymers have had serious limitations in specific applications such as transistors and memories. For example, the performance of field-effect transistors (FETs) based on conducting polymers cannot rival that of FETs based on single-crystalline inorganic semiconductors, such as Si and Ge, which have charge carrier mobilities that are about three orders of magnitude higher [1]. The mobilities of FETs based on solution-processed conducting polymers [e.g., poly(3-hexylthiophene)] are generally found to be in the range of 0.1 cm^2 V^{-1} s^{-1}, similar to those of amorphous silicon FETs [2]. A mobility of 10.5 cm^2 V^{-1} s^{-1} has been the best reported thus far [3]. Conducting polymer FETs are not practically suitable for use in applications requiring high switching speeds.

Another example of conducting polymer applications involves electrochromic devices for smart windows and flexible displays. Conducting polymers are one of the most attractive electrochromic materials because of advantages such as high coloration efficiency, rapid switching ability, and diverse colors [4]. The switching time is one of the important parameters in display technology. The electrochromism of conducting polymers is based on reversible redox reactions accompanying ion exchange, and is completely different from the operating mechanism of FETs. The switching time is mainly affected by redox reaction efficiency, which depends on the ion diffusion rate and conductivity. Notably, it was recently reported that poly(3,4-ethylenedioxythiophene) (PEDOT) nanotubes with wall thicknesses of 10–20 nm exhibited fast switching speeds of less than 10 ms, though the color contrast was low [5]. However, electrochromic displays based on conducting polymers have not yet been considered for commercialization. Stability, rapid response times, and efficient color changes are still critical parameters that need improvement.

Conducting polymers have also been used for sensor applications as a signal transducer. There are several important parameters in sensor technology, such as sensitivity, selectivity, and response time [6,7]. In most cases, a response time on the order of seconds is enough for human recognition. Thus, it is easier to meet the requirements for response time than for other parameters. The sensing mechanisms of conducting polymers can involve redox reactions, ion adsorption and desorption, volume and weight changes, chain conformational changes, or charge transfer and screening. Compared to the inorganic counterparts, conducting polymers have an advantage in achieving high sensitivity and selectivity by virtue of their chemical and structural diversity. Conducting polymers also share the strengths of polymers over other materials, including low-temperature synthesis and processing, large-area manufacture, flexibility, and cost effectiveness. As a result, conducting polymers can be competitive in sensor applications and sensors are therefore considered to be one of the most practical applications of conducting polymers.

Conducting polymers have mostly been synthesized in the form of powder and film using chemical and electrochemical polymerization methods, respectively. It is important to precisely control the structure and morphology of conducting polymers during the synthesis process, since most of them are insoluble in common solvents and not thermoplastic. There have been remarkable advances in the synthesis of nanomaterials over the past several decades. Particularly, a variety of metallic and inorganic semiconductor nanostructures have been fabricated, which have exhibited unique electrical, optical, and chemical properties. Nanostructured materials feature high surface-to-volume ratio and small dimensions, which are very beneficial for sensor applications. The high surface area facilitates enhanced interaction between the materials and analytes, which leads to high sensitivity, and the small dimensions enable fast adsorption/desorption kinetics for analytes in the material, which allows a rapid response time. Accordingly, recent years have witnessed a shift in sensor technology towards more sensitive recognition elements, highly sophisticated architectures, and miniaturization due to the emergence of nanomaterials and nanotechnology. Conducting polymer nanomaterials also have strong potential for yielding enhanced sensor performance compared to their bulk counterparts [8]. However, polymers are unstable at the nanometer scale due to the nature of covalent bonds, which makes their nanostructures unstable as well. Because of this, progress in the synthesis of polymer nanomaterials has been relatively slow so far, and only limited research has been conducted on what fascinating properties polymer nanomaterials possess, in contrast to inorganic nanomaterials [9].

There are several conducting polymers that have been extensively investigated for practical applications, which are summarized in Table 1. All of them possess high conductivity and good environmental stability, and their polymerization reactions are not only straightforward but also proceed with high yield. Polypyrrole (PPy) features low oxidation potential and good biocompatibility, and the advantages of polyaniline (PANI) include that the monomer is very inexpensive. Polythiophene (PTh) has many useful derivatives, one of which is PEDOT developed by Bayer AG (Leverkusen, Germany). PEDOT features optical transparency and can be soluble in water with polystyrene sulfonate (PSS). Owing to these desirable characteristics, PPy, PANI, PTh and their derivatives have become leading materials in various applications fields. This review addresses research efforts to fabricate and manipulate nanostructures mainly consisting of the representative conducting polymers for sensor applications, and highlights remarkable recent examples with a focus on materials functionalization, transduction mechanisms, and device characteristics in sensor applications.

Table 1. Representative conducting polymers.

Name	Structure
Polypyrrole (PPy)	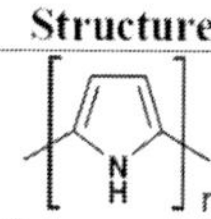
Polyaniline (PANI)	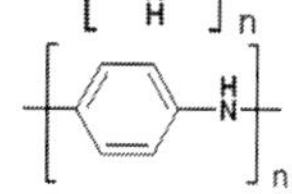
Polythiophene (PTh)	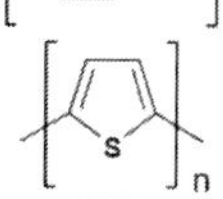
Poly(3,4-ethylenedioxythiophene) (PEDOT)	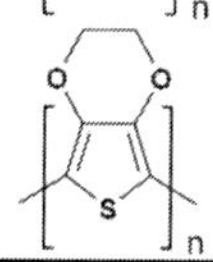

2. FABRICATION OF CONDUCTING POLYMER NANOMATERIALS

Precise control over the size and morphology of conducting polymers at the nanoscale is essential to improving the performance of related sensors. As mentioned, polymers are highly unstable at the nanometer scale, which is one of the greatest obstacles in building polymer nanoarchitectures. Nevertheless, numerous efforts have been made to fabricate polymer nanomaterials with well-defined size and morphology, and various types of conducting polymer nanostructures have been fabricated in a controlled fashion.

Conducting polymers have traditionally been synthesized via chemical or electrochemical oxidation polymerization. The overall process includes the oxidation of monomers, followed by the coupling of the charged monomers to produce the polymer chains. Chemical polymerization is advantageous for large-scale production at low cost, while electrochemical polymerization offers the possibility of in-situ formation, such as on an electrode for a sensor device. Conducting polymers can be obtained in the presence of various oxidizing agents. Oxidation polymerizations with acid or peroxide initiators result in insulating materials that require a post-doping process. Metal salts that can act as both oxidizing and doping agents are used to conduct the oxidation polymerization, which directly yields polymers in a conductive state. For example, ferric salts including $FeCl_3$ and $Fe(ClO_4)_3$ are widely employed. Owing to their electrical conductivity, conducting polymers can grow electrochemically without oxidizing agents on an electrode. It is possible to tailor the polymer thickness by controlling the applied potential, polymerization time, and electrolyte.

Conducting polymer nanostructures have been fabricated with the aid of templates during the polymerization process. The synthetic routes are traditionally classified into three classes depending on the kind of template: hard template synthesis, soft template synthesis, and template-free synthesis. However, it is sometimes ambiguous to distinguish the hard and soft templates. For example, track-etched polymer membranes and polymer nanofibers are both soft, but they are categorized as a hard template and a soft template, respectively. Therefore, the templates are classified here as solid-phase and molecular templates as a new standard. Table 2 summarizes the templates frequently used for fabricating nanostructures and the possible resulting products.

Table 2. Template synthesis methods.

Products	Templates			
	Solid-Phase	Ref.	Molecular	Ref.
Nanospheres	Nanospheres	[10]	Spherical micelles	[13]
Nanocapsules	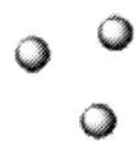	[11,12]	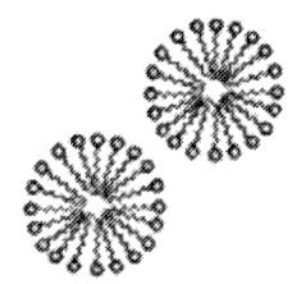	[14]
Nanorods/fibers	Porous matrices	[15–17]	Rod-like micelles	[20,21]
Nanotubes	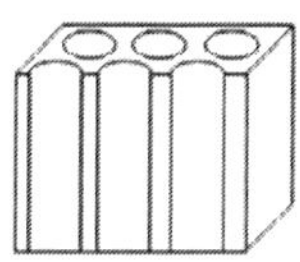	[18,19]		[22,23]
Nanofibers	Nanorods	[24,25]	Liquid crystals	[27–29]
Nanotubes		[26]	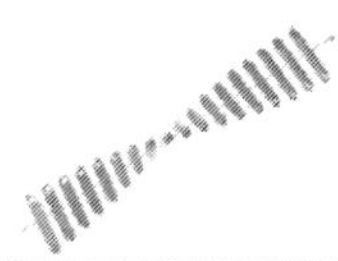	

2.1. Solid-Phase Template Synthesis

Organic/inorganic nanoparticles, anodic alumina membranes, track-etched polymer membranes, and mesoporous silica are examples of solid-phase templates. Anodic alumina and track-etched polymer membranes with parallel cylindrical nanopores have often been chosen for the production of nanotubes or nanorods [30]. The porous alumina membranes are formed by anodizing high-purity alumina disks in an acidic electrolyte, and the track-etched polymer membranes are produced by irradiating the membrane with high-energy heavy ions, followed by chemical etching. The use of such solid-phase templates is advantageous for simply tailoring the dimensions of nanomaterials, and it has thus been extensively studied for synthetic routes to obtain 1D nanostructures of organic and inorganic materials. Conventional electrodeposition and electrophoresis techniques can be simply applied to the templates to yield nanostructures. Because the geometry and morphology of the resulting nanomaterials are endowed by the template itself, precise control of the diameter and length is possible with the solid-phase templates. However, it is very hard to completely remove the template without degradation or irreversible aggregation of the resulting nanomaterials. Scale-up for commercial applications is also highly difficult due to the complicated synthetic process and high cost. Therefore, the solid-phase template synthesis might be suitable for fabricating high-value products in the form of composites without the removal of template. A good example was suggested by Sommerdijk et al. [31]. A glucose sensor was developed using a track-etched membrane, in which PEDOT was coated onto the inner walls of the cylindrical pores. Since the PEDOT nanotube-containing membrane was easy to handle, it could be directly integrated into a sensing cell as a working electrode to detect glucose. This approach fully utilized the structural advantages of the PEDOT nanotubes in a sensor application, and thus offers a model for a successful way of using solid-phase templates.

Electrospinning is a simple and scalable technique that can create continuous and aligned polymer nanofibers and microfibers under a high electric field. Electrospun nanofibers are made from various polymer melts and solutions, which can serve as templates to direct the formation of conducting polymer nanofibers and nanotubes [32]. In a recent study [33], ultrathin poly(methyl methacrylate) (PMMA) nanofibers with an average diameter of 60 nm were obtained through electrospinning for use as a template (Figure 1a), and then immersed in ferric chloride solution for the adsorption of ferric ions on the PMMA nanofibers (Figure 1b). Subsequently, PEDOT was coated onto the PMMA nanofibers via vapor deposition polymerization at controlled temperatures and pressures to yield core-shell structured nanofibers (Figure 1c,e,g). Unique surface substructures such as nanonodules and nanorods could be grown on the nanofibers by adjusting major kinetic factors. The removal of the PMMA core can be done by selective solvent etching, which results in nanotubular structures (Figure 1d,f,h). The multidimensional nanofibers and nanotubes with surface substructures will have significant advantages when used as transducers for sensors.

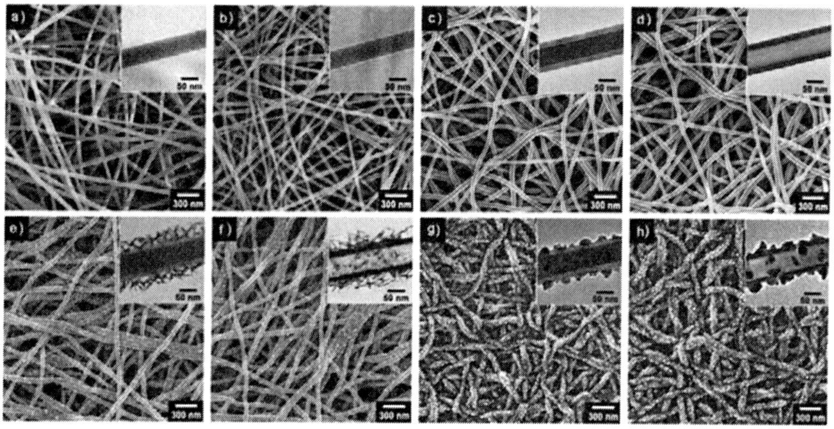

Figure 1. Multidimensional poly(3,4-ethylenedioxythiophene) (PEDOT) nanostructures with unique surface substructures. (**a–h**) The poly(methyl methacrylate) (PMMA) nanofibers function as a template and substrate for the growth of PEDOT under different synthetic conditions (temperature and pressure). The morphologies of the resulting nanomaterials were characterized by scanning electron microscopy (SEM) and transmission electron microscopy (TEM) (right top inset images): PMMA nanofibers (**a**) before and (**b**) after ferric ion adsorption; PMMA/PEDOT nanofibers with smooth layer surface (**c**) before and (**d**) after core etching; PMMA/PEDOT nanofibers with nanorod surface (**e**) before and (**f**) after core etching; PMMA/ PEDOT nanofibers with nanonodule surface (**g**) before and (**h**) after core etching. With permission from [33]; Copyright 2012, American Chemical Society.

2.2. Molecular Template Synthesis

Molecular template synthesis has strengths compared to solid-phase template synthesis. The most widely used molecular templates include surfactants, liquid crystals, and polyelectrolytes. Because molecular template synthesis is comparatively straightforward and cost-effective, it is suitable for large-scale production. However, the molecular templates are not robust, static entities, leading to considerable difficulty in obtaining the desired nanostructures. Surfactant templating is a typical example of molecular template synthesis. Surfactants have the ability to self-assemble into ordered molecular structures called micelles. The micelles exist in various forms, and their dimensions generally range from a few to a few tens of nanometers, making them very suitable as templates. Microemulsions consisting of thermodynamically stable micelles with size less than 50 nm have been used in polymerization techniques to prepare polymer nanoparticles. However, micelles are highly sensitive to the surrounding environment, which makes it very difficult to achieve stable microemulsion systems for polymerization, preventing their widespread utilization for industrial products. One of the strategies for making stable microemulsion systems is to use co-surfactants like long-chain alcohols. A notable example is described in Figure 2a [13]. A cationic surfactant, dodecyltrimethylammonium bromide (DTAB) was used to obtain spherical micelles reinforced with decanol in aqueous solution. The decanol can play a role in retarding the diffusion of monomers through the aqueous phase. Consequently, monodisperse PPy nanoparticles (60 nm in diameter) were obtained on a gram scale, which is a very large quantity in laboratory-scale synthesis. Figure 2b shows a TEM image of the PPy nanoparticles and the inset shows a photo of a Petri dish containing 12 g of the nanoparticles.

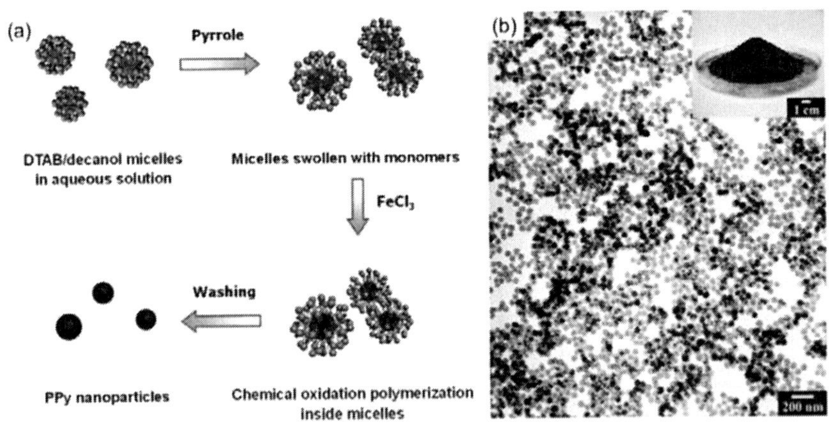

Figure 2. (**a**) Schematic illustration of the preparation of PPy nanoparticles in a cationic surfactant (DTAB)/co-surfactant (decanol) emulsion system; (**b**) TEM image of monodisperse PPy nanoparticles prepared through micelle templating (inset: photograph showing a Petri dish containing 12 g of PPy nanoparticles obtained in a single polymerization reaction). With permission from [13]; Copyright 2005, Wiley-VCH Verlag GmbH & Co. KGaA.

This example indicates potential for use of the microemulsion technique for the efficient mass production of nanoparticles. However, the technique requires high surfactant concentration, which is problematic in terms of cost and environmental pollution. Another study was reported for

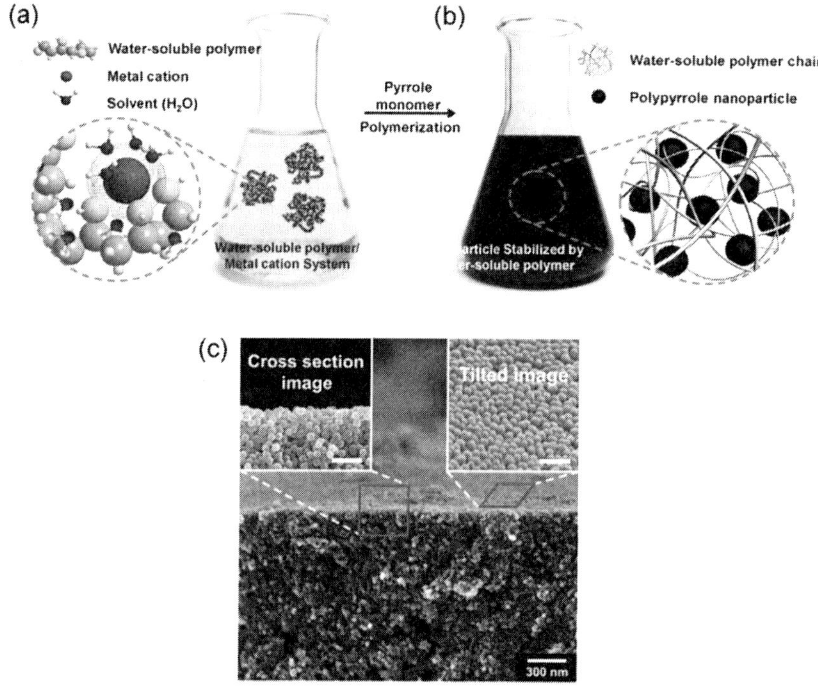

Figure 3. Schematic illustration of the formation of PPy nanoparticles in an aqueous dispersion of water-soluble polymer (PVA)/metal cation (ferric ion) complexes, and SEM images of the resulting nanoparticles. (**a**) Hydroxyl groups of PVA chains coordinate with ferric ions by an ion-dipole interaction in an aqueous medium; (**b**) The ferric ions act as the oxidizing agent for chemical oxidation of pyrrole monomers. After polymerization, the resulting PPy nanoparticles are stabilized by PVA chains; (**c**) Tilted and cross-section SEM images of the PPy nanoparticles stacked on a substrate (scale bar: 100 nm). With permission from [34]; Copyright 2010, Wiley-VCH Verlag GmbH & Co. KGaA.

producing PPy nanoparticles using a kind of dispersion polymerization, where water-soluble polymers provide steric stability. It is known that dispersion polymerization produces polymer microbeads. Jang et al. were able to produce PPy nanoparticles just by using a polymeric stabilizer, polyvinyl alcohol (PVA), in aqueous solution [34]. Figure 3a,b represent the synthesis process in which ferric ions were anchored on PVA chains via coordination bonding. The polymerization of pyrrole monomers proceeds only with ferric ions on PVA chains, which allows enough steric stability to yield nanometer-sized particles, as illustrated in Figure 3b. Figure 3c exhibits the SEM images of the resulting PPy nanoparticles, in which the size of the nanoparticles was surprisingly as small as ca. 25 nm in diameter. Considering that previous studies were limited to the synthesis of micrometer-sized particles with steric stabilizers, this approach is very innovative.

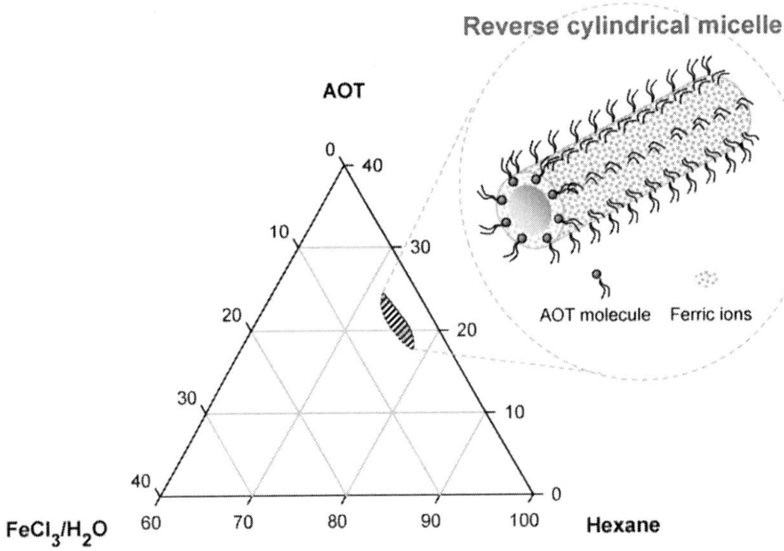

Figure 4. Partial ternary phase diagram for the hexane/AOT/aqueous $FeCl_3$ solution system determined at 15 °C. The molar concentration of the aqueous $FeCl_3$ solution was 16.2 M. The marked area corresponds to the AOT reverse cylindrical micelle phase region. With permission from [22]; Copyright 2005, American Chemical Society.

Aş shown in these examples, many examples of the formation of spherical nanoparticles can be found. However, it is hard to fabricate one-dimensional (1D) nanostructures such as nanofibers and nanotubes using molecular templates, and only a few studies have been reported [35]. 1D nanostructures have advantages over spherical or film-type counterparts in sensor applications, such as anisotropic physical properties and easy integration with two-terminal microelectrodes. Jang and Yoon reported the formation of PPy nanotubes in a reverse micelle system, which was the first notable example of fabricating conducting polymer nanotubes with a molecular template [22,23]. Sodium bis(2-ethylhexyl) sulfosuccinate (AOT) molecules were assembled with ferric ions to reverse cylindrical micelles in an apolar solvent, where ferric ions were adsorbed on anionic AOT headgroups. Figure 4 shows a partial ternary phase diagram of a hexane/AOT/aqueous ferric chloride solution. AOT has hydrophobic twin tail groups, and its sulfonate headgroup forms hydrogen bonds with water, making the AOT reverse micelles highly stable in an oil phase. Pyrrole is hydrophobic. Therefore, the oxidation polymerization of pyrrole monomers proceeds by ferric ions onto the reverse cylindrical micelles, leading to 1D nanostructures with hollow interiors.

Several different types of PPy nanowires have been obtained using surfactants, although their formation mechanisms have not been clarified. Hence, PPy has been a focus of many studies regarding the preparation of nanostructures. PPy can have a cross-linked molecular structure during polymerization, which would enhance its dimensional stability at the nanometer scale and make it possible to form relatively stable nanostructures. PANI and PEDOT nanostructures have been also prepared with the assistance of surfactants. PANI nanowire networks were synthesized using a cationic surfactant, hexadecyltrimethylammonium bromide, and an organic diacid, oxalic acid, in aqueous solution. The PANI nanowires had diameters of 35–100 nm depending on the polymerization conditions [36]. Unique clip-like nanostructures of PPy, PANI, and PEDOT could be even fabricated in bulk quantities using an anionic oxidant/cationic surfactant complex as a template [37]. The judicious combination of main parameters, such as surfactants, oxidizing/doping agents, pH, temperature, and other structure-directing agents, provides infinite possibilities of fabricating nanostructures with desirable morphology.

2.3. Template-Free Synthesis

Template-free synthesis, which is based on the self-assembly of molecular building blocks, is naturally very straightforward. However, extensive efforts are required to design and synthesize building blocks that are able to assemble into nanostructures under certain conditions. Fortunately, several species (precursors) can spontaneously form nanostructures without such artificial efforts. Typically, PANI has intrinsically 1D morphology at the nanometer scale when borne in aqueous solution, and thus, it has been intensively investigated to fabricate 1D PANI nanostructures in the absence of templates [38,39]. The formation of PANI nanofibrils has been studied extensively. The growth control of PANI nanofibrils in the presence of a steric stabilizer was reported, in addition to their morphology-dependent electrochemical properties [40]. Wan et al. have actively reported the formation of PANI nanotubes under various

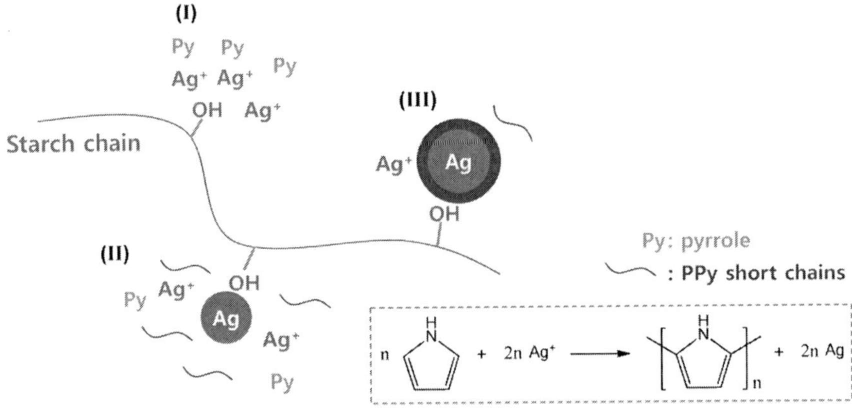

Figure 5. Schematic illustration of the formation mechanism of Ag-PPy nanoparticles: the reaction process could be divided into three stages (I, II, and III). Right bottom: the scheme describing the chemical reaction between pyrrole and silver cation. With permission from [47]; Copyright 2012, American Chemical Society.

synthesis conditions without templates [41,42,43]. It was found that subtle changes in polymerization parameters often result in drastic differences in the morphology of the resulting PANI product.

Tseng et al. developed a site-specific electrochemical method for the fabrication of individually addressable PPy, PANI, and PEDOT nanowire frameworks on microelectrode junctions [44,45]. Numerous intercrossing polymer nanowire (50 to 200 nm in diameter) networks were formed between the two microelectrodes (2-μm gap). Shi et al. simply synthesized Au-PEDOT core-shell nanocables by the one-step interfacial polymerization of EDOT (in organic phase) and $HAuCl_4$ (in aqueous phase) [46]. The nanocables have lengths of a few micrometers, outer diameters of around 50 nm, and central cores around 30 nm. The dissolution of the Au core using iodine solution gave rise to hollow PEDOT nanotubes. Similar research was also conducted to synthesize Ag-PPy core-shell nanoparticles with an average core diameter of 36 nm and a shell thickness of 13 nm via a simple one-pot synthesis, as represented in Figure 5 [47]. A silver precursor, silver nitrate, acted as an oxidizing agent for the chemical polymerization of pyrrole. The pyrrole monomers were oxidized by silver cations, yielding PPy and silver atoms concurrently (see the reaction scheme in Figure 5). In the initial stages I and II, the hydroxyl groups of soluble starch provide nucleation sites where silver cations are reduced to atomic silver by the pyrrole monomers, whereas pyrrole monomers are oxidized to radical cations that lead to the generation of PPy short chains. While the silver nanoseeds are being formed in stage II, additional silver cations are adsorbed onto the nanoseed surface due to a sort of common ion adsorption effect. The silver nanoseed surfaces become the active sites for oxidation of the surrounding pyrrole monomers and PPy short chains, finally resulting in the formation of a PPy shell. These bottom-up approaches offer an efficient and simple route for the fabrication of a nanostructured metal/conducting polymer complex.

3. SENSOR APPLICATIONS

A variety of sensors have been formulated using conducting polymers in different transduction modes. The transduction modes can be divided into five main classes based upon the operating principle into conductometric, potentiometric, amperometric, colorimetric, and gravimetric modes [48,49]. The conductometric mode uses changes in electrical conductivity in response to an analyte interaction. The conductivity of a conducting polymer material bridging the gap between two adjacent electrodes is commonly measured as a function of analyte concentration, and it can be also monitored with a fixed potential in solution.

Potentiometric sensing mode is based on analyte-induced changes in the chemical potential of a system when no current is flowing. The change in the open-circuit potential of the system is monitored, which is mostly proportional to the logarithm of the concentration of analyte. The chemical and diffusion processes have to be at equilibrium conditions in the potentiometric mode for a thermodynamically accurate signal to yield.

Amperometric mode refers to either single-potential amperometry or variable-potential amperometry. The principle of amperometric sensing is to measure the current generated by the redox reaction of an analyte at a sensing (working) electrode, where the current is subject to Faraday's law and a dynamic reaction achieving steady-state conditions in the system.

Voltammetric mode is also a variant of amperometric mode. It monitors the change in current while varying the applied potential. Colorimetric sensors quantitate changes in optical absorption characteristics, which depend on the local electronic structure. The sensitivity of the bandgap of conducting polymers to analyte-induced changes provides a useful means to create this kind of sensor.

Lastly, gravimetric mode takes advantage of a weight change in a conducting polymer as a result of analyte-polymer interaction. Minute weight changes in the polymer can be normally monitored using a quartz crystal microbalance. Numerous conducting polymer sensors based on these transduction mechanisms have been devised in order to detect various chemical and biological species.

3.1. Chemical Sensors

There are many chemical species of concern that must be detected, including toxic gases, volatile organic compounds, alcohol, and humidity. Various conducting polymer nanostructures have been used to detect them. Chemically synthesized polymer nanoparticles suspended in solvents can be simply deposited on a prefabricated electrode by drop casting to construct a sensor substrate. Kwon et al. investigated the gas sensing properties of PPy nanoparticles with diameters of 20, 60, and 100 nm. They found that decreasing the nanoparticle size produced a noticeable increase in the sensitivity toward ammonia gas. It was anticipated that the increased sensitivity can be attributed to the higher surface-to-volume ratio of the 20-nm nanoparticles compared to the 60 and 100-nm nanoparticles [50]. The use of conducting polymer nanoparticles obtained via micelle templating and electrospinning in sensor applications has also been reported. Highly sensitive chemiresistive sensors based on PPy nanotubes were devised to discriminate volatile organic compounds and toxic gases [51]. PPy nanotubes were transferred onto a polydimethylsiloxane substrate using a dry-transfer method and then micro-patterned gold electrodes were deposited onto the nanotubes by thermal evaporation. Other conducting polymer nanostructures such as PPy nanoparticles, PEDOT nanorods, and PEDOT nanotubes were also integrated into electrode substrates to induce discriminative responses toward individual analytes. The lowest detectable concentration of the sensors was ca. 0.01 ppm for ammonia, selectively.

Electrochemical polymerization allows controllable polymer deposition on an electrode. Recently, PPy nanowires grown electrochemically on a microelectrode substrate were used as conductometric transducer to detect hydrogen gas at room temperature (see Figure 6a) [52]. The diameters of the nanowires were in the range of 40–90 nm and their lengths were on the order of several tens of micrometers. It was found that the PPy nanowires were stacked at high density (Figure 6b), and a number of the nanowires bridged the gap between the electrodes (Figure 6c). The sensor was exposed to different concentrations of hydrogen gas at room temperature in a closed chamber, during which the resistance of the PPy nanowire electrode was monitored. Upon exposure to hydrogen gas, a drop in resistance was observed. The resistance of the PPy nanowires is expected to decrease when exposed to hydrogen, which is a reducing gas. The sensors had a linear detection range of ca. 600–2500 ppm for hydrogen gas, and their sensitivities were found to be dependent of the amount of the deposited nanowires. On the other hand, the sensors showed a loss in sensitivity due to carbon monoxide gas interference, which has remained a critical issue to be solved. The detection of hydrogen at low concentrations is very important, because it is explosive in the presence of oxygen. Interestingly, it has been found that PANI has the ability to interact with hydrogen. Kaner and Weiller demonstrated that hydrogen interacts directly with doped PANI nanofibers to induce a small change in the conductivity of the nanofibers [53]. The direct mass uptake of hydrogen by PANI nanofibers was also observed using a quartz crystal microbalance (ca. 3% relative to the nanofiber mass). A plausible mechanism of the hydrogen/PANI interaction involves hydrogen interacting with doped PANI at the charged amine nitrogen sites, followed by the dissociation of hydrogen with the formation of new N–H bonds at the amine nitrogen of the PANI chain [53,54]. Subsequently, charge transfer between adjacent amine nitrogens returns the PANI back to its original doped, emeraldine-salt form with a release of hydrogen.

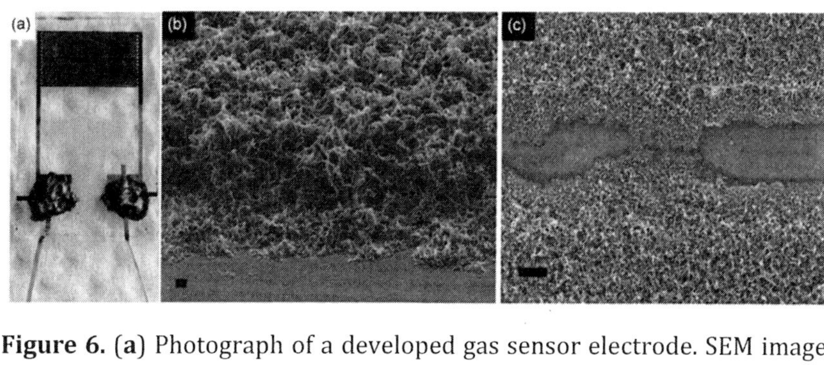

Figure 6. (**a**) Photograph of a developed gas sensor electrode. SEM images of PPy nanowires deposited on the electrode substrate: (**b**) 65° tilted view (scale bar = 1 μm) and (**c**) top view showing PPy nanowires bridging the insulating gap between the gold electrodes (scale bar = 10 μm). With permission from [52]; Copyright 2012, American Chemical Society.

An integrated electrochemical and electrical nanosensor for the sensitive and selective detection of nitroaromatic explosives vapors (TNT) was demonstrated recently [55]. The integrated sensor chip consists of two main parts with a thin layer coating of BMIM-PF_6: (i) PEDOT nanojunctions

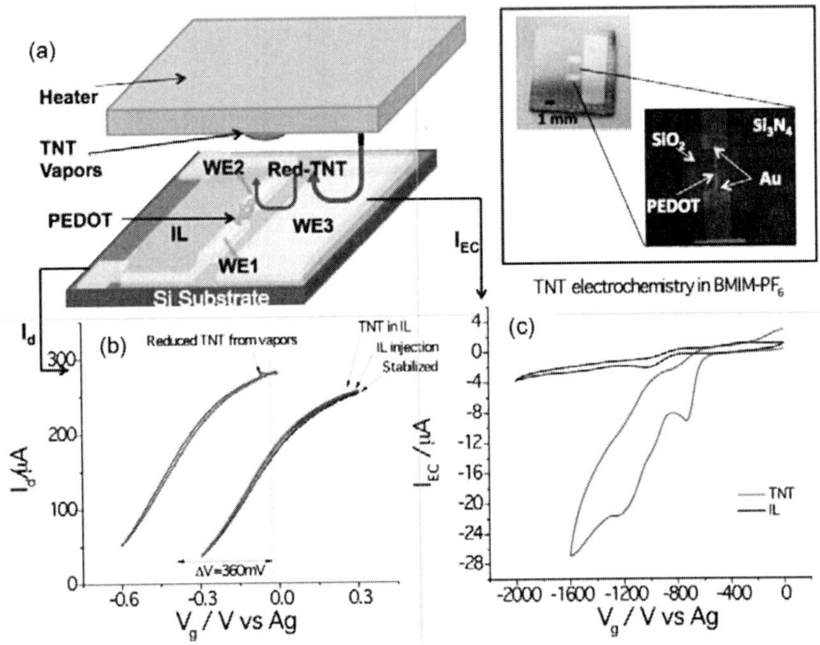

Figure 7. (**a**) A hybrid nanosensor consisting of a conducting polymer nanojunction (polymer bridged between WE1 and WE2) and a working electrode (WE3) on a Si chip. The chip is covered with a thin layer of ionic liquid (BMIM-PF_6) serving as an electrolyte and preconcentration medium. Upon heating TNT particulates (to 60 °C), TNT vapor is generated which is collected by the ionic liquid layer. The analyte is reduced and detected electrochemically on WE3, and the reduction products are detected by the polymer nanojunctions. Inset: Optical micrograph of the sensor chip used and an SEM image of the PEDOT nanojunction; (**b**) Current (I_d) via the PEDOT nanojunction plotted as a function of WE1 potential before and after exposure to TNT; (**c**) Cyclic voltammograms of a blank BMIM-PF_6 solution (black) and 4 ppm TNT in BMIM-PF_6 (red). The large reduction current in the latter case is due to the reduction of TNT. Note that an Ag wire quasi-reference electrode and a Pt counter electrode are used. With permission from [55]; Copyright 2010, American Chemical Society.

for conductometric detection and (ii) electrodes for electrochemical detection (Figure 7). The sensor utilizes both the electrochemical signatures of the reduction of the analytes, and the specific interactions of the reduction products with the polymer nanojunction to achieve excellent selectivity. BMIM-PF_6 is used not only as a stable electrolyte, but also as a preconcentration medium for nitroaromatic compounds. Because TNT has a low vapor pressure (4.8×10^{-6} Torr) at room temperature, TNT residues are expected to last for a long time under ambient conditions, which makes it possible for them to be collected and detected in vapor form upon heating (Figure 7a). Upon exposure to the TNT vapor with no potential applied to WE3, the characteristic I_d–V_g curve of the PEDOT nanojunction remains unchanged (cyan curve in Figure 7b). In contrast, when a negative enough potential (−1.8 to −1.2 V) is applied to the electrode, the cyclic voltammogram (red curve in Figure 7c) shows multiple peaks, due to the reduction of TNT. The total amount of the potential shift reached ca. −0.36 V within 2 min (red curve in Figure 7b), which provided sensitive detection of TNT. The sensor was capable of detecting ppt-level TNT within 1–2 min in the presence of various common interferents found in ambient air.

Conducting polymers can be complexed with many species to create composite materials that are sensitive to certain chemical agents. PANI nanofibers treated with $CuCl_2$ showed the ability to detect hydrogen sulfide with change in resistance by 4 orders of magnitude [56]. The $CuCl_2$ in the PANI nanofibers reacts with hydrogen sulfide to yield HCl, finally resulting in the further doping of the nanofibers. This reaction scheme has been extended to the detection of other specific analytes such as phosgene [57] and arsine [58]. Phosgene, a colorless toxic gas with a permissible exposure level of 0.1 ppm, was used as a chemical weapon during World War I. Arsine is also a highly toxic gas used in the semiconductor industry, with a permissible exposure level of 50 ppb. Water-soluble amines such as ethylenediamine, phenylenediamine, and metanilic acid were found to make PANI nanofibers sensitive to phosgene. In addition, water-soluble metal salts were be readily incorporated into PANI nanofibers, such as $CuCl_2$, $CuBr_2$, CuF_2, $Cu(O_2CCH_3)_2$, $Cu(NO_3)_2$, $EuCl_2$, $NiCl_2$, $FeCl_3$, and $CoCl_2$. The salts were screened to find the best candidates for detecting arsine. The $CuBr_2$/PANI nanofiber composite gave the best response with a change in resistance of more than an order of magnitude upon exposure to arsine. One of the notorious toxic gases is sarin, which is an organophosphorous compound. Flexible nerve agent sensors based on hydroxylated PEDOT nanotubes (HPNTs) with unique surface substructures were recently demonstrated [59]. Axially aligned HPNTs were produced via electrospinning process under a magnetic field, and then integrated into a conductometric substrate. Dimethyl methylphosphonate (DMMP) is commonly used as a simulant for sarin. Upon exposure to DMMO, the HPNTs showed a sharp increase in resistance. The reaction time was as fast as ca. 1 s, and the recovery time was as good as 3–25 s. The lowest detectable concentration was as low as 10 ppt. Figure 8 provides information on the selectivity of the HPNTs toward similar organophosphorus compounds such as trimethyl phosphate (TMP), methyl dichlorophosphate (MDCP), and trichlorophosphate (TCP), used as nerve gas simulants. The hydroxyl side chain of PEDOT was found to interact with the nerve agent simulants through hydrogen bonding. The hydrogen-bond strength was calculated to increase in the order TCP $<$ MDCP $<$ sarin $\approx$ DMMP $<$ TMP (Figure 8b), which would lead to different response intensities with the same tendency (Figure 8a). The response of HPNTs against 15 volatile organic compounds, chosen as potential interferents, was examined and compared to responses from other sensing materials using principal components analysis. The first three principal component scores are plotted in Figure 8c. Unique signatures of the analytes could be observed, denoted by their segregation into separate regions of the plot, allowing identification of individual analytes by their responses. Particularly, DMMP exhibited clearly differentiable components, which validated the selective recognition ability of HPNTs. The response of the HPNTs was found to be durable even under mechanical deformation, which provided the possibility of fabricating a wearable nerve agent sensor.

Optical modulation of the major properties of conducting polymers has offered attractive strategies for detecting various target species. Debarnot et al. recently developed a "multilayer integrated optical sensor (MIOS)" using PANI as a sensitive material and demonstrated the detection of ammonia gas [60]. Figure 9 schematically illustrates the structural difference between the conventional waveguide sensor and the MIOS. In the MIOS structure, a thin passive film of PMMA is deposited on a small section of the planar waveguide known as the interaction length (L_I), and then the sensitive material PANI is coated onto the passive film. The planar waveguide consists of a glycidyl ether of bisphenol A, which is a promising candidate for simply fabricating optical waveguides due to the excellent optical transparency and easy processing [61]. PANI is opaque and has a high refractive index and absorption coefficient. PANI interacts with ammonia gas in such a way that the adsorption and desorption of ammonia give rise to deprotonation and protonation of PANI, respectively, leading to variation of the optical properties of PANI. The PMMA film is transparent with a refractive index lower than that of the waveguide to satisfy the total internal reflection condition for light propagation, and importantly, it reduces the optical losses induced by just evanescent wave/sensitive material coupling. A significant change was observed in the guided light output power after the sensor was exposed to ammonia gas, due to the variation in PANI absorption characteristics. Compared to the conventional transmission technique, the planar waveguide provides a much longer light propagation distance due to the multiple reflections of the guided light, leading to higher sensitivity. The sensor exhibited a logarithmic linear optical response within a range of ammonia concentration of ca. 90–4600 ppm.

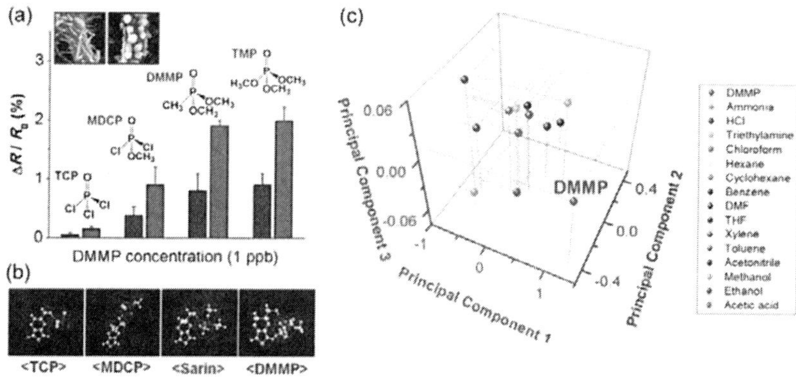

Figure 8. Sensing performance of chemical nerve agent sensor based on hydroxylated PEDOT nanotubes (HPNTs). (**a**) Histogram showing the response of HPNTs toward similar organophosphorus compounds at 1 ppb (TCP, MDCP, DMMP, TMP); (**b**) 3D graphics showing the formation of hydrogen bonds between nerve agent stimulant molecules and HEDOT; (**c**) Principal components analysis plot using response intensity inputs from four different conducting polymer nanomaterials (two different HPNTs, pristine PEDOT nanotubes, and PPy nanotubes) to the 16 analytes (including DMMP): each analyte concentration was fixed at around 4 ppm. With permission from [59]; Copyright 2012, American Chemical Society.

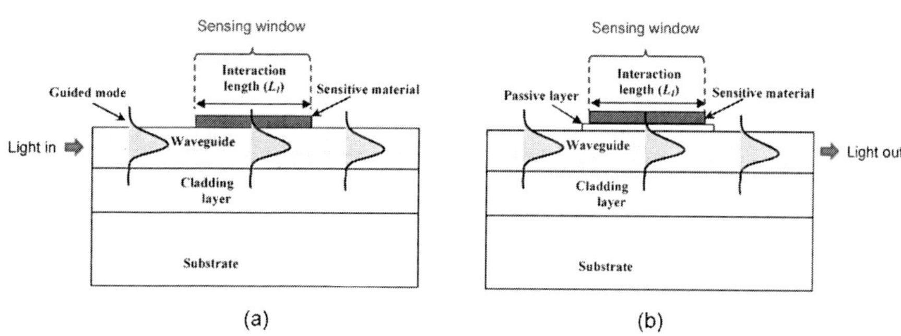

Figure 9. Schematic structures of the two types of waveguide sensors: (**a**) conventional waveguide sensor; (**b**) multilayer integrated waveguide sensor. With permission from [60]; Copyright 2008, American Chemical Society.

3.2. Biosensors

Biosensing is commonly conducted in a solution phase to maintain the biological activity of target species. Accordingly, it is critical to immobilize transducer materials on sensing platforms for continuously obtaining reliable signals. Compared to metals and ceramics, conducting polymers are more compatible with biological systems. Inorganic nanomaterials have been readily integrated into biosensor platforms using lithography and focused ion beam techniques. However, the integration of conducting polymer nanomaterials into biosensors has been limited due to their incompatibility with the traditional microfabrication processes. Chemical, thermal, and kinetic damage can possibly degrade conducting polymers during the microfabrication process. An alternative strategy was developed to circumvent this issue by Jang and Yoon [62]. Covalent linkages between PPy nanotubes and a microelectrode substrate were made to achieve reliable electrical contact in solution. PPy nanotubes with carboxyl groups were prepared and the carboxyl groups were chemically coupled with the surface amino group of the electrode substrate (Figure 10a). A liquid, ion-gated field-effect transistor (FET) sensor could be successfully fabricated using this method (Figure 10b,c). The structure of the FET sensor is similar to that of the normal metal-oxide-semiconductor FET, except for the gate, which incorporates the means of transduction from a chemical event to a voltage [63]. In the liquid ion-gated FET configuration, two metal electrodes, called the source and drain are deposited on a substrate. The gate electrode is electrically isolated from the two electrodes and immersed in an electrolyte, as illustrated in Figure 10b. The gate potential affects the density of charge carriers in the semiconductor channel. The channel is normally modified with molecular or polymeric receptors for selectively recognizing the analyte of interest. The recognition of the analyte by the receptor can alter the gate potential being applied on the channel, which modulates the source-drain current. With this operating principle, many different types of FET sensors have been devised in order to detect glucose [64], odorants [65], tastants [66], hormones [67], and proteins [68]. Several critical parameters determining the FET sensor response have been also investigated. For example, Mulchandani et al. fabricated single PPy nanowire-based FET sensors for real-time pH monitoring and examined how the diameter of the nanowire affects the sensor performance [69]. PPy nanowires with three different diameters (ca. 60, 80, and 200 nm) were anchored on a pair of gold electrodes with different gap lengths (1 and 4 μm). The FET sensors had higher sensitivity with lower diameter and higher length.

Electrochemical deposition can also be an excellent approach to fabricating biosensor platforms based on conducting polymer nanomaterials.

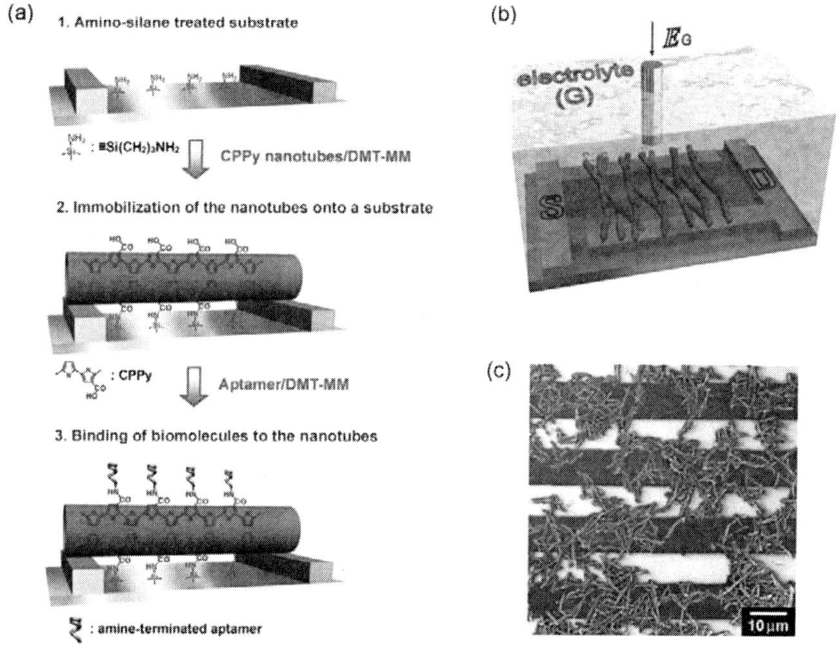

Figure 10. Schematic illustration of (**a**) reaction steps for the fabrication of a sensor platform based on carboxylated PPy nanotubes; and (**b**) a liquid ion-gated FET sensor; (**c**) SEM image of carboxylated PPy nanotubes that are deposited on the interdigitated microelectrode substrate. With permission from [62]; Copyright 2008, Wiley-VCH Verlag GmbH & Co. KGaA.

Mulchandani and Myung reported a single-step electrodeposition technique for the formation of PPy and PANI nanowires to bridge the gap between two electrodes on a silicon wafer [70,71]. The deposition and growth of the nanowires are based on electrochemical oxidation polymerization. The nanowires grew from the cathode to the anode through an electrolyte channel when an electric potential was applied, and the dimensions of the nanowires were determined by the width of the channel and the distance between electrodes. Since this process is biomolecule-friendly, receptors can be introduced into the polymer during the electropolymerization. For example, avidin protein could be entrapped in the PPy nanowire, which allowed the detection of a biotin-conjugated DNA [72]. A similar electrochemical approach was developed by the same research group [73]. They used dielectrophoretic alignment to achieve a single PPy nanowire connection between a pair of microelectrodes, and then performed maskless selective gold electrodeposition on the microelectrodes to anchor the nanowire ends (Figure 11a,b). The nanowire surface with the secondary amine groups was functionalized with an antibody using carbodiimide linker chemistry to detect a cancer antigen. The detection of cancer antigen CA 125 in human blood plasma was demonstrated, showing the utility of the sensor for real samples (Figure 11c).

Another important issue in biosensors is related to how receptors are coupled with transducers. Versatile receptors including enzymes, antibodies, nucleic acids, and cells have been introduced into conducting polymers through adsorption, entrapment, or covalent binding. A widely known example involves enzyme-based glucose detection for diagnosing and managing diabetes. Traditionally, amperometric glucose detection has been achieved with glucose oxidase (GOx) enzyme electrodes. The GOx has been immobilized on electrodes by chemical cross-linking, eletrodeposition or electrostatic interactions [74]. Recent studies have proposed new alternatives to the conventional enzyme electrode [75,76]. Conducting polymer nanomaterials have mainly been employed as the conductive matrix with redox properties. A FET-based sensor was developed by using GOx enzyme-attached PPy nanotubes as the conductive channel [64]. The PPy nanotubes were prepared by the chemical polymerization of pyrrole-3-carboxylic acid inside the cylindrical pores of an alumina membrane. GOx contains terminal amino groups on its lysine residues. Thus, GOx can be covalently attached to the nanotubes via condensation reaction between the carboxylic group and the amino group. Cai et al. took advantage of the covalent attachment of GOx to PANI nanofibers for glucose detection [77]. They used a well-known EDC/NHS chemistry to attach the enzyme to the nanofibers. GOx kept its natural structure and electroactivity, and consequently, the nanofiber electrode showed a wide linear range of 0.01 to 1 mM with a detection limit of 0.5 μM for glucose. In contrast, a conducting polymer nanojunction sensor described by Tao et al. just used a simple adsorption of GOx on polyaniline/poly(acrylic acid) nanojunction to build up an enzyme electrode [78]. The nanojunction sensor showed specific and rapid responses toward glucose with the detection limit of micromolar levels. A new type of amperometric glucose sensor developed by Wu and Yin used PANI-wrapped boron nitride nanotubes decorated with Pt nanoparticles as the electrode material [79]. GOx was just mixed with the Pt/PANI/boron nitride nanotubes, which resulted in a cotton-like enzyme nano-hybrid. Surprisingly, the enzyme nano-hybrid electrode had good acid stability and heat resistance.

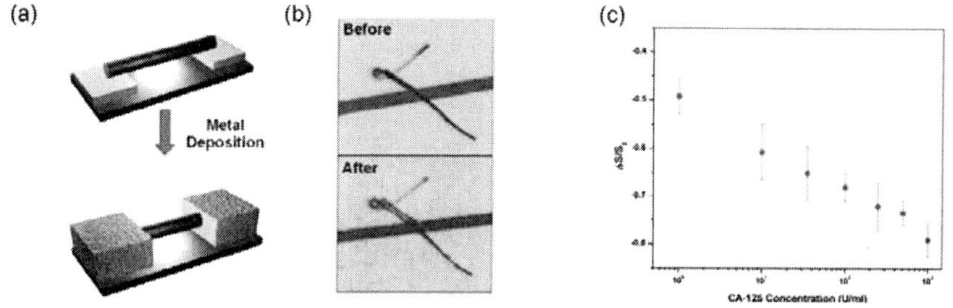

Figure 11. (**a**) Schematics of maskless electrodeposition of gold along with (**b**) optical images of before and after selective gold electrodeposition on gold microelectrodes separated by a 3 μm gap connected with a single PPy nanowire; (**c**) A calibration curve in terms of normalized conductance change of a single PPy nanowire biosensor in spiked human blood plasma suggesting the utility of this sensor for real sample measurements. With permission from [73]; Copyright 2009, American Chemical Society.

Molecularly imprinted sensors that require no receptors have also been fabricated using conducting polymers. For example, a simple method for the photoelectrochemical sensing of microcystin-LR (MC-LR) was achieved by using PPy/titania nanotubes with MC-LR recognition sites [80]. MC-LR is a chemical species that may cause structural and functional disturbances of the liver, and is a potential cancer threat [81]. Owing to its ubiquity and high toxicity, MC-LR is an element of concern in water-purity control and environmental monitoring. Optical transduction can facilitate the detection of electrochemically inert species such as MC-LR. Titania nanotubes have high surface area and photoelectric activity. The recognition sites of MC-LR were imprinted in the interior PPy layer of titania nanotubes. The photocatalytic oxidation of MC-LR under a certain bias potential caused changes in the photocurrent, and PPy played a role in further enhancing the photoelectric conversion efficiency. MC-LR could be determined with a linear range of 0.5 to 100 μg L^{-1}, and the detection limit was 0.1 μg L^{-1}. In another example, Kan et al. reported an electrochemical sensor based on molecularly imprinted PPy for epinephrine recognition [82]. Epinephrine, also known as adrenaline, is an important neurotransmitter in the mammalian central nervous system, and is present in human serum at nanomolar levels. Pyrrole monomers were electrochemically polymerized with epinephrine molecules on the surface of an electrode coated with silica nanoparticles and carbon nanotubes. Epinephrine can be imprinted inside PPy via hydrogen bonding during the polymerization. After the removal of silica nanoparticles and epinephrine, a molecularly imprinted PPy/carbon nanotube film with high porosity could be produced on the electrode. The amperometric response of the PPy/carbon nanotube electrode was recorded by the addition of epinephrine in phosphate buffer solution. Good linearity between the epinephrine concentration and the current response was observed in the range of 3×10^{-7} to 1×10^{-3} M and the detection limit was found to be 3×10^{-8} M.

3.3. Others

Mechanical sensors based on conducting polymers that transduce dynamic responses into measurable signals are receiving growing attention [83]. As a notable example, a strain sensor was developed by using electrospun PEDOT/PSS/PVA nanofibers [84]. PEDOT/PSS/PVA nanofibers were collected on a polyimide (Kapton) film with two patterned contact electrodes and then annealed at 80 °C for 15 min. The nanofiber film was packaged with a thin layer of polydimethylsiloxane to prevent contamination, and its current change was monitored under external strains. Figure 12a shows the time-resolved current response of the sensor to repeated exposure to different tensile strains. The current decreased with tensile strain, and increased with compressive strain. The current was fully recovered to the original state when the strain was relieved, indicating that the sensor had high reproducibility and good stability in response. Additionally, the behavior of the sensor in response to human actions was examined, as shown in Figure 12b. The sensor was affixed to an index finger, and the current response during bending and unbending actions were recorded. The sensor gave an immediate response to the actions of the finger and the current response was highly reversible and reproducible.

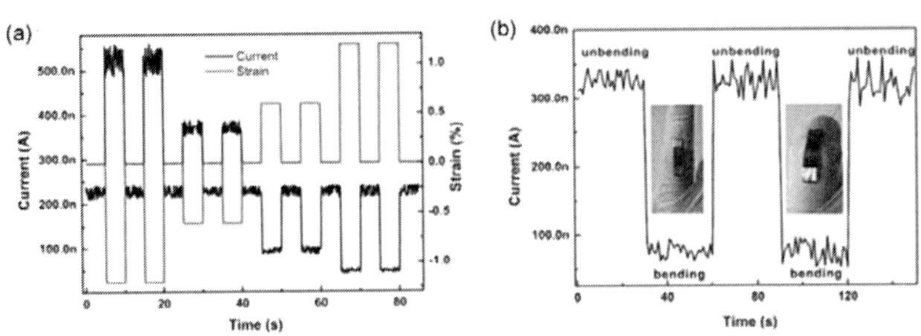

Figure 12. (**a**) The change in resistance of the strain sensor under different external strains; (**b**) The current response of the strain sensor with cyclical bending and unbending actions of the finger. With permission from [84]; Copyright 2011, Royal Chemical Society.

4. SUMMARY AND OUTLOOK

A number of studies have demonstrated that conducting polymer nanomaterials are promising candidates for building state-of-the-art sensors, due to their unique advantages over other materials. The most important parameters that determine the sensor performance include response/recovery time, sensitivity, selectivity, and stability. A great deal of effort has been directed toward enhancing these parameters over the past decades. The response and recovery times and the sensitivity have experienced impressive improvements with great advances in nanotechnology. However, selectivity is still a challenge. Detecting target species in a complex environment remains a difficult task, and is hindering the widespread application of conducting polymer-based sensors. Most sensors have been demonstrated only under controlled laboratory conditions, while actual samples comprise a range of interferent chemicals and substances. A conducting polymer by itself lacks the specificity or selectivity toward target species, and thus, it is crucial to judiciously functionalize the polymer with appropriate receptors. Another concern is that conducting polymers may degrade over time, even in dry, oxygen-free environments. New efforts should be aimed at improving the stability of the sensor response.

Sensors play an ever-increasing role in environmental monitoring, medical diagnosis, industrial safety control, security, and so forth. Conducting polymer nanomaterials are believed to have much unexplored potential for sensor applications. Thus, future research into conducting-polymer nanomaterials-based sensors will offer great potential for the construction of next-generation sensor devices. In particular, it is anticipated that wearable or flexible high-performance sensors will be developed using conducting polymer nanomaterials in the near future.

ACKNOWLEDGMENTS

This work was supported by the Basic Science Research Program (2012R1A1A1042024) and the Global Research Laboratory (K20903002024l2E010004010) through the National Research Foundation of Korea (NRF), funded by the Ministry of Science, ICT and Future Planning.

CONFLICTS OF INTEREST

The authors declare no conflict of interest.

REFERENCES

1. Taur, Y.; Ning, T.H. Fundamental of Modern VLSI Devices; Cambridge University Press: New York, NY, USA, 1998; p. 11.

2. Dimitrakopoulos, C.D.; Malenfant, P.R.L. Organic thin film transistors for large area electronics. Adv. Mater. 2002, 14, 99–117.

3. Li, J.; Zhao, Y.; Tan, H.S.; Guo, Y.; Di, C.A.; Yu, G.; Liu, Y.; Lin, M.; Lim, S.H.; Zhou, Y.; et al. A stable solution-processed polymer semiconductor with record high-mobility for printed transistors. Sci. Rep. 2012, 2.

4. Cho, S.I.; Lee, S.B. Fast electrochemistry of conductive polymer nanotubes: Synthesis, mechanism, and application. Acc. Chem. Res. 2008, 41, 699–707.

5. Cho, S.I.; Kwon, W.J.; Choi, S.J.; Kim, P.; Park, S.A.; Kim, J.; Son, S.J.; Xiao, R.; Kim, S.H.; Lee, S.B. Nanotube-based ultrafast electrochromic display. Adv. Mater. 2005, 17, 171–175.

6. Rahman, M.A.; Kumar, P.; Park, D.S.; Shim, Y.B. Electrochemical sensors based on organic conjugated polymers. Sensors 2008, 8, 118–141.

7. Bai, H.; Shi, G.Q. Gas sensors based on conducting polymers. Sensors 2007, 7, 267–307.

8. Yoon, H.; Jang, J. Conducting-polymer nanomaterials for high-performance sensor applications: Issues and challenges. Adv. Funct. Mater. 2009, 19, 1567–1576.

9. Yoon, H.; Choi, M.J.; Lee, K.A.; Jang, J.S. Versatile strategies for fabricating polymer nanomaterials with controlled size and morphology. Macromol. Res. 2008, 16, 85–102.

10. Shin, S.; Yoon, H.; Jang, J. Polymer-encapsulated iron oxide nanoparticles as highly efficient Fenton catalysts. Catal. Commun. 2008, 10, 178–182.

11. Cheng, D.M.; Zhou, X.D.; Xia, H.B.; Chan, H.S.O. Novel method for the preparation of polymeric hollow nanospheres containing silver cores with different sizes. Chem. Mater. 2005, 17, 3578–3581.

12. Hao, L.Y.; Zhu, C.L.; Jiang, W.Q.; Chen, C.N.; Hu, Y.; Chen, Z.Y. Sandwich Fe2O3@SiO2@PPy ellipsoidal spheres and four types of hollow capsules by hematite olivary particles. J. Mater. Chem. 2004, 14, 2929–2934.

13. Jang, J.; Yoon, H. Multigram-scale fabrication of monodisperse conducting polymer and magnetic carbon nanoparticles. Small 2005, 1, 1195–1199.

14. Bajpai, V.; He, P.G.; Dai, L.M. Conducting-polymer microcontainers: Controlled syntheses and potential applications. Adv. Funct. Mater. 2004, 14, 145–151.

15. Liu, R.; Lee, S.B. MnO2/Poly(3,4-ethylenedioxythiophene) coaxial nanowires by one-step coelectrodeposition for electrochemical energy storage. J. Am. Chem. Soc. 2008, 130, 2942–2943.

16. Callegari, V.; Gence, L.; Melinte, S.; Demoustier-Champagne, S. Electrochemically template-grown multi-segmented gold-conducting polymer nanowires with tunable electronic behavior. Chem. Mater. 2009, 21, 4241–4247.

17. Li, G.T.; Bhosale, S.; Wang, T.Y.; Zhang, Y.; Zhu, H.S.; Fuhrhop, K.H. Gram-scale synthesis of submicrometer-long polythiophene wires in mesoporous silica matrices. Angew. Chem. Int. Ed. 2003, 42, 3818–3821.

18. Xiao, R.; Cho, S.I.; Liu, R.; Lee, S.B. Controlled electrochemical synthesis of conductive polymer nanotube structures. J. Am. Chem. Soc. 2007, 129, 4483–4489.

19. Han, M.G.; Foulger, S.H. 1-dimensional structures of poly(3,4-ethylenedioxythiophene) (PEDOT): A chemical route to tubes, rods, thimbles, and belts. Chem. Commun. 2005, 24, 3092–3094.

20. Yoon, H.; Chang, M.; Jang, J. Formation of 1D poly(3,4-ethylenedioxythiophene) nanomaterials in reverse microemulsions and their application to chemical sensors. Adv. Funct. Mater. 2007, 17, 431–436.

21. Zhang, X.T.; Zhang, J.; Song, W.H.; Liu, Z.F. Controllable synthesis of conducting polypyrrole nanostructures. J. Phys. Chem. B 2006, 110, 1158–1165.

22. Jang, J.; Yoon, H. Formation mechanism of conducting polypyrrole nanotubes in reverse micelle systems. Langmuir 2005, 21, 11484–11489.

23. Jang, J.; Yoon, H. Facile fabrication of polypyrrole nanotubes using reverse microemulsion polymerization. Chem. Commun. 2003, 720–721.

24. Zhang, X.Y.; Goux, W.J.; Manohar, S.K. Synthesis of polyaniline nanofibers by "nanofiber seeding". J. Am. Chem. Soc. 2004, 126, 4502–4503.

25. Laforgue, A.; Robitaille, L. Deposition of ultrathin coatings of polypyrrole and poly(3,4-ethylenedioxythiophene) onto electrospun nanofibers using a vapor-phase polymerization method. Chem. Mater. 2010, 22, 2474–2480.

26. Zhang, X.Y.; Manohar, S.K. Narrow pore-diameter polypyrrole nanotubes. J. Am. Chem. Soc. 2005, 127, 14156–14157.

27. Huang, L.M.; Wang, Z.B.; Wang, H.T.; Cheng, X.L.; Mitra, A.; Yan, Y.S. Polyaniline nanowires by electropolymerization from liquid crystalline phases. J. Mater. Chem. 2002, 12, 388–391.

28. Goto, H. Electrochemical polymerization of pyrrole in cholesteric liquid crystals: Morphology and optical properties. J. Polym. Sci. Polym. Chem. 2007, 45, 1377–1387.

29. Hulvat, J.F.; Stupp, S.I. Anisotropic properties of conducting polymers prepared by liquid crystal templating. Adv. Mater. 2004, 16, 589–592.

30. Martin, J.; Maiz, J.; Sacristan, J.; Mijangos, C. Tailored polymer-based nanorods and nanotubes by "template synthesis": From preparation to applications. Polymer 2012, 53, 1149–1166.

31. Kros, A.; van Hovell, W.F.M.; Sommerdijk, N.A.J.M.; Nolte, R.J.M. Poly(3,4-ethylenedioxythiophene)-based glucose biosensors. Adv. Mater. 2001, 13, 1555–1557.

32. Greiner, A.; Wendorff, J.H. Electrospinning: A fascinating method for the preparation of ultrathin fibres. Angew. Chem. Int. Ed. 2007, 46, 5670–5703.

33. Kwon, O.S.; Park, S.J.; Park, H.W.; Kim, T.; Kang, M.; Jang, J.; Yoon, H. Kinetically controlled formation of multidimensional poly(3,4-ethylenedioxythiophene) nanostructures in vapor-deposition polymerization. Chem. Mater. 2012, 24, 4088–4092.

34. Hong, J.Y.; Yoon, H.; Jang, J. Kinetic study of the formation of polypyrrole nanoparticles in water-soluble polymer/metal cation systems: A light-scattering analysis. Small 2010, 6, 679–686.

35. Long, Y.Z.; Li, M.M.; Gu, C.Z.; Wan, M.X.; Duvail, J.L.; Liu, Z.W.; Fan, Z.Y. Recent advances in synthesis, physical properties and applications of conducting polymer nanotubes and nanofibers. Prog. Polym. Sci. 2011, 36, 1415–1442.

36. Zhong, W.B.; Deng, J.Y.; Yang, Y.S.; Yang, W.T. Synthesis of large-area three-dimensional polyaniline nanowire networks using a "soft template". Macromol. Rapid. Commun. 2005, 26, 395–400.

37. Liu, Z.; Zhang, X.Y.; Poyraz, S.; Surwade, S.P.; Manohar, S.K. Oxidative template for conducting polymer nanoclips. J. Am. Chem. Soc. 2010, 132, 13158–13159.

38. Tran, H.D.; D'Arcy, J.M.; Wang, Y.; Beltramo, P.J.; Strong, V.A.; Kaner, R.B. The oxidation of aniline to produce "polyaniline": A process

yielding many different nanoscale structures. J. Mater. Chem. 2011, 21, 3534–3550.

39. Ding, H.J.; Shen, J.Y.; Wan, M.X.; Chen, Z.J. Formation mechanism of polyaniline nanotubes by a simplified template-free method. Macromol. Chem. Phys. 2008, 209, 864–871.

40. Park, H.W.; Kim, T.; Huh, J.; Kang, M.; Lee, J.E.; Yoon, H. Anisotropic growth control of polyaniline nanostructures and their morphology-dependent electrochemical characteristics. ACS Nano 2012, 6, 7624–7633.

41. Du, J.M.; Zhang, J.L.; Han, B.X.; Liu, Z.M.; Wan, M.X. Polyaniline microtubes synthesized via supercritical CO2 and aqueous interfacial polymerization. Synth. Met. 2005, 155, 523–526.

42. Wan, M.X. Some issues related to polyaniline micro-/nanostructures. Macromol. Rapid Commun. 2009, 30, 963–975.

43. Zhang, L.J.; Wan, M.X. Chiral polyaniline nanotubes synthesized via a self-assembly process. Thin Solid Films 2005, 477, 24–31.

44. Wang, J.; Chan, S.; Carlson, R.R.; Luo, Y.; Ge, G.L.; Ries, R.S.; Heath, J.R.; Tseng, H.R. Electrochemically fabricated polyaniline nanoframework electrode junctions that function as resistive sensors. Nano Lett. 2004, 4, 1693–1697.

45. Alam, M.M.; Wang, J.; Guo, Y.Y.; Lee, S.P.; Tseng, H.R. Electrolyte-gated transistors based on conducting polymer nanowire junction arrays. J. Phys. Chem. B 2005, 109, 12777–12784.

46. Lu, G.W.; Li, C.; Shen, J.Y.; Chen, Z.J.; Shi, G.Q. Preparation of highly conductive gold-poly(3,4-ethylenedioxythiophene) nanocables and their conversion to poly(3,4-ethylenedioxythiophene) nanotubes. J. Phys. Chem. C 2007, 111, 5926–5931.

47. Chang, M.; Kim, T.; Park, H.W.; Kang, M.; Reichmanis, E.; Yoon, H. Imparting chemical stability in nanoparticulate silver via a conjugated polymer casing approach. ACS Appl. Mater. Interfaces 2012, 4, 4357–4365.

48. Bakker, E.; Qin, Y. Electrochemical sensors. Anal. Chem. 2006, 78, 3965–3983.

49. Stetter, J.R.; Li, J. Amperometric gas sensors—A review. Chem. Rev. 2008, 108, 352–366.

50. Kwon, O.S.; Hong, J.Y.; Park, S.J.; Jang, Y.; Jang, J. Resistive gas sensors based on precisely size-controlled polypyrrole nanoparticles: Effects of particle size and deposition method. J. Phys. Chem. C 2010, 114, 18874–18879.

51. Kwon, O.S.; Park, S.J.; Yoon, H.; Jang, J. Highly sensitive and selective chemiresistive sensors based on multidimensional polypyrrole nanotubes. Chem. Commun. 2012, 48, 10526–10528.

52. Al-Mashat, L.; Debiemme-Chouvy, C.; Borensztajn, S.; Wlodarski, W. Electropolymerized polypyrrole nanowires for hydrogen gas sensing. J. Phys. Chem. C 2012, 116, 13388–13394.

53. Virji, S.; Kaner, R.B.; Weiller, B.H. Hydrogen sensors based on conductivity changes in polyaniline nanofibers. J. Phys. Chem. B 2006, 110, 22266–22270.

54. Fowler, J.D.; Virji, S.; Kaner, R.B.; Weiller, B.H. Hydrogen detection by polyaniline nanofibers on gold and platinum electrodes. J. Phys. Chem. C 2009, 113, 6444–6449.

55. Aguilar, A.D.; Forzani, E.S.; Leright, M.; Tsow, F.; Cagan, A.; Iglesias, R.A.; Nagahara, L.A.; Amlani, I.; Tsui, R.; Tao, N.J. A hybrid nanosensor for TNT vapor detection. Nano Lett. 2010, 10, 380–384.

56. Virji, S.; Fowler, J.D.; Baker, C.O.; Huang, J.X.; Kaner, R.B.; Weiller, B.H. Polyaniline manofiber composites with metal salts: Chemical sensors for hydrogen sulfide. Small 2005, 1, 624–627.

57. Virji, S.; Kojima, R.; Fowler, J.D.; Villanueva, J.G.; Kaner, R.B.; Weiller, B.H. Polyaniline nanofiber composites with amines: Novel materials for phosgene detection. Nano Res. 2009, 2, 135–142.

58. Virji, S.; Kojima, R.; Fowler, J.D.; Kaner, R.B.; Weiller, B.H. Polyaniline nanofiber-metal salt composite materials for arsine detection. Chem. Mater. 2009, 21, 3056–3061.

59. Kwon, O.S.; Park, S.J.; Lee, J.S.; Park, E.; Kim, T.; Park, H.W.; You, S.A.; Yoon, H.; Jang, J. Multidimensional conducting polymer nanotubes for ultrasensitive chemical nerve agent sensing. Nano Lett. 2012, 12, 2797–2802.

60. Airoudj, A.; Debarnot, D.; Beche, B.; Poncin-Epaillard, F. Design and sensing properties of an integrated optical gas sensor based on a multilayer structure. Anal. Chem. 2008, 80, 9188–9194.

61. Pelletier, N.; Beche, B.; Gaviot, E.; Camberlein, L.; Grossard, N.; Polet, F.; Zyss, J. Single-mode rib optical waveguides on SOG/SU-8 polymer and integrated Mach-Zehnder for designing thermal sensors. IEEE Sens. J. 2006, 6, 565–570.

62. Yoon, H.; Kim, J.H.; Lee, N.; Kim, B.G.; Jang, J. A novel sensor platform based on aptamer-conjugated polypyrrole nanotubes for label-free electrochemical protein detection. ChemBioChem 2008, 9, 634–641.

63. Lee, C.S.; Kim, S.K.; Kim, M. Ion-sensitive field-effect transistor for biological sensing. Sensors 2009, 9, 7111–7131.

64. Yoon, H.; Ko, S.; Jang, J. Field-effect-transistor sensor based on enzyme-functionalized polypyrrole nanotubes for glucose detection. J. Phys. Chem. B 2008, 112, 9992–9997.

65. Yoon, H.; Lee, S.H.; Kwon, O.S.; Song, H.S.; Oh, E.H.; Park, T.H.; Jang, J. Polypyrrole nanotubes conjugated with human olfactory receptors: High-performance transducers for FET-type bioelectronic noses. Angew. Chem. Int. Ed. 2009, 48, 2755–2758.

66. Song, H.S.; Kwon, O.S.; Lee, S.H.; Park, S.J.; Kim, U.K.; Jang, J.; Park, T.H. Human taste receptor-functionalized field effect transistor as a human-like nanobioelectronic tongue. Nano Lett. 2013, 13, 172–178.

67. Kwon, O.S.; Ahn, S.R.; Park, S.J.; Song, H.S.; Lee, S.H.; Lee, J.S.; Hong, J.Y.; Lee, J.S.; You, S.A.; Yoon, H.; et al. Ultrasensitive and selective recognition of peptide hormone using close-packed arrays of hPTHR-conjugated polymer nanoparticles. ACS Nano 2012, 6, 5549–5558.

68. Yoon, H.; Jang, J. A field-effect-transistor sensor based on polypyrrole nanotubes coupled with heparin for thrombin detection. Mol. Cryst. Liq. Cryst. 2008, 491, 21–31.

69. Shirale, D.J.; Bangar, M.A.; Chen, W.; Myung, N.V.; Mulchandani, A. Effect of aspect ratio (length:diameter) on a single polypyrrole nanowire FET device. J. Phys. Chem. C 2010, 114, 13375–13380.

70. Ramanathan, K.; Bangar, M.A.; Yun, M.H.; Chen, W.F.; Mulchandani, A.; Myung, N.V. Individually addressable conducting polymer nanowires array. Nano Lett. 2004, 4, 1237–1239.

71. Yun, M.H.; Myung, N.V.; Vasquez, R.P.; Lee, C.S.; Menke, E.; Penner, R.M. Electrochemically grown wires for individually addressable sensor arrays. Nano Lett. 2004, 4, 419–422.

72. Ramanathan, K.; Bangar, M.A.; Yun, M.; Chen, W.; Myung, N.V.; Mulchandani, A. Bioaffinity sensing using biologically functionalized conducting-polymer nanowire. J. Am. Chem. Soc. 2005, 127, 496–497.

73. Bangar, M.A.; Shirale, D.J.; Chen, W.; Myung, N.V.; Mulchandani, A. Single conducting polymer nanowire chemiresistive label-free immunosensor for cancer biomarker. Anal. Chem. 2009, 81, 2168–2175.

74. Wang, J. Electrochemical glucose biosensors. Chem. Rev. 2008, 108, 814–825.

75. Yoon, H.; Ahn, J.H.; Barone, P.W.; Yum, K.; Sharma, R.; Boghossian, A.A.; Han, J.H.; Strano, M.S. Periplasmic binding proteins as optical modulators of single-walled carbon nanotube fluorescence: Amplifying a nanoscale actuator. Angew. Chem. Int. Ed. 2011, 50, 1828–1831.

76. Barone, P.W.; Yoon, H.; Ortiz-Garcia, R.; Zhang, J.Q.; Ahn, J.H.; Kim, J.H.; Strano, M.S. Modulation of single-walled carbon nanotube photoluminescence by hydrogel swelling. ACS Nano 2009, 3, 3869–3877.

77. Zhao, M.; Wu, X.M.; Cai, C.X. Polyaniline nanofibers: Synthesis, characterization, and application to direct electron transfer of glucose oxidase. J. Phys. Chem. C 2009, 113, 4987–4996.

78. Forzani, E.S.; Zhang, H.Q.; Nagahara, L.A.; Amlani, I.; Tsui, R.; Tao, N.J. A conducting polymer nanojunction sensor for glucose detection. Nano Lett. 2004, 4, 1785–1788.

79. Wu, J.M.; Yin, L.W. Platinum nanoparticle modified polyaniline-functionalized Boron Nitride nanotubes for amperometric glucose enzyme biosensor. ACS Appl. Mater. Interfaces 2011, 3, 4354–4362.

80. Chen, K.; Liu, M.C.; Zhao, G.H.; Shi, H.J.; Fan, L.F.; Zhao, S.C. Fabrication of a novel and simple microcystin-LR photoelectrochemical sensor with high sensitivity and selectivity. Environ. Sci. Technol. 2012, 46, 11955–11961.

81. Xia, Y.T.; Deng, J.L.; Jiang, L. Simple and highly sensitive detection of hepatotoxin microcystin-LR via colorimetric variation based on polydiacetylene vesicles. Sensors Actuat. B 2010, 145, 713–719.

82. Zhou, H.; Xu, G.L.; Zhu, A.H.; Zhao, Z.; Ren, C.C.; Nie, L.L.; Kan, X.W. A multiporous electrochemical sensor for epinephrine recognition and detection based on molecularly imprinted polypyrrole. RSC Adv. 2012, 2, 7803–7808.

83. Alici, G.; Spinks, G.M.; Madden, J.D.; Wu, Y.Z.; Wallace, G.G. Response characterization of electroactive polymers as mechanical sensors. IEEE-ASME T. Mech. 2008, 13, 187–196.

84. Liu, N.S.; Fang, G.J.; Wan, J.W.; Zhou, H.; Long, H.; Zhao, X.Z. Electrospun PEDOT:PSS-PVA nanofiber based ultrahigh-strain sensors with controllable electrical conductivity. J. Mater. Chem. 2011, 21, 18962–18966.

9 CHAPTER

TRANSPARENT CONDUCTIVE FILMS FABRICATED FROM POLYTHIOPHENE NANOFIBERS COMPOSITED WITH CONVENTIONAL POLYMERS

Borjigin Aronggaowa [1], Yuriko Toda [1], Noriyuki Ito [1], Kazuhiro Shikinaka [2] and Takeshi Shimomura [2,*]

[1] *Graduate School of Bio-Applications and Systems Engineering, Tokyo University of Agriculture and Technology, Tokyo 184-8588, Japan*

[2] *Department of Organic and Polymer Materials Chemistry, Graduate School of Engineering, Tokyo University of Agriculture and Technology, Koganei, Tokyo 184-8588, Japan*

ABSTRACT

Transparent, conductive films were prepared by compositing poly(3-hexylthiophene) (P3HT) nanofibers with poly(methyl methacrylate) (PMMA). The transparency, conductivity, atmospheric stability, and mechanical strength of the resulting nanofiber composite films when doped with $AuCl_3$ were evaluated and compared with those of P3HT nanofiber mats. The conductivity of the nanofiber composite films was 4.1 $S \cdot cm^{-1}$, which is about seven times less than that which was previously reported for a nanofiber mat with the same optical transmittance (~80%) reported by

Aronggaowa et al. The time dependence of the transmittance, however, showed that the doping state of the nanofiber composite films in air was more stable than that of the nanofiber mats. The fracture stress of the nanofiber composite film was determined to be 12.3 MPa at 3.8% strain.

Keywords: conducting polymer; nanofiber; transparent conductive film; poly(3-hexylthiophene); composite film

1. INTRODUCTION

Materials that are both optically transparent and electrically conductive have attracted significant interest due to their numerous applications, such as in solar cells, flat-panel displays, touch-sensitive screens, light-emitting diodes, and photovoltaic cells. Indium tin oxide (ITO) has become the standard material for such applications. Commercially-available ITO films, with thicknesses in the range of 150–200 nm, have greater than 90% transmittance at visible wavelengths and a sheet resistance of <8 Ω/□. The high cost of indium, however, has made ITO increasingly expensive and, in addition, it is not suitable for flexible devices due to its brittle nature. A metal nanowire network [1,2] was recently thoroughly investigated as a substitute for ITO in certain promising applications.

Films made from conducting polymers such as polyaniline (PANI) composites [3,4,5,6,7], poly(3,4-ethylenedioxythiophene)/poly(styrenesulfonate) (PEDOT/PSS) [8,9,10], and poly(3-alkylthiophene) (P3AT) [11] have also been studied. Such polymers have the advantages of being lightweight, highly flexible, and chemically stable. Unfortunately, conducting polymers typically exhibit absorption in the visible wavelength range and, thus, they are considered unsuitable for applications that require high-conductivity throughout relatively thick films.

We have recently reported a transparent, conductive, polymeric nanofiber mat made from poly(3-hexylthiophene) (P3HT), which exhibits superior transparency and conductivity originating from the presence of highly effective conducting pathways, a large void fraction generated by a bulky nanofiber network, and a high dopant concentration due to a large surface area [12]. Compared to materials currently used in actual touch panels, however, the conductivity of this nanofiber mat, at a transmittance of over 88%, is an order of magnitude less than that required. Furthermore, the stability of the doping state was not sufficient and the mat was quite brittle mechanically. Therefore, some of the problems associated with this material must be mitigated prior to its application.

On the other hand, conductive nanofiller/polymer composite materials have attracted great academic and industrial attention, as they offer high conductivity in a conventional flexible polymeric film [13,14,15,16,17], and various new materials have been developed to satisfy market needs, exhibiting for example, specific thermal [18,19,20,21,22,23,24,25], mechanical [26,27], and electrochemical properties [28,29,30,31,32,33,34,35,36] in addition to conductivity. There are several different types of conductive nanofillers, including carbon nanotubes [18,19,20,21,22,23,24,28,29,30,31,32,33,34,35,36], carbon fibers [26], metal particles [25], and conductive polymers [37,38]. Our aim in the present work was to solve the problems associated with the P3HT nanofiber mat as a transparent, conductive film by using the nanocomposite technique.

Qiu et al. have reported that P3HT can be recrystallized into nanofiber form, in combination with conventional non-crystalline polymers such as poly(methyl methacrylate) (PMMA) and polystyrene (PS) to form nanofiber composite films [38,39,40]. These nanofiber composite films exhibit semiconducting properties and can be used to fabricate a flexible field-effect transistor (FET) simply by adding electrodes. We have also observed that an effective nanofiber network was percolated over these films by use of a Kelvin-probe microscope (KFM) [41]. If these composite films can be doped, they represent potential flexible films with considerable mechanical strength and both high transparency (due to the transparent matrix) and high conductivity (due to the well-developed P3HT nanofiber network). Furthermore, since the nanofibers are embedded in the matrix, these composite films are expected to show high environmental stability.

In this study, we prepared doped P3HT nanofiber composite films embedded in PMMA, and investigated the environmental stability, transparency, and conductivity of these films in comparison to a standard P3HT nanofiber mat.

2. EXPERIMENTAL SECTION

2.1. Composite Film Preparation

Regioregular poly(3-hexylthiophene) (P3HT) for use as a conducting polymeric nanofiber was purchased from the Sigma-Aldrich Co., Inc. (St Louis, MO, USA), and was used without further purification. We chose poly(methyl methacrylate) (PMMA) to form the matrix of the high transparency composite film, as reported previously [38,39,40]. PMMA (M_w 120,000) was also purchased from Sigma-Aldrich and was used without further purification. P3HT nanofiber composite films were prepared via the following simple procedure [38]. P3HT and PMMA powders were added into a solvent mixture composed of chloroform (as a good solvent for P3HT) and anisole (as a poor solvent for P3HT), such that the concentration of the polymer blend was 1 wt %, with stirring at 60 °C [38]. P3HT/PMMA solutions were prepared with P3HT/PMMA weight ratios of 0.1:99.9, 0.3:99.7, 1:99, 2:98, 3:97, 5:95, 10:90 and 20:80, in a fixed 5:5 (v/v) solvent mixture of chloroform/anisole. In addition, 5:95 weight ratio P3HT/PMMA solutions were prepared in chloroform/anisole mixed solvents made in ratios of 10:0, 8:2, 6:4, 5:5, 4:6, 3:7, 2:8 and 0:10 (v/v). Each solution was cooled gradually to 20 °C at a rate of 25 °C/h without stirring. The formation of nanofibers was indicated by a change in the solution color from a transparent yellow to a turbid reddish brown, which became evident after the solution sat undisturbed, in air, for one week.

Glass substrates 1 mm thick and 12 × 26 mm^2 in size were washed in 10 min intervals in an alkaline cleaner, chloroform, acetone, methanol, and deionized water under sonication and dried with nitrogen. Each polymer solution was subsequently spin-coated onto the surface of a glass substrate at either 3000 or 5000 rpm for 60 s. The thickness of each film was controlled by the number of droplets of the polymer solution applied during the spin-coating process.

$AuCl_3$ and $FeCl_3$ were used as the dopants and anhydrous acetonitrile was chosen as the doping solvent. $AuCl_3$ (99.0%) was purchased from Sigma-Aldrich, $FeCl_3$ was purchased from Wako Pure Chemical Industries (Osaka, Japan) and anhydrous acetonitrile was obtained from Kanto Chemical Industries (Tokyo, Japan). All of these chemicals were used without further purification. $FeCl_3$ (81.0 mg) or $AuCl_3$ (151.7 mg) was dissolved in acetonitrile (5 mL) to form an acceptor solution (0.1 M) [42]. The film-coated glass substrates were then immersed in the acceptor solution for 60 s to stimulate doping, after which they were washed three times with acetonitrile for 30 s, each time, and dried in a vacuum chamber.

Samples for the measurement of tensile strength were prepared by the following method. An aqueous poly(vinyl alcohol) (PVA) solution (1 wt %) was spin-coated uniformly onto a glass substrate and dried completely under vacuum for 15 min. Composite films were formed on these PVA-coated substrates using the same approach described above for the preparation of composite films for optical and electrical measurements. The substrate was then immersed in hot water and the composite film was peeled from the substrate as the PVA dissolved.

2.2. Measurement Methods

The film thickness measurements and surface observations of composite films were performed by scanning probe microscopy (SPM; CP-II microscope, Veeco Instruments Inc., Plainview, NY, USA) in the scanning force microscope (SFM) mode. The instrument was equipped with a commercial silicon cantilever (Veeco Instruments Inc., Plainview, NY, USA) with a spring constant and resonant frequency of about 40 N/m and 300 KHz, respectively. Transmittance spectra were obtained using an ultraviolet-visible spectrometer (V-630, JASCO Co. Ltd., Tokyo, Japan). The electrical conductivities of films were measured with an impedance analyzer (LORESTA GP, Mitsubishi Chemical Co., Ltd., Tokyo, Japan). Electrical conductivity measurements of non-doped samples with high resistances were performed using a Keithley model 236 source measure unit (SMU) (Keithley Instruments, Inc., Cleveland, OH, USA) incorporating two Pt electrodes with a 20 μm gap fabricated by sputter deposition.

The tensile stress–strain measurements were carried out using a Tensilon Universal Tester (Tensilon RTC-1250, Orientec Co., Tokyo, Japan) at 25 °C, testing sample specimens 2 mm wide, 20 mm long and typically 0.02 mm thick. A position, 5 mm past the edge of each long end of the sample, was clamped using polyimide tape to prevent the sample from slipping in the holder. The tensile stress–strain responses were measured using a load cell (UR-10L-A, Orientec Co., Tokyo, Japan) and tests were performed at a crosshead speed of 10 mm/min.

3. RESULTS AND DISCUSSION

3.1. P3HT Nanofiber Formation in Conventional Polymer Films

Films were made by spin-coating polymer solutions with P3HT/PMMA weight ratios of 0.1:99.9, 0.3:99.7, 1:99, 2:98, 3:97, 5:95, 10:90, and 20:80 in 5:5 (v/v) chloroform/anisole and SFM observations were performed. Figure 1 shows the SFM topological images of films with P3HT/PMMA ratios of (a) 0.3:99.7; (b) 1:99; (c) 3:97; (d) 5:95; (e) 10:90 and (f) 20:80. We could see nanofibers even with a P3HT/PMMA ratio of 0.3:99.7 in the SFM image, demonstrating that a nanofiber composite film could be made. The nanofibers, however, did not connect with one another in films with a P3HT content of less than 0.3%, and so a nanofiber network did not develop (Figure 1a). The nanofiber networks began to form with P3HT contents above 1%, so this seems to be a percolation limit based on SFM observations. In the cases of the (e) 10:90 and (f) 20:80 solutions, the nanofiber network was so aggregated that there were some areas over which individual nanofibers could not be discriminated in the SFM image. The nanofibers were observed to be a bit shorter in the composite films than in the homogeneous nanofiber mat. This could be explained by the inhibition of π–π stacking by the PMMA matrix, making it difficult to form an ordered structure of P3HT nanofibers. This tendency is a disadvantage of the nanofiber composite films.

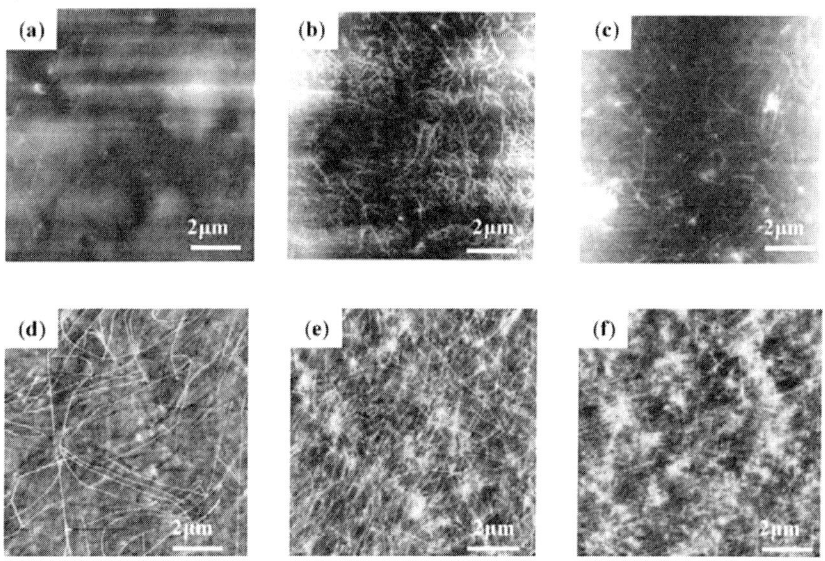

Figure 1. SFM images of P3HT/PMMA nanofiber composite films made with P3HT/PMMA in weight ratios of (**a**) 0.3:99.7; (**b**) 1:99; (**c**) 3:97; (**d**) 5:95; (**e**) 10:90; and (**f**) 20:80.

The film thicknesses were measured by SFM. The films were first scratched in the manipulation mode by the cantilever and the height difference between the resulting groove and a surrounding scratch-free flat area of the film was then measured in the SFM intermittent-contact mode [12]. Figure 2 shows the average film thickness as a function of the number of droplets of 1 wt % polymer solution (using 5:95 P3HT/PMMA in 5:5 chloroform/anisole) during the spin-coating process at 5000 rpm for 30 s. This figure also includes the film thickness as a function of number of droplets in the case of a nanofiber mat made using a 0.05 wt % P3HT solution [12]. In both nanofiber composite films and nanofiber mats, an almost linear increase in film thickness with the number of drops was observed. In addition, the film thickness of the nanofiber composite film increased more rapidly than that of the nanofiber mat. This is simply due to the difference in the polymer concentration between 1 wt % in the case of the P3HT/PMMA blend and 0.05 wt % in the case of the P3HT alone. In the trials detailed below, the film thicknesses of nanofiber composite films (from 5:95 P3HT/PMMA) were estimated by this dependence.

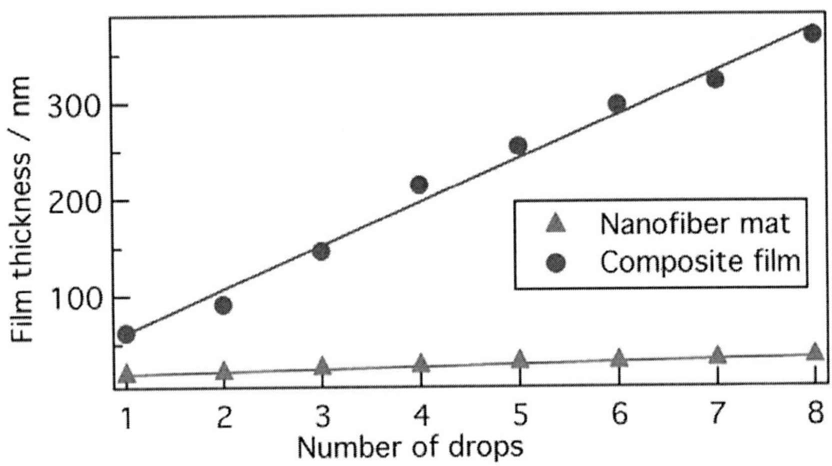

Figure 2. Film thicknesses of nanofiber mats [12] and nanofiber composite films (from 5:95 P3HT/PMMA) as a function of number of droplets of polymer solution.

In addition, Figure 3 shows the average film thickness as a function of the P3HT weight ratio in the P3HT/PMMA solution in 5:5 chloroform/anisole, based on placing a drop of 1 wt % polymer solution on a substrate during spin-coating at 5000 rpm for 30 s. With increasing PMMA ratios, the film thickness was increased, and this tendency may be explained by the increase in the viscosity of the polymer solutions.

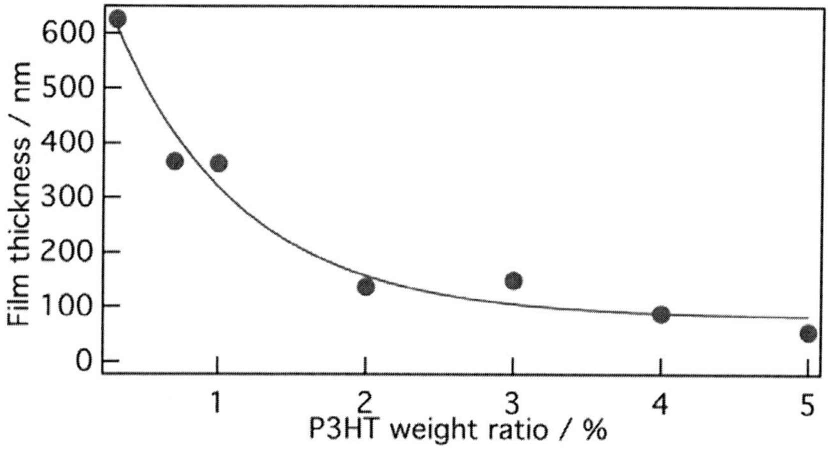

Figure 3. The average film thickness as a function of P3HT/PMMA composition.

We previously reported that P3HT nanofibers with different degrees of crystallinity can be made from solvent mixtures of chloroform and anisole with different compositions [43]. When increasing the volume ratio of chloroform (acting as a good solvent) to 70%, the resulting nanofibers had higher internal ordering and showed larger field-effect mobility, while, when using a volume ratio of chloroform above 80%, we could not observe any nanofibers. It is possible to see a similar tendency in the composite films produced in the present study.

P3HT/PMMA solutions (at a 5:95 ratio) in mixed solvents with varying solvent ratios were spin-coated onto glass substrates at 5000 rpm for 30 s. Figure 4 shows the SFM topological images of films made from solvents with chloroform/anisole ratios of 10:0, 8:2, 6:4, 5:5, 4:6, 3:7, 2:8, and 0:10. Under the good solvent (chloroform)-rich condition, i.e., the 10:0 ratio, a sea-island structure in a smooth surface was observed without any nanofibers. The nanofiber structures formed when the chloroform volume ratio was less than 80%. As the chloroform ratio required for nanofiber formation was larger than that without PMMA, these results indicate that PMMA exhibits poor compatibility with P3HT. However, the observed trend in which an increasing ratio of the poor solvent anisole induced π–π stacking was essentially the same as was observed in the absence of PMMA.

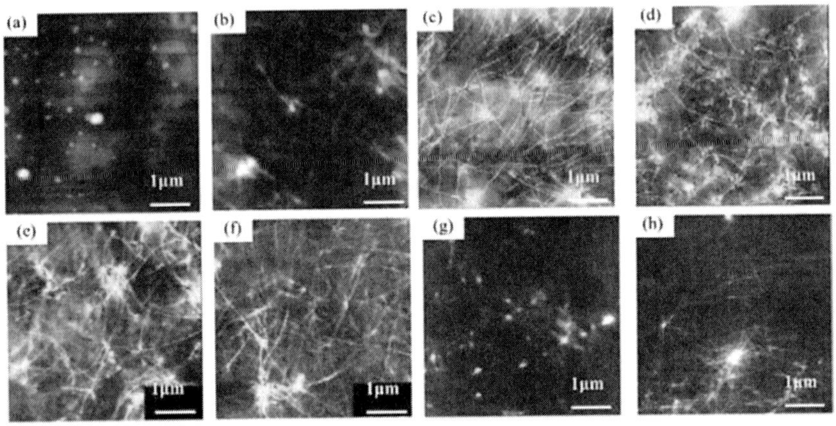

Figure 4. SFM images of P3HT/PMMA nanofiber composite films made from a mixed solvent of chloroform/anisole in volume ratios of (**a**) 10:0; (**b**) 8:2; (**c**) 6:4; (**d**) 5:5; (**e**) 4:6; (**f**) 3:7; (**g**) 2:8; and (**h**) 0:10.

Table 1 shows the conductivity of P3HT/PMMA films (at a 5:95 ratio) made from solvents with different chloroform/anisole ratios. The film thicknesses were kept relatively constant by controlling the number of drops applied for spin-coating. The film made from a 5:5 solvent mixture had the largest conductivity, and thus this condition seems to be the most suitable.

Table 1. Conductivity of composite films produced from different solvent ratios.

Chloroform/anisole (*v/v*)	Nanofibers [a]	Conductivity/S·cm^{-1}
10:0	−	$<10^{-12}$
8:2	+	2.1×10^{-10}
7:3	+	7.4×10^{-9}
6:4	+	8.5×10^{-9}
5:5	+	1.8×10^{-8}
4:6	+	6.3×10^{-9}
2:8	+	9.7×10^{-9}
0:10	+	8.6×10^{-11}

Notes: [a] Nanofibers; +: formed; −: did not form.

3.2. Optical Transparency of Nanofiber Composite Films

A nanofiber composite film was prepared from 5:95 P3HT/PMMA cast from 5:5 chloroform/anisole and the UV transmittance spectrum of this film is shown in Figure 5. The main absorption peaks were observed at 520 and 550 nm, and the formation of nanofibers was confirmed by the shoulder at 550–600 nm, attributed to H aggregates formed by π–π interaction in the same manner as which occurs in the nanofiber mat. Doping was performed by dipping the nanofiber film in an $AuCl_3$ solution and, after doping, the transmittance in the visible light range increased while the transmittance of near-infrared light decreased. This change is due to the formation of polaron or bipolaron bands, as also takes place with typical conducting polymers, including P3HT [44]. This result thus indicates that the nanofiber composite film can be effectively doped in the same manner as the P3HT nanofiber mat.

Figure 6 shows the relationship between transmittance at the peak absorbance wavelength and film thickness of P3HT/PMMA nanofiber composite films with a 5:95 P3HT/PMMA ratio as cast from 5:5 chloroform/anisole, before and after doping. The transmittance decreased linearly with the film thickness, and this tendency was similar to that of the nanofiber mat.

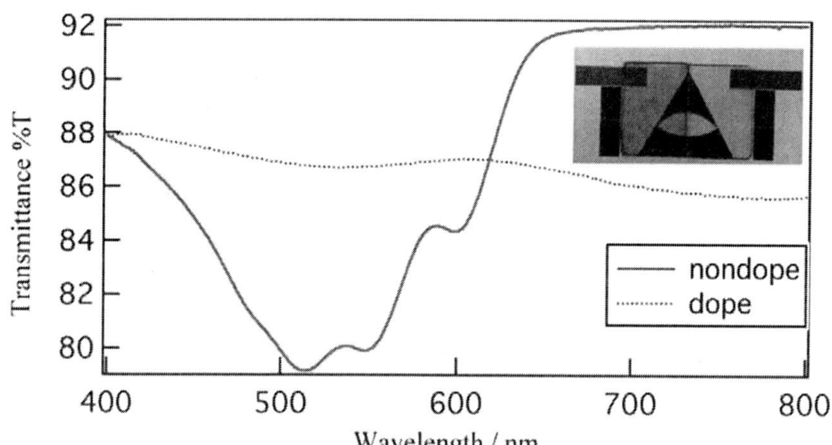

Figure 5. Transmittance spectra and photographic images of a P3HT/PMMA composite film before and after doping with $AuCl_3$. In the photographs, the red tinted left side is prior to doping and the transparent right side is after doping.

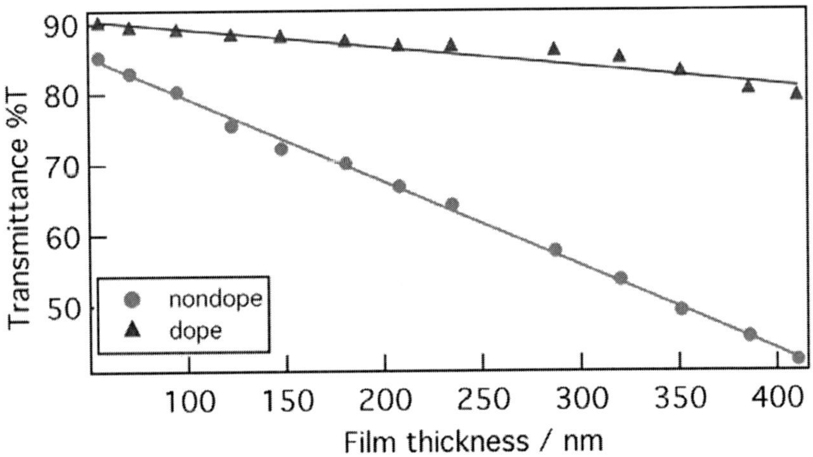

Figure 6. Transmittance dependence on film thickness of P3HT/PMMA nanofiber composite films at the peak absorbance wavelength (514.0 nm) before and after doping with $AuCl_3$.

3.3. Doping Stability of Nanofiber Composite Films

In order to research the doping stability of nanofiber composite films in air at room temperature, the time dependence of the transmittance at the peak absorbance wavelength of P3HT/PMMA nanofiber composite films (obtained from 5:95 P3HT/PMMA cast from 5:5 chloroform/anisole) doped by (a) $FeCl_3$ and (b) $AuCl_3$ was assessed, as shown in Figure 7. Transmittance spectra doped by (a) $FeCl_3$ and (b) $AuCl_3$ were shown in Figures S1 and S2, respectively. In addition, the time dependence of transmittance of nanofiber mats doped by (a) $FeCl_3$ and (b) $AuCl_3$ was measured. The nanofiber composite films and nanofiber mats were prepared so as to have nearly the same transmittance before doping.

Immediately following the doping process, the transmittance of samples doped with $FeCl_3$ decreased almost monotonically with time. The nanofiber mat doped with $FeCl_3$ decreased, particularly, rapidly, from 87% to 82%, during 38 h. The transmittance of the composite films decreased more gradually than that of the P3HT nanofiber mat. These results indicate that the composite films were more stable in air than the nanofiber mats.

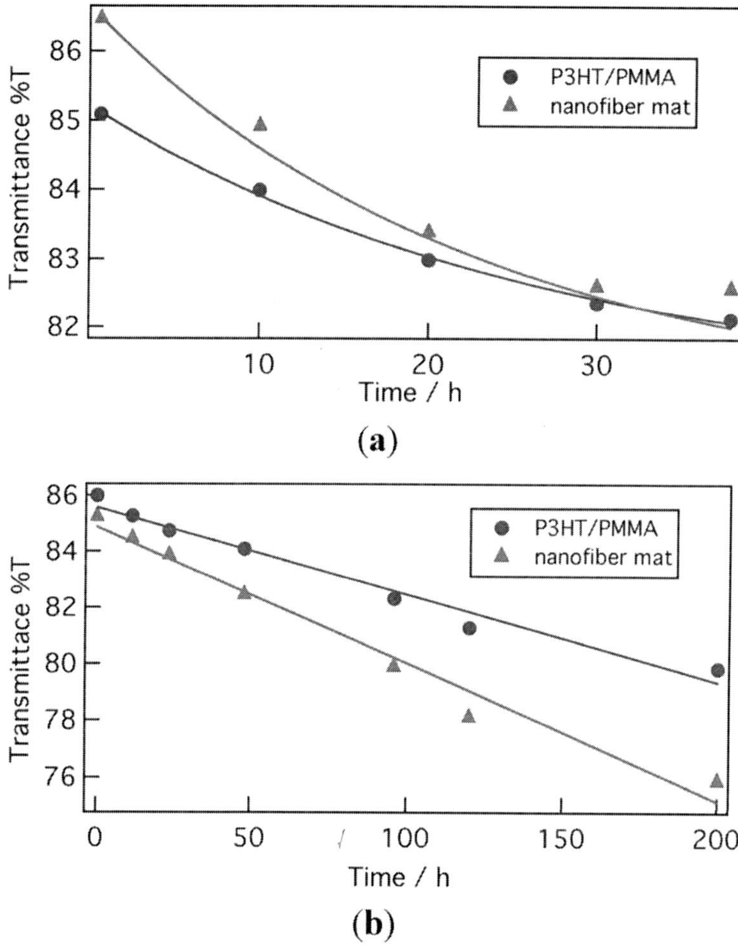

Figure 7. Time dependence of the transmittance at the peak absorbance wavelength (nanofiber mat: 554.5 nm, nanofiber composite film: 514 nm) of a nanofiber mat and a P3HT/PMMA nanofiber composite film before and after doping by (**a**) $FeCl_3$ and (**b**) $AuCl_3$. Measurements were carried out, in air, at room temperature.

In the case of doping with $AuCl_3$, less significant decreases in transmittance were observed immediately following the doping process. Following 200 h, the transmittance decreases of the P3HT/PMMA nanofiber composite films and P3HT nanofiber mats were 7.1% and 10.9%, respectively. Although doping by $AuCl_3$ thus produced considerably more stable films, the doping state of the nanofiber mat was again more unstable than that of the composite films, just as in the case of doping with $FeCl_3$.

3.4. Relation between Transmittance and Conductivity of Nanofiber Composite Films

Figure 8 shows the conductivity of P3HT/PMMA nanofiber composite films, made from 5:95 P3HT/PMMA ratio cast from 5:5 chloroform/anisole and doped by $AuCl_3$, as a function of transmittance. The conductivity of the nanofiber mat doped by $AuCl_3$ as a function of transmittance was also added. Since we measured the conductivity by the 4-pin probe method, the contact resistance was almost negligible and the conductivity should be essentially independent of the film thickness, and, thus, of the transparency of the films. However, the results show that the conductivity depends slightly on the transparency and, by extension, the film thickness. Various factors have been suggested as explanations for this phenomenon, such as surface scattering, the skin effect, discontinuity of the film and inhomogeneous doping [45,46,47].

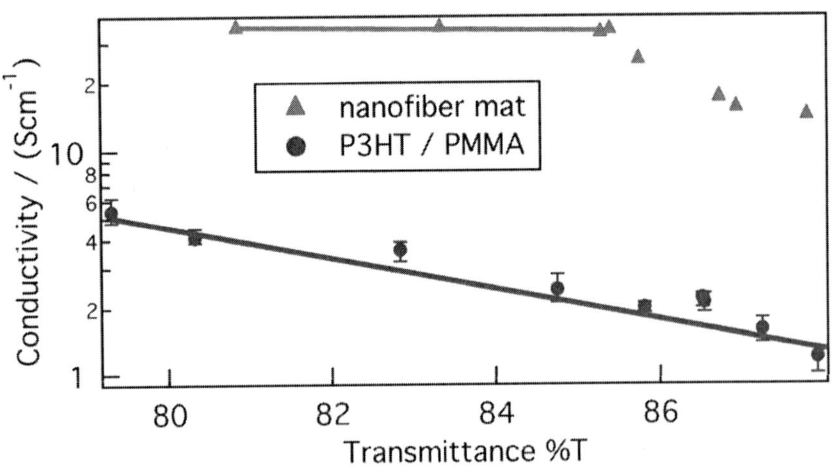

Figure 8. The conductivity of P3HT/PMMA nanofiber composite films and nanofiber mats doped by $AuCl_3$ as a function of transmittance.

The conductivity of the P3HT nanofiber mat at a transmittance value of 80.0% was 35 S·cm^{-1}, while that of the P3HT/PMMA nanofiber composite film was 4.1 S·cm^{-1}. Consequently, the conductivity of the nanofiber mat was 7–9 times larger than that of the nanofiber composite film. If many of the voids in the nanofiber mat were filled by the PMMA matrix, the nanofiber composite film should have nearly the same conductivity as the nanofiber mat, and so the low conductivity of the nanofiber composite films was attributed to the lower content of nanofiber structures.

3.5. Mechanical Strength of Nanofiber Composite Films

In general, conducting polymer including P3HT are too brittle to use as the sole components of flexible devices, while nanofiber mats are sufficiently flexible. Nanofibers were therefore embedded in a conventional polymer film to allow their use in flexible devices. For mechanical measurements, we prepared a self-standing film of the P3HT/PMMA nanofiber composite as follows. A 10 wt % PVA solution was spin-coated onto a glass substrate, after which a solution of 5:95 P3HT/PMMA in 5:5 chloroform/anisole was spin-coated uniformly onto the PVA film. After drying under vacuum, the film was allowed to spontaneously peel from the substrate in hot water and was then washed with additional water. The peeled film (having a thickness of 20 μm) was dried completely under vacuum and a 2 × 20 mm^2 membrane was cut from the film for evaluation of the mechanical properties. The tensile stress–strain response of the membrane was measured and the tensile strength curve is shown in Figure 9.

The breaking strength of the P3HT/PMMA composite film was calculated as 12.3 ± 0.3 MPa with a strain of 3.8% ± 0.4%. This experimental value was lower than the typical breaking strength of 48–76 MPa and elongation of 2%–10% of a PMMA film [48], and was not sufficient for fabrication of the intended devices. However, these mechanical properties are improved compared to the values previously reported for the homogeneous nanofiber mat, which cannot be used for practical applications due to its brittleness. Furthermore, it is well known that there is a trade-off between mechanical strength and conductivity and transparency. In order to improve these films and simultaneously achieve high conductivity, high transparency, and acceptable mechanical strength, we will have to overcome this three-way deadlock in future work.

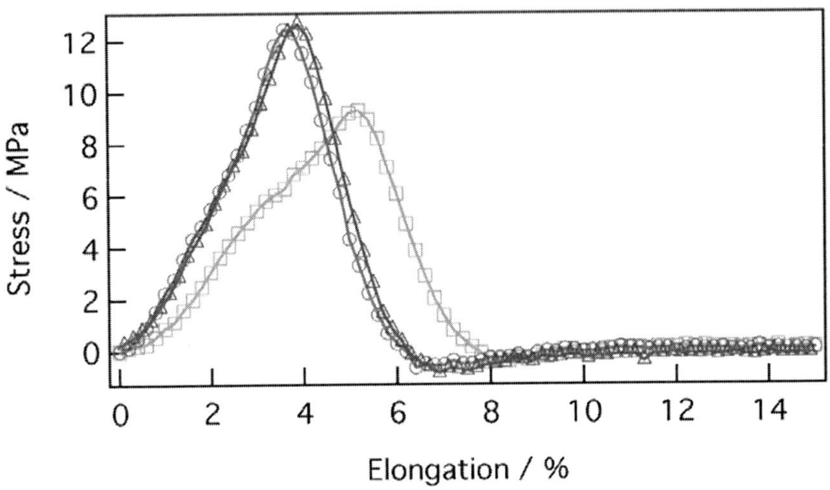

Figure 9. Stress–strain curve of a P3HT/PMMA composite film. Red and blue lines show standard measurement data, while the green line was obtained from a sample, which exhibited partial rupture during testing.

4. CONCLUSIONS

Nanofiber composite films made of P3HT/PMMA doped with $AuCl_3$ showed higher atmospheric stability, with their transmittance decreasing by only 7.1% after 200 h. They also exhibited improved fracture stress of 12.3 MPa at 3.8% strain compared to nanofiber mats, while their conductivity was about seven times less (4.1 $S \cdot cm^{-1}$) than that of a nanofiber mat with the same transmittance of about 80%. To overcome the issues identified in this study, the nanofiber density in these materials should be increased to allow their application as transparent, conductive, and flexible films.

ACKNOWLEDGMENTS

This work was supported in part by a Grant-Aid for Scientific Research (B) (No. 20350102) from the Ministry of Education, Culture, Sports, Science and Technology of Japan, and an Adaptable and Seamless Technology Transfer Program through Target-Driven R&D (AS231Z04000D) from the Japan Science and Technology Agency.

CONFLICTS OF INTEREST

The authors declare no conflict of interest.

REFERENCES

1. Lee, J.Y.; Connor, S.T.; Cui, Y.; Peumans, P. Solution-processed metal nanowire mesh transparent electrodes. Nano Lett. 2008, 8, 689–692.

2. De, S.; Higgins, T.M.; Lyons, P.E.; Doherty, E.M.; Nirmalraj, P.N.; Blau, W.J.; Boland, J.J.; Coleman, J.N. Silver nanowire networks as flexible, transparent, conducting films: Extremely high DC to optical conductivity ratios. ACS Nano 2009, 3, 1767–1774.

3. Zhang, H.; Li, C. Chemical synthesis of transparent and conducting polyanilinepoly(ethylene terephthalate) composite films. Synth.Met. 1991, 44, 143–146.

4. Cao, Y.; Treacy, G.M.; Smith, P.; Heeger, A.J. Optical-quality transparent conductive polyaniline films. Synth. Met. 1993, 57, 3526–3531.

5. Cao, Y.; Treacy, G.M.; Smith, P.; Heeger, A.J. Solution-cast films of polyaniline: Optical-quality transparent electrodes. Appl. Phys. Lett. 1992, 60, 2711–2713.

6. Byun, S.W.; Im, S.S. Degradation kinetics of electrical conductivity in transparent polyaniline-nylon 6 composite films. Synth. Met. 1995, 69, 219–220.

7. Wan, M.X.; Li, M.; Li, J.C.; Liu, Z.X. Transparent and conducting coatings of polyaniline composite. Thin Solid Films 1995, 259, 188–193.

8. Jonas, F.; Krafft, W.; Muys, B. Poly(3,4-ethylenedioxythiophene): Conductive coatings, technical applications and properties. Macromol. Symp. 1995, 100, 169–173.

9. Groenendaal, L.B.; Jonas, F.; Freitag, D.; Pielartzik, H.; Reynolds, J.R. Poly(3,4-ethylenedioxythiophene) and its derivatives: Past, present, and future. Adv. Mater. 2000, 12, 481–494.

10. Yan, H.; Jo, T.; Okuzaki, H. Highly conductive and transparent poly(3,4-ethylenedioxythiophene)/poly(4-styrenesulfonate) thin films. Polym. J. 2009, 41, 1028–1029.

11. Apperloo, J.J.; van Haare, J.A.E.H.; Janssen, R.A.J. Transparent highly oxidized conjugated polymer films from solution. Synth. Met. 1999, 101, 417–420.

12. Aronggaowa, B.; Kawasaki, M.; Shimomura, T. Thin, transparent conductive films fabricated from conducting polymer nanofibers. Polym. J. 2013.

13. Breuer, O.; Sundararaj, U. Big returns from small fibers: A review of polymer/carbon nanotube composites. Polym. Comp. 2004, 25, 630–645.

14. Winey, K.I.; Vaia, R.A. Polymer nanocomposites. MRS Bull. 2007, 32, 314–322.

15. Maruyama, B.; Alam, H. Carbon nanotubes and nanofibers in composite materials. SAMPE J. 2002, 38, 59–70.

16. Ong, B.S.; Wu, Y.L.; Liu, P.; Gardner, S. High-performance semiconducting polythiophenes for organic thin-film transistors. J. Am. Chem. Soc. 2004, 126, 3378–3379.

17. McCulloch, I.; Heeney, M.; Bailey, C.; Genevicius, K.; Macdonald, I.; Shkunov, M.; Sparrowe, D.; Tierney, S.; Wagner, R.; Zhang, W.M.; et al. Liquid-crystalline semiconducting polymers with high charge-carrier mobility. Nat. Mater. 2006, 5, 328–333.

18. Narkis, M.; Tobolsky, A.V. Chemically crosslinked polyethylene: Modulus–temperature relations and heat stability. J. Appl. Polym. Sci. 1969, 13, 2257–2263.

19. Meyer, J. Glass transition temperature as a guide to selection of polymers suitable for PTC materials. Polym. Eng. Sci. 1973, 13, 462–468.

20. Gubbels, F.; Jér^ome, R.; Vanlathem, E.; Deltour, R.; Blacher, S.; Brouers, F. Kinetic and thermodynamic control of the selective localization of carbon black at the interface of immiscible polymer blends. Chem. Mater. 1998, 10, 1227–1235.

21. Zhang, C.; Yi, X.S.; Yui, H.; Asai, S.; Sumita, M.J. Morphology and electrical properties of short carbon fiber-filled polymer blends: High-density polyethylene/poly(methyl methacrylate). Appl. Polym. Sci. 1998, 69, 1813–1819.

22. Lagrève, C.; Feller, J.F.; Linossier, I.; Levesque, G. Poly(butylene terephthalate)/poly(ethylene-co-alkyl acrylate)/carbon black conductive composites: Influence of composition and morphology on electrical properties. Polym. Eng. Sci. 2001, 41, 1124–1132.

23. Feller, J.F. Conductive polymer composites: Influence of extrusion conditions on positive temperature coefficient effect (PTC) of poly(butylene terephthalate)/poly(olefin)-carbon black blends. J. Appl. Polym. Sci. 2004, 91, 2151–2157.

24. Cheah, K.; Forsyth, M.; Simon, G.P. Conducting composite using an immiscible polymer blend matrix. Synth. Met. 1999, 102, 1232–1233.

25. Boiteux, G.; Fournier, J.; Issotier, D.; Seytre, G.; Marichy, G. Conductive thermoset composites: PTC effect. Synth. Met. 1999, 102, 1234–1235.

26. Weber, I.; Schwartz, P. Monitoring bending in carbon-fibre/epoxy composite strands: A comparison between mechanical and resistance techniques. Comp. Sci. Technol. 2001, 61, 849–853.

27. lcakova, J.; Saha, P.; Kresalek, V.; Quadrat, O. Pre-exponential factor and activation energy of electrical conductivity in polyester resin/carbon fibre composites. Synth. Met. 2000, 113, 83–88.

28. Severin, E.J.; Doleman, B.J.; Lewis, N.S. An investigation of the concentration dependence and response to analyte mixtures of carbon black/insulating organic polymer composite vapor detectors. Anal. Chem. 2000, 72, 658–668.

29. Munoz, B.C.; Steinthal, G.; Sunshine, S. Conductive polymer-carbon black composites-based sensor arrays for use in an electronic nose. Sens. Rev. 1999, 19, 300–305.

30. Lundberg, B.; Sundquist, B.J. Resistivity of a composite conducting polymer as a function of temperature, pressure, and environment: Application as a pressure and gas concentration transducer. Appl. Phys. 1986, 60, 1074–1079.

31. Chen, J.; Tsubokawa, N. Novel gas sensor from polymer-grafted carbon black: Vapor response of electric resistance of conducting composites prepared from poly(ethylene-block-ethylene oxide)-grafted carbon black. J. Appl. Polym. Sci. 2000, 77, 2437–2447.

32. Srivastava, S.; Tchoudakov, R.; Narkis, M.A. Preliminary investigation of conductive immiscible polymer blends as sensor materials. Polym. Eng. Sci. 2000, 40, 1522–1528.

33. Narkis, M.; Srivastava, S.; Tchoudakov, R.; Breuer, O. Sensors for liquids based on conductive immiscible polymer blends. Synth. Met. 2000, 113, 29–34.

34. Chen, X.D.; Jiang, Y.D.; Wu, Z.M.; Li, D.; Yang, J.D. Morphology and gas-sensitive properties of polymer based composite films. Sens. Actuators B Chem. 2000, 66, 37–39.

35. Chen, J.; Tsubokawa, N. A Novel gas sensor from polymer-grafted carbon black: Responsiveness of electric resistance of conducting composite from LDPE and PE-b-PEO-grafted carbon black in various vapors. Polym. Adv. Technol. 2000, 11, 101–107.

36. Burl, M.C.; Sisk, B.C.; Vaid, T.P.; Lewis, N.S. Classification performance of carbon black-polymer composite vapor detector arrays as a function of array size and detector composition. Sens. Actuators B Chem. 2002, 87, 130–149.

37. Lin, C.W.; Hwang, B.G.; Lee, C.R. Characteristics and sensing behavior of electrochemically codeposited polypyrrole-poly(vinyl alcohol)thin film exposed to ethanol vapors. J. Appl. Polym. Sci. 1999, 73, 2079–2087.

38. Qiu, L.Z.; Lim, J.A.; Wang, X.H.; Lee, W.H.; Hwang, M.; Cho, K. Versatile use of vertical-phase-separation-induced bilayer structures in organic thin-film transistors. Adv. Mater. 2008, 20, 1141–1145.

39. Qiu, L.; Lee, W.H.; Wang, X.; Kim, J.S.; Lim, J.A.; Kwak, D.; Cho, K. Organic thin-film transistors based on polythiophene nanowires embedded in insulating polymer. Adv. Mater. 2009, 21, 1349–1353.

40. Lim, J.A.; Kim, J.H.; Qiu, L.Z.; Lee, W.H.; Lee, H.S.; Kwak, D.; Cho, K. Inkjet-printed single-droplet organic transistors based on semiconductor nanowires embedded in insulating polymers. Adv. Funct. Mater. 2010, 20, 3292–3297.

41. Yoshida, K.; Kawasaki, M.; Toda, Y.; Yamashita, S.; Shimomura, T. Microscopic conduction pathways of poly(3-hexylthiophene) nanofibers embedded in polymer film. Polym. J. 2012, 44, 371–374.

42. Ahlskog, M. Stability studies on AuCl3-doped poly(3-hexylthiophene): Comparison with FeCl3 doping. Synth. Met. 1995, 72, 197–200.

43. Shimomura, T.; Takahashi, T.; Ichimura, Y.; Nakagawa, S.; Noguchi, K.; Heike, S.; Hasuizume, T. Relationship between structural coherence and intrinsic carrier transport in an isolated poly(3-hexylthiophene) nanofiber. Phys. Rev. B 2011, 83, 115314:1–115314:10.

44. Chung, T.C.; Kaufman, J.H.; Heeger, A.J.; Wudl, F. Charge storage in doped poly(thiophene): Optical and electrochemical studies. Phys. Rev. B 1981, 30, 702–710.

45. Maissel, L.I. Electrical Properties of Metallic Thin Films. In Handbook of Thin Film Technology; Maissel, L.I., Glang, R., Eds.; McGraw-Hill: New York, NY, USA, 1970; Chapter 13.

46. Kivelson, S.; Heeger, A.J. First order phase transition to a metallic state in PA: A strong-coupling polaronic metal. Phys. Rev. Lett. 1985, 55, 308–311.

47. Kim, J.; Sohn, D.; Sung, Y.; Kim, E.-R. Fabrication and characterization of conductive polypyrrole thin film prepared by in situ vapor-phase polymerization. Synth. Met. 2003, 132, 309–313.

48. Hsu, S.L. Poly(methyl methacrylate). In Polymer Data Handbook; Mark, J.E., Ed.; Oxford University Press: Oxford, UK, 1999; p. 656.

INDEX

M

N

O

P

R

S

T